AF594972

Groundwater Hydrological Model Simulation

Groundwater Hydrological Model Simulation

Editor

Cristina Di Salvo

MDPI • Basel • Beijing • Wuhan • Barcelona • Belgrade • Manchester • Tokyo • Cluj • Tianjin

Editor
Cristina Di Salvo
Italian National Research
Council,
Italy

Editorial Office
MDPI
St. Alban-Anlage 66
4052 Basel, Switzerland

This is a reprint of articles from the Special Issue published online in the open access journal *Water* (ISSN 2073-4441) (available at: https://www.mdpi.com/journal/water/special_issues/Hydrological_Model).

For citation purposes, cite each article independently as indicated on the article page online and as indicated below:

LastName, A.A.; LastName, B.B.; LastName, C.C. Article Title. *Journal Name* **Year**, *Volume Number*, Page Range.

ISBN 978-3-0365-7142-3 (Hbk)
ISBN 978-3-0365-7143-0 (PDF)

Contents

About the Editor . . . . **vii**

Cristina Di Salvo
Groundwater Hydrological Model Simulation
Reprinted from: *Water* **2023**, *15*, 822, doi:10.3390/w15040822 . . . . **1**

Davide Sartirana, Chiara Zanotti, Marco Rotiroti, Mattia De Amicis, Mariachiara Caschetto, Agnese Redaelli, Letizia Fumagalli and Tullia Bonomi
Quantifying Groundwater Infiltrations into Subway Lines and Underground Car Parks Using MODFLOW-USG
Reprinted from: *Water* **2022**, *14*, 4130, doi:10.3390/w14244130 . . . . **5**

Kaouther Ncibi, Micòl Mastrocicco, Nicolò Colombani, Gianluigi Busico, Riheb Hadji, Younes Hamed and Khan Shuhab
Differentiating Nitrate Origins and Fate in a Semi-Arid Basin (Tunisia) via Geostatistical Analyses and Groundwater Modelling
Reprinted from: *Water* **2022**, *14*, 4124, doi:10.3390/w14244124 . . . . **29**

Matia Menichini, Linda Franceschi, Brunella Raco, Giulio Masetti, Andrea Scozzari and Marco Doveri
Groundwater Modeling with Process-Based and Data-Driven Approaches in the Context of Climate Change
Reprinted from: *Water* **2022**, *14*, 3956, doi:10.3390/w14233956 . . . . **47**

Eric D. Morway, Daniel T. Feinstein and Randall J. Hunt
Simulation of Heat Flow in a Synthetic Watershed: The Role of the Unsaturated Zone
Reprinted from: *Water* **2022**, *14*, 3883, doi:10.3390/w14233883 . . . . **73**

Luca Alberti, Matteo Antelmi, Gabriele Oberto, Ivana La Licata and Pietro Mazzon
Evaluation of Fresh Groundwater Lens Volume and Its Possible Use in Nauru Island
Reprinted from: *Water* **2022**, *14*, 3201, doi:10.3390/w14203201 . . . . **97**

Rudy Rossetto, Alberto Cisotto, Nico Dalla Libera, Andrea Braidot, Luca Sebastiani, Laura Ercoli and Iacopo Borsi
ORGANICS: A QGIS Plugin for Simulating One-Dimensional Transport of Dissolved Substances in Surface Water
Reprinted from: *Water* **2022**, *14*, 2850, doi:10.3390/w14182850 . . . . **117**

Daniel T. Feinstein, Randall J. Hunt and Eric D. Morway
Simulation of Heat Flow in a Synthetic Watershed: Lags and Dampening across Multiple Pathways under a Climate-Forcing Scenario
Reprinted from: *Water* **2022**, *14*, 2810, doi:10.3390/w14182810 . . . . **133**

Elisabetta Preziosi, Nicolas Guyennon, Anna Bruna Petrangeli, Emanuele Romano and Cristina Di Salvo
A Stepwise Modelling Approach to Identifying Structural Features That Control Groundwater Flow in a Folded Carbonate Aquifer System
Reprinted from: *Water* **2022**, *14*, 2475, doi:10.3390/w14162475 . . . . **157**

Julla Kabeto, Dereje Adeba, Motuma Shiferaw Regasa and Megersa Kebede Leta
Groundwater Potential Assessment Using GIS and Remote Sensing Techniques: Case Study of West Arsi Zone, Ethiopia
Reprinted from: *Water* **2022**, *14*, 1838, doi:10.3390/w14121838 . . . . **173**

Alireza Asadi and Kushal Adhikari
Minimizing Errors in the Prediction of Water Levels Using Kriging Technique in Residuals of the Groundwater Model
Reprinted from: *Water* **2022**, *14*, 426, doi:10.3390/w14030426 . . . . **203**

Cristina Di Salvo
Improving Results of Existing Groundwater Numerical Models Using Machine Learning Techniques: A Review
Reprinted from: *Water* **2022**, *14*, 2307, doi:10.3390/w14152307 . . . . **219**

About the Editor

Cristina Di Salvo

Cristina Di Salvo, PhD, is a researcher working at CNR-Institute of Environmental Geology and Geoengineering. She graduated in Geological Sciences and holds a PhD in Hydrogeology with a dissertation on 3D numerical models in urban areas. Her research topics include: groundwater modelling, urban hydrology, groundwater mapping, and hydrogeological risks. She has been the research leader of research projects involving the hydrogeological risks in urban areas, with particular focus on archaeological areas.

MDPI

Editorial

Groundwater Hydrological Model Simulation

Cristina Di Salvo

CNR-Institute of Environmental Geology and Geoengineering, Area della Ricerca di Roma 1—Montelibretti Via Salaria km 29,300, 00015 Monterotondo, Italy; cristina.disalvo@igag.cnr.it

1. Introduction

The management of groundwater resources commonly involves challenges and complexities, which are taken on by researchers using a variety of different strategies. In particular, groundwater numerical modelling is a widely-used and effective approach to simulating and analysing groundwater dynamics under varying conditions. Models are set up to investigate particular features of the groundwater system that need to be better understood; they are generally implemented in order to test conceptual hypotheses arising from field observations of the system. For this reason, numerical models need to be supported by proper field data acquisition and elaboration, a correct conceptualization of the natural system, optimal selection of the computer code and solver, and an effective calibration process. Despite their wide use, each model is different from the others, and modellers must find the best technique to solve specific problems and meet specific objectives. One area that has especially promoted innovation in modelling technique is the study of climate change—it poses new challenges and requires investigation techniques to adapt to new needs.

This Special Issue aims to gather contributions emphasising different aspects of groundwater modelling, focusing on the latest developments and applications for water resources management, including innovative applications of traditional models, the implementation of new open source platforms for groundwater modelling, and the use of artificial intelligence to explore data and expedite the calibration process.

The Special Issue comprises 10 articles and 1 review paper, with contributions from over 47 authors. Geographically, the case studies concern 5 countries extending over 4 continents (United States of America (North America); Ethiopia and Tunisia (Africa); Italy (Europe); Nauru (Oceania)), with very different features (e.g., urban environment, carbonate mountain areas, coastal aquifers).

Specifically, the topics covered by the contributions collected in this SI include:

- the interactions between groundwater and the underground infrastructures in urban areas;
- the use of groundwater models to determine the origin of groundwater contamination;
- the testing of modelling approaches to simulate the impact of climate change;
- the testing of modelling techniques to optimize groundwater management;
- the development of open source software and tools to manage groundwater models;
- application of geostatistical tools to reduce model error and improve predictions;
- comparative studies among numerical models and machine learning techniques.

The SI offers a wide overview of recent applications of groundwater modelling harnessing a variety of techniques. The common goal of all the studies is to test methodologies that can be used to find optimal solutions for supporting stakeholders in adopting proper measures to manage groundwater. Depending on the case study, these measures are aimed at: reducing the damage due to flooding of urban structures; ensuring water supply while guaranteeing a sustainable water balance; evaluating and managing the effect of climate change on sensitive ecosystems; preventing the degradation of good-quality resources; reducing the threat of saltwater intrusion.

Citation: Di Salvo, C. Groundwater Hydrological Model Simulation. *Water* **2023**, *15*, 822. https://doi.org/10.3390/w15040822

Received: 31 January 2023
Accepted: 13 February 2023
Published: 20 February 2023

2. Overview of the Contributions of the Special Issue

The paper "Quantifying Groundwater Infiltrations into Subway Lines and Underground Car Parks Using MODFLOW-USG" [1] investigates the interaction between groundwater and underground infrastructures in a portion of urban Milan (Italy). The authors developed a steady-state MODFLOW-USG model which combined the use of Wall (HFB) and Drain (DRN) packages, for simulating underground infrastructures (i.e., subway lines and public car parks). The model was calibrated against a condition of high water levels. The quantification of groundwater infiltration shows agreement with historical information about submerged structures, giving confidence that the model can be used to predict infiltration related to water table oscillation and, thus, be a support in the design of dewatering systems or other proposed solutions to secure urban structures from potential infiltration damages.

In the contribution "Differentiating Nitrate Origins and Fate in a Semi-Arid Basin (Tunisia) via Geostatistical Analyses and Groundwater Modelling" [2], a MODFLOW-2005 groundwater flow model and a MODPATH advective particle tracking model have been combined with geostatistical analyses based on data from hydrochemical and hydrogeological characterization. Modelling is applied to a multi-aquifer groundwater flow system to verify the hypothesis of geogenic origin of NO_{3-} in the semi-confined aquifer. While the uppermost unconfined aquifer is contaminated by NO_{3-} by anthropic activities, models result show that the leakage of NO_{3-} through the aquitard is negligible. Authors conclude that the high NO_{3-} concentration in the deepest aquifer is associated with pre-Triassic evaporite dissolution and, thus, has a natural origin. These findings based on the model application should help guide proper management of the contaminated aquifers.

The paper "Groundwater Modelling with Process-Based and Data-Driven Approaches in the Context of Climate Change" [3] investigates the application of alternative modelling approaches (process-based, data-driven, and integrated data-driven/process-based) to simulate the effects of different climate scenario on three porous aquifers. Results distinguish key characteristics for each aquifer, such as the ability of storage capacity to mitigate the effects of dry climate conditions or the dramatic sensitivity of a system to climate extremes. In general, the study highlights that choosing the modelling approach based on the specific aquifer features is fundamental to obtaining a modelling tool efficient in supporting groundwater management actions aimed at mitigation of the effects of climate change.

In the paper "Simulation of heat flow in a synthetic watershed: The role of the unsaturated zone" [4], the authors applied a coupled flow (MODFLOW-NWT) and transport (MT3D-USGS) model for simulating unsaturated/saturated heat transport due to atmospheric warming via a synthetic three-dimensional representative watershed. An important novelty of the research is the focus on the unsaturated zone (UZ) and the effect of variable depth–to–water table on heat flow to the water table and surface-water features. The approach is computationally efficient and gives rise to a flexible tool for evaluating the temperature response to warming and trends of heat transport across the watershed. The research highlighted that: (1) the heat flow forcing function is the product of infiltration temperatures and infiltration rates; (2) the UZ has a strong damping effect on the warming signal; (3) the warming is buffered also at discharge points, where shallow and deep flow converge; (4) the stream baseflow response to heat forcing is influenced by the lateral extent of the riparian zone. The authors conclude that explicit representation of the UZ in models is important to realistically evaluating the impacts of climate change on fragile ecosystems such as riparian zones or stream habitats.

The paper "Evaluation of Fresh Groundwater Lens Volume and Its Possible Use in Nauru Island" [5] presents a particular case study concerning the groundwater system characterization and modelling of the Nauru Atoll Island (Pacific Ocean). After a large-scale study for detecting the location of freshwater lenses, a local-scale study was made, aimed at quantifying the freshwater lens thickness and volume for supply uses through a geoelectrical tomography survey, and a 3D density-dependent numerical model implemented in SEAWAT. The main scientific finding is that freshwater in small islands can unexpectedly accumulate right along the seashore and not in the centre of the island. Furthermore, the calibrated model can be used to design sustainable groundwater exploitation systems that avoid the exacerbation of saltwater intrusion.

The paper "ORGANICS: A QGIS Plugin for Simulating One-Dimensional Transport of Dissolved Substances in Surface Water" [6] describes the development and testing of a QGIS plugin, which simulates the concentration of a contaminant along the profile of the watercourse. Attempting to embed surface water solute transport modelling into GIS by inputting the entry point concentration and the average speed of surface water, this tool allows GIS experts to perform first level yet fast simulations of the concentration of the pollutant in surface water bodies. The code is open source and free, which facilitates the reproducibility of the run analyses.

The paper "Simulation of Heat Flow in a Synthetic Watershed: Lags and Dampening across Multiple Pathways under a Climate-Forcing Scenario" [7] is a continuation of companion research already presented in [4]. The processes of overland flow, infiltration through an unsaturated zone (UZ) and groundwater flow discharge to a surface-water network are simulated by a synthetic flow and transport watershed model under a 30-year warming signal. Quantitative results for the transient distribution of heat flow conditions demonstrate the dampening effect of the UZ in the warming transferred to the water table (about 40% of the warming applied to watershed infiltration) and the dampening effect of the aquifer on the heat discharged to the stream network (about 10% of the original warm-up signal). Despite the subsurface lag and storage effects, simulated temperatures in surface waters increase due to the addition of heat by storm runoff which bypasses the UZ. The relevance of this study lies in the fact that provides a possible workflow for climate-change modelling application, allowing for a detailed analysis of warming trends at the groundwater/surface water interface, which are areas of great importance for aquatic ecosystems.

In the paper "A Stepwise Modelling Approach to Identifying Structural Features That Control Groundwater Flow in a Folded Carbonate Aquifer System" [8], the authors set up a procedure to test a numerical modelling technique in a carbonate aquifer characterized by a complex geological structure that constitutes a source of good quality water for human consumption. Three models were implemented by gradually adding complexity to the model grid using an equivalent porous medium approach: single layer (2D), three layers (quasi-3D), and five layers (fully 3D). This was done in order to find the best match with the observed aquifer outflow to the river. The Newton–Raphson formulation for MODFLOW-2005 was used to solve numerical instabilities. Results demonstrated that folded and faulted geological structure control groundwater flow dynamics, and thus need to be adequately represented by a full-3D model. These findings are relevant in applications involving the management of groundwater in corrugated carbonate, which are often exploited for water supply.

In the paper "Using GIS and Remote Sensing Techniques: Case Study of West Arsi Zone, Ethiopia" [9], remote sensing data and geographic information system tools are used to evaluate the groundwater potential of the study area. By means of a chain of GIS tools, parameters influencing groundwater were extracted, mapped, and elaborated in a GIS environment; the procedure was validated by means of borehole data. Results show that the method provides a fast and accurate technique to detect the groundwater potential of an area, furnishing a tool for optimize the planning of groundwater exploitation.

The paper "Minimizing Errors in the Prediction of Water Levels Using Kriging Technique in Residuals of the Groundwater Model" [10] describes an application of the kriging geostatistical tool to the groundwater level residuals of a MODFLOW model developed in the Edwards–Trinity (Plateau) aquifer (Texas), aimed at improving predictions at unsampled locations. The average absolute model error was reduced from 31 m to 5 m, while the average residual standard error decreased from 9.7 to 4.7 m. The authors argue that their procedure makes model results more reliable, allowing design of more informative monitoring systems, and ultimately leading to more efficient management of groundwater resources.

The paper "Improving Results of Existing Groundwater Numerical Models Using Machine Learning Techniques: A Review" [11] presents a review of papers comparing the use of numerical and machine learning methods for groundwater level modelling. The review highlights the advantages or disadvantages of both techniques, depending on the objectives of the model. A promising strategy is to use both methods as complementary to each other: machine learning techniques can improve the calibration of numerical models whereas process-based numerical models are suitable to understand the physical system and, on turn, select proper input variables for machine learning models. Furthermore, machine learning models can provide rapid and effective solutions for groundwater management and are computationally efficient tools to correct head error prediction of numerical models.

The approaches and techniques featured in this SI are a sample of the many innovations being applied to groundwater modelling in order improve water management and to respond to short- and long-term threats to water supply.

Conflicts of Interest: The author declare no conflict of interest.

References

1. Sartirana, D.; Zanotti, C.; Rotiroti, M.; De Amicis, M.; Caschetto, M.; Redaelli, A.; Fumagalli, L.; Bonomi, T. Quantifying Groundwater Infiltrations into Subway Lines and Underground Car Parks Using MODFLOW-USG. *Water* **2022**, *14*, 4130. [CrossRef]
2. Ncibi, K.; Mastrocicco, M.; Colombani, N.; Busico, G.; Hadji, R.; Hamed, Y.; Shuhab, K. Differentiating Nitrate Origins and Fate in a Semi-Arid Basin (Tunisia) via Geostatistical Analyses and Groundwater Modelling. *Water* **2022**, *14*, 4124. [CrossRef]
3. Menichini, M.; Franceschi, L.; Raco, B.; Masetti, G.; Scozzari, A.; Doveri, M. Groundwater Modeling with Process-Based and Data-Driven Approaches in the Context of Climate Change. *Water* **2022**, *14*, 3956. [CrossRef]
4. Morway, E.D.; Feinstein, D.T.; Hunt, R.J. Simulation of heat flow in a synthetic watershed: The role of the unsaturated zone. *Water* **2022**, *14*, 3883. [CrossRef]
5. Alberti, L.; Antelmi, M.; Oberto, G.; La Licata, I.; Mazzon, P. Evaluation of Fresh Groundwater Lens Volume and Its Possible Use in Nauru Island. *Water* **2022**, *14*, 3201. [CrossRef]
6. Rossetto, R.; Cisotto, A.; Dalla Libera, N.; Braidot, A.; Sebastiani, L.; Ercoli, L.; Borsi, I. ORGANICS: A QGIS Plugin for Simulating One-Dimensional Transport of Dissolved Substances in Surface Water. *Water* **2022**, *14*, 2850. [CrossRef]
7. Feinstein, D.T.; Hunt, R.J.; Morway, E.D. Simulation of Heat Flow in a Synthetic Watershed: Lags and Dampening across Multiple Pathways under a Climate-Forcing Scenario. *Water* **2022**, *14*, 2810. [CrossRef]
8. Preziosi, E.; Guyennon, N.; Petrangeli, A.B.; Romano, E.; Di Salvo, C. A Stepwise Modelling Approach to Identifying Structural Features That Control Groundwater Flow in a Folded Carbonate Aquifer System. *Water* **2022**, *14*, 2475. [CrossRef]
9. Kabeto, J.; Adeba, D.; Regasa, M.S.; Leta, M.K. Groundwater Potential Assessment Using GIS and Remote Sensing Techniques: Case Study of West Arsi Zone, Ethiopia. *Water* **2022**, *14*, 1838. [CrossRef]
10. Asadi, A.; Adhikari, K. Minimizing Errors in the Prediction of Water Levels Using Kriging Technique in Residuals of the Groundwater Model. *Water* **2022**, *14*, 426. [CrossRef]
11. Di Salvo, C. Improving Results of Existing Groundwater Numerical Models Using Machine Learning Techniques: A Review. *Water* **2022**, *14*, 2307. [CrossRef]

Article

Quantifying Groundwater Infiltrations into Subway Lines and Underground Car Parks Using MODFLOW-USG

Davide Sartirana *, Chiara Zanotti , Marco Rotiroti , Mattia De Amicis, Mariachiara Caschetto, Agnese Redaelli, Letizia Fumagalli and Tullia Bonomi

Department of Earth and Environmental Sciences, University of Milano-Bicocca, Piazza della Scienza 1, 20126 Milan, Italy

* Correspondence: d.sartirana1@campus.unimib.it

Abstract: Urbanization is a worldwide process that recently has culminated in wider use of the subsurface, determining a significant interaction between groundwater and underground infrastructures. This can result in infiltrations, corrosion, and stability issues for the subsurface elements. Numerical models are the most applied tools to manage these situations. Using MODFLOW-USG and combining the use of Wall (HFB) and DRN packages, this study aimed at simulating underground infrastructures (i.e., subway lines and public car parks) and quantifying their infiltrations. This issue has been deeply investigated to evaluate water inrush during tunnel construction, but problems also occur with regard to the operation of tunnels. The methodology has involved developing a steady-state groundwater flow model, calibrated against a maximum groundwater condition, for the western portion of Milan city (Northern Italy, Lombardy Region). Overall findings pointed out that the most impacted areas are sections of subway tunnels already identified as submerged. This spatial coherence with historical information could act both as validation of the model and a step forward, as infiltrations resulting from an interaction with the water table were quantified. The methodology allowed for the improvement of the urban conceptual model and could support the stakeholders in adopting proper measures to manage the interactions between groundwater and the underground infrastructures.

Keywords: urban hydrogeology; rising groundwater levels; shallow aquifer; 3D geodatabase; horizontal flow barrier; Milan; Italy

Citation: Sartirana, D.; Zanotti, C.; Rotiroti, M.; De Amicis, M.; Caschetto, M.; Redaelli, A.; Fumagalli, L.; Bonomi, T. Quantifying Groundwater Infiltrations into Subway Lines and Underground Car Parks Using MODFLOW-USG. *Water* **2022**, *14*, 4130. https://doi.org/10.3390/w14244130

Academic Editor: Craig Allan

Received: 11 November 2022
Accepted: 15 December 2022
Published: 19 December 2022

1. Introduction

Urban hydrogeology is a specific branch of research [1,2] that has been constantly developed in recent years as a consequence of rapid urbanization phenomena that have been witnessed in most parts of the world [3]. Considering that 70% of the world population is expected to live in urban areas by 2050 [4], urbanization can be defined as a world-wide process [5]. Thus, it is reasonable to think that in the next few years a huge effort will be allocated to research into urban hydrogeology [6].

Overexploitation and deterioration of urban water resources act as the main consequences of this rapid urbanization [7]. To put a brake on urban sprawl, a vertical urban development has occurred, determining an augmented use of urban underground [8–12]. However, the construction of ever-deeper structures [13] can impact groundwater (GW) with regards to flow, quality, and thermal issues [5,14,15].

With respect to GW flow, different cities around the world have observed rising water table levels, as a consequence of the deindustrialization process, that have generated some interference between GW and underground infrastructures (UIs) such as basements, car parks, and subway lines [16–23]. Numerical GW flow modeling was widely adopted as the main tool to evaluate the barrier effect of UIs to flow patterns, GW budget [14], and the possible side effects on the underground elements (i.e., corrosion and stability issues).

Concerning engineering issues, GW inflow into tunnels has been predicted in urban areas by adopting analytical solutions [24], synthetic modeling [25], and steady-state numerical modeling on real cases [26,27] to properly design the tunnel drainage system during the construction phase. In fact, water inrush is a challenging issue to face, causing negative impacts on tunnel stability, generating subsidence damage [25,28], heavy financial losses, and losing construction time [26,29].

At the same time, the problem of damages in operating tunnels, as water seepage or lining cracking, requires consideration [30–33]. Despite the lower water amounts penetrating inside the UIs over a long period, GW could determine severe issues, such as temporary unusability, which require waterproofing works and lead to economic losses. Thus, quantifying infiltrations could help to assess proper mitigation strategies [34], supporting the stakeholders in the complex task of urban GW management. To do so, among the different approaches applied in the literature, groundwater infiltrations into subsurface elements have been evaluated by modeling the underground infrastructures by means of the DRN package [34–36]. Recently, a single model layer was developed by Golian et al. [37] to restore groundwater levels after tunneling. In this work, an unsealed and a sealed underground tunnel were modeled using RIV and HFB packages, respectively. The latter has been applied in various fields of groundwater modeling: from coastal areas to model slurry walls containing seawater intrusion [38,39], to geophysical modeling to simulate faults [40,41], to urban contexts in industrial sites [42], or to evaluate the impact of underground infrastructures on groundwater levels [43,44].

The existence of 3D geodatabases, gathering information on underground structures [45,46] and frequently scattered over many institutions and stakeholders [1,47–49], could support the adoption of these packages to properly model UIs. In this way, it should be possible to precisely define their relationship with the water levels, thus improving the urban conceptual model.

Based on these assumptions, the aim of this study has been to quantify GW infiltrations into different categories of UIs (i.e., subway lines and underground car parks), considering different UI conditions (i.e., intact, saturated, and leaky walls). The methodology that has been applied involves developing a local 3D GW numerical flow model for the western area of Milan metropolitan city (Lombardy, Northern Italy). Through this model, the most critical portions of the subsurface network suffering from GW infiltrations have been evaluated. Interactions with the water table and possible infiltrations in subway line M4 (to be inaugurated in 2023) and two public car parks that are currently under construction were also analyzed.

By means of this model, the stakeholders would be able to design management solutions to secure the infrastructures from being flooded in the future. The model has been realized as steady-state with MODFLOW-USG [50] and calibrated using a trial and error approach against a GW maximum condition that was defined in a previous work as documented by Sartirana et al. [51]. HFB and DRN packages have been coupled to model the UIs, reproducing their geometries and volumes through the adoption of grid refinement, contributing to the quantification of GW infiltrations into subsurface elements. In particular, the top and the bottom of the UIs were modeled through the HFB package; to the best of the authors' knowledge, this application of the HFB package could represent an improvement in modeling the UIs. In fact, the relation between GW and the UIs along the vertical sides of a model cell could be thus considered. Moreover, as for Milan city, this is the first time that car parks have been considered in a 3D GW numerical flow model, while being studied for the adoption of GW-level time-series clustering to suggest targeted guidelines for the construction of new underground public car parks [51].

The methodology presented here could be implemented for other urban realities, serving as a way of managing a documented interaction between GW and the UIs that may lead to a planned subsurface infrastructure development with possibly great potential for an integrated management strategy.

2. Urban Conceptual Model of the Study Area

The study area covers 100 km^2 inside the Milan metropolitan area (Figure 1). Human activities have always characterized this area, especially through industrial and agricultural activities that are still conducted in the western and southern areas of Milan [52,53]. The city hosts 1.4 million inhabitants [54] and is currently undergoing an important urban transformation [55]. It is located in the middle of the Po Valley, whose hydrogeologic structure has been deeply examined both in the past [56] and recently [57]. Three main hydro structures were identified: a shallow hydro structure (ISS), an intermediate (ISI), and a deep (ISP) hydro structure. Within the model domain, an ISS has a medium thickness of 40 m with a bottom surface ranging from 100 m above sea level (a.s.l.) (to the north) to about 60 m a.s.l. (to the south). It hosts a shallow aquifer (Figure 2) (i.e., Aquifer Group A1, Regione Lombardia and ENI Divisione AGIP 2002 [56]), where all the underground infrastructures are located. This aquifer is not exploited for drinking needs. Sands and gravels mainly characterize this hydro structure. The same lithologies, but with an increasing presence of silty and clayey horizons, constitute the ISI, that mostly corresponds to Aquifer Groups A2 and B of Regione Lombardia and ENI Divisione AGIP 2002 [56]. An ISP, having a more uncertain lithological composition, was not modeled within this study.

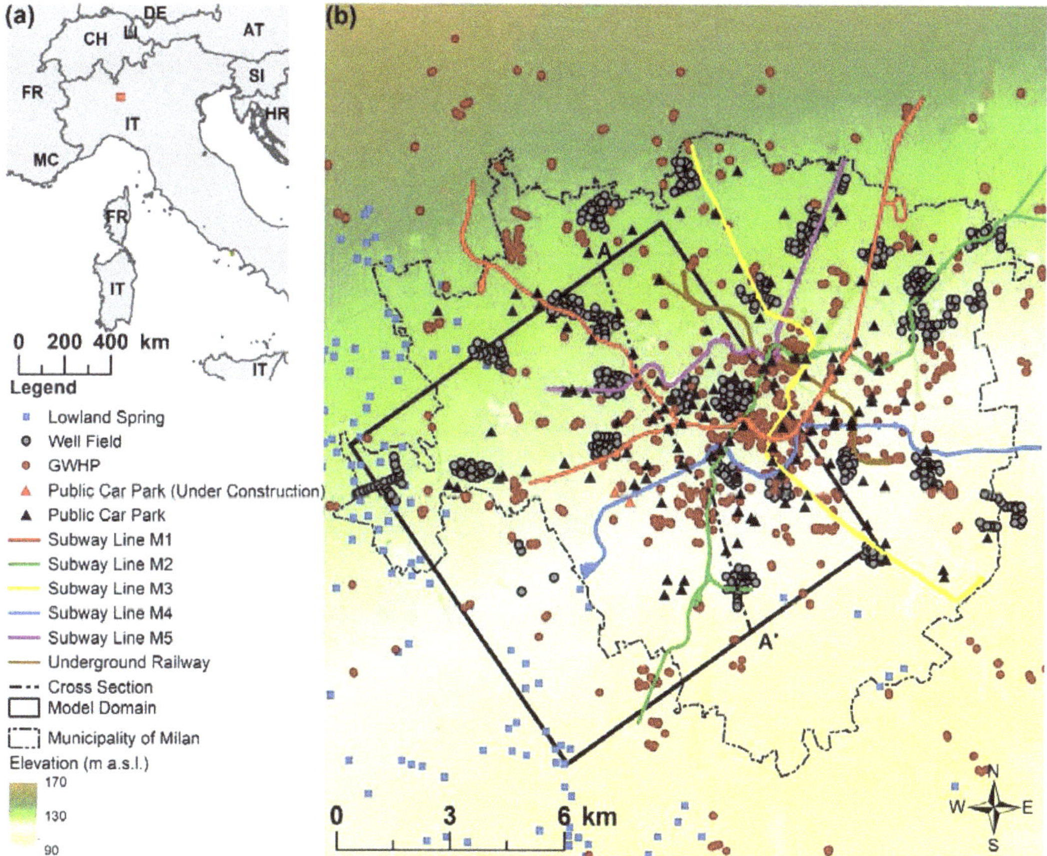

Figure 1. (**a**) Geographical setting of the study area; (**b**) main hydrogeologic features (lowland springs) Color coding for the subway lines respects the color coding used by the subway managing company. Public car parks have been represented as triangles to differentiate them from wells. (Image readapted from Sartirana et al. [51]).

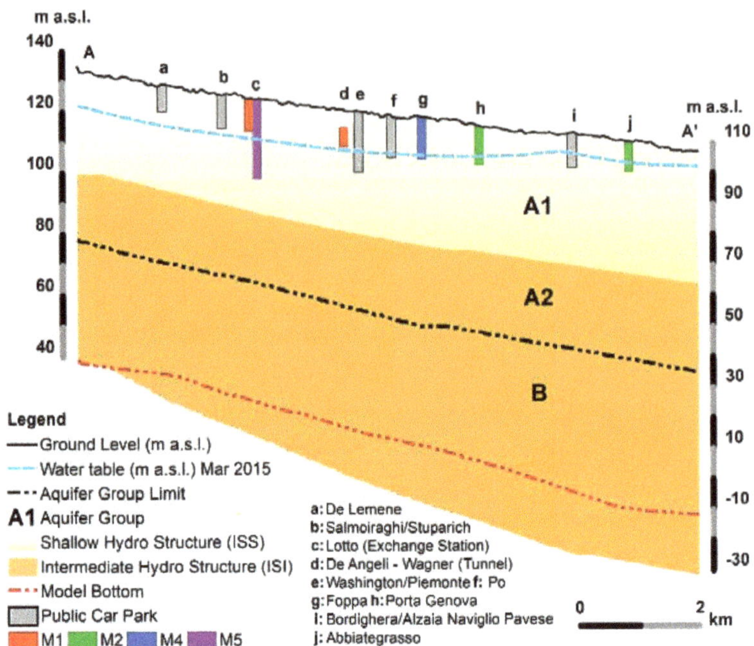

Figure 2. Hydrogeologic schematic cross sections AA' (N–S) of the study area, showing the location of some UIs and their relationship with the groundwater condition of Mar 2015 [51]. For their location on map, please refer to Figure 1.

Industrial needs triggered an extensive groundwater withdrawal since the early 1960s. Consequently, the water table reached its maximum depth of more than 30 m in the northern part of the city around 1975, thus determining the minimum GW levels due to significant water exploitation [58,59]. During the same time frame, some UIs (car parks, subway lines M1 and M2) were built, sometimes without proper lining methods, without consideration for a possible future GW level rise. Subsequently, since the beginning of the 1990s, the decommissioning of many industrial sites, mainly located in the northern sector of the city, generated a rise in GW level, determining flooding episodes for these oldest and shallowest subway lines and for some underground car parks built starting from the middle of the 1980s [60,61]. Consequently, the most recent and deepest subway lines (M3, M4 to be inaugurated in 2023, and M5) have been designed with lining systems. As for underground car parks, 126 public car parks are now listed in the city [51]: 65 out of 126 are located in the model domain. The construction of two new underground car parks (Figure 1b) is currently taking place close to the Gelsomini and Frattini stations of subway line M4. These car parks are named Brasilia (placed just northward of the stations) and Scalabrini (to the south of the stations), respectively; both have been designed to be two floors deep (i.e., 8 m depth as calculated by Sartirana et al, 2020).

The water table rise occurred differently among different areas of the town, with a maximum rise of about 10–15 m in the north, and a more dampened effect in the other sectors [51]. Particularly, a low significant rising trend was evidenced in the west and south, respectively, due to local geological conditions and the hydraulic gradient that constrains the water table close to the ground level, thus reducing the water table oscillations.

In the downtown area, an increasing presence of open-loop groundwater heat pumps (GWHPs) for geothermal needs (Figure 1b), together with the presence of a huge number of UIs, could induce an anthropogenic control on water table rising; due to extraction and injection wells systems, the water withdrawn is usually returned to the shallow aquifer, thus determining a non-consumptive use of the resource [62]. These systems sometimes discharge exploited water to surface water bodies to control the GW rise.

Public-supply well fields withdraw water used for drinking needs, and have screens to tap the semi-confined and confined aquifer units. A total of 261 wells, belonging to 13 well fields, are located inside the considered domain.

The construction of new underground car parks takes place in the framework of the adoption of the Plan of Government for the Territory (PGT) [63], that regulates further subsurface occupation as a measure against excessive soil consumption. In this context, numerical modeling, possibly combined with the application of other techniques aimed at better understanding the urban conceptual model [51], could represent a valid tool to coordinate urban underground development, thus supporting stakeholders in their decision-making process.

3. Materials and Methods

The numerical model was built considering an already-existing urban conceptual model [51], integrating its contents, when possible, with Open Data information [64]. The core of the methodology was the modeling of the UIs (see Underground Infrastructures Modeling) to evaluate GW infiltrations. Different scenarios of conductance were realized to quantify infiltrations simulating different wall conditions; the results have been examined in order to discuss possible strategies to manage GW/UI interactions.

3.1. Numerical Model

A steady-state numerical flow model was developed using MODFLOW-USG [50], and Groundwater Vistas 8 [65] was used as the graphical user interface.

3.1.1. Grid Discretization

The model grid (Figure 3) was composed of 1,668,348 cells and was horizontally structured by applying a quadtree refinement: cell dimension ranges were from 100 m in the peripheral areas, up to 12.5 m around subway lines and public car parks (i.e., fourth level of refinement); in proximity to public car parks currently under construction, a fifth quadtree level of refinement was applied (i.e., 6.25 m) (Figure 3a). The grid was rotated by 35° from the offset (X = 1,509,407, Y = 5,026,235, Monte Mario Italy 1; ESPG: 3003) to be perpendicular to the general NW–SE groundwater flow direction of the domain [58,66].

The vertical discretization (Figure 3c) consisted of 18 layers. The first 8 layers, with an average thickness of three meters, included all the UIs lying in the shallow aquifer (layers 1–10, ISS/Aquifer Group A1); layers 11 to 14 had a medium thickness of seven meters to model the first portion of the ISI (Aquifer Group A2). Layers 15 to 17 (with a medium thickness of 6 m) were adopted to represent the aquitard (AQ), while the last layer, with a medium thickness of 20 m, aimed at modeling the final portion of the ISS (Aquifer Group B).

3.1.2. Boundary Conditions

Boundary conditions (Figure 4), used to outline the hydrogeologic system, were represented through Neumann and Cauchy conditions:

- General Head Boundary (GHB) was used to model the initial heads along the borders around the study area, at their real distance from the analyzed domain. As for their hydraulic head values, the initial information was taken from a piezometric map of March 2015 (Mar15) for the study area [51]. In addition, the main quarries located inside the domain (Figure 4) have been represented as GHBs.
- WELL (WEL) was used to model the 261 public wells and 785 groundwater heat pumps (GWHPs) described in Section 2. Information on well discharge was readapted from De Caro et al. [61] with regard to public wells, and from Regione Lombardia [64] for GWHPs. Finally, a further 384 private wells fell within the analyzed domain; as

their well discharge was mostly unknown, a discharge value of -432 m3/d was initially attributed to these wells.

- Recharge (RCH): 5 zones, based on land use, were identified from the geographic database Dusaf 6.0 [67]; their values were calculated as the contribution of precipitations, irrigations, and runoff. The initial values for each zone were calculated starting from the precipitation data of Paderno Dugnano rain gauge (located just northward of the city of Milan), monitored by the regional environmental protection agency [68]. Precipitations amounted to 1496.2 mm/yr for the twelve months before Mar15, the period chosen for model calibration. Absence of infiltration was considered for urban areas and for surface water elements (i.e., quarries), while 20% of infiltration was attributed to the other recharge areas; moreover, an additional contribution from recharge infiltration was attributed to irrigational areas.

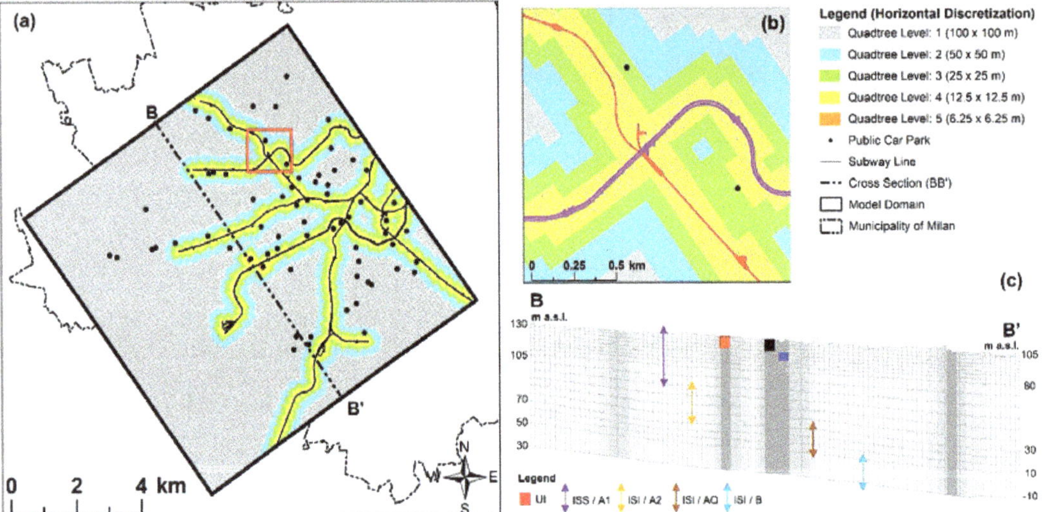

Figure 3. (**a**) Grid horizontal discretization; the red rectangle points to the sector area represented in (**b**); (**b**) example of quadtree refinement close to Lotto exchange station (see Figure 2); (**c**) grid vertical discretization. Please note that for (**b**), the same color coding of Figure 1b has been used for subway lines.

Underground Infrastructures Modeling

The underground railway (Figure 1b) occupies only a small portion of the northwestern area; thus, it was not considered within the study. All the UIs (Figure 4) were conceptualized and modeled by coupling the Wall (HFB) [69] and the DRN [70] packages. The capabilities of both packages were combined to properly simulate and evaluate the exchange between the UIs and the surrounding aquifer. HFB offers the ability to isolate individual components to consider how water is passed between an engineered element such as a subway line and the aquifer. On the other hand, the DRN package enables the modeler to assign a head inside the engineered structures. In this case, the DRN package was adopted to simulate a fictitious water-collection system within the UIs. Combining the capabilities of these two packages is the core of the proposed methodology. To support and validate the adopted methodology, the model domain was discretized into 187 zones: the aquifer of interest, and all the UIs' elements. Thus, water exchanged between neighboring zones, based on the MODFLOW solution [71,72], was quantified.

To properly model the depth of the UIs, information regarding the UIs' bottom and the diameter of the subway tunnels has been obtained from an already-existing 3D GDB of the subsurface elements for the study area [46]. Subsequently, the following rule was adopted as the main constraint to model the UIs: if an UI occupied a layer of more than 50% of its thickness, the UI was then represented inside that layer; otherwise, if this constraint was not respected, the UI was then modeled in the overlying layer.

The wall usually goes along any of the four horizontal sides of each cell, but in MODFLOW-2005 there is no option to specify a vertical barrier. Notwithstanding, the adoption of MODFLOW-USG allowed for wider flexibility in using the HFB package, as the barrier could be aligned along any face of the unstructured grid [50]; thus, HFB cells could also be placed at the intersection between two nodes sharing the same X and Y coordinates, in contiguous layers. This enabled the reproduction not only of the lateral sides, but also the top and the bottom of all the subsurface elements. To do so, the initial information about the lateral sides of the UIs was integrated by "manually" compiling the HFB package, adding the position of the top/bottom of the UIs.

The drainage network was placed inside the UI and positioned at the bottom layer of each section of the UI. The drain head (i.e., drain elevation) was fixed as equal to the bottom elevations of the UI. In this way, the possible groundwater inflow into the UIs could be drained, quantifying the amount of water to be withdrawn to dry the infrastructure. The conceptual model of the adopted approach to simulate the underground infrastructures network is represented in Figure 5.

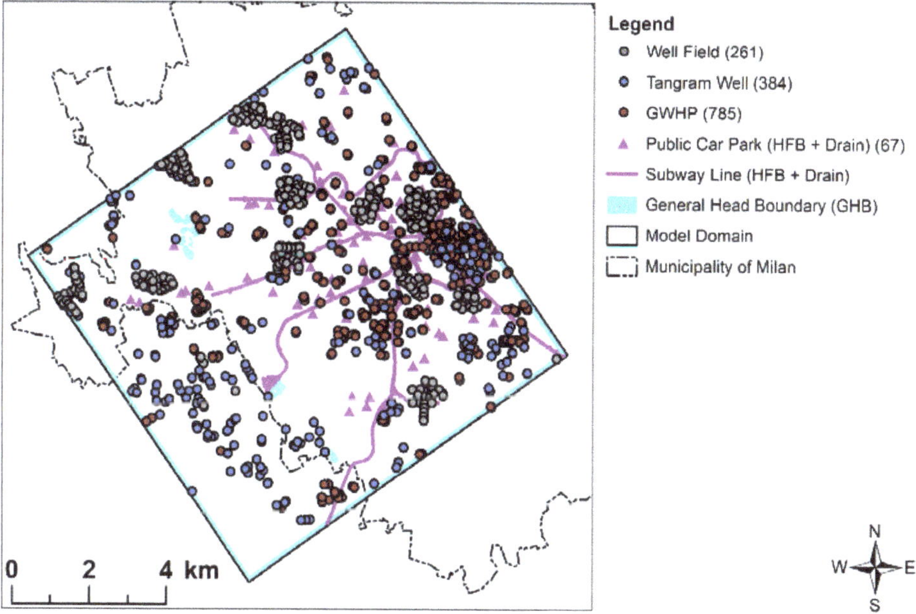

Figure 4. Model boundary conditions. GHBs' distance from the model area has been indicated. Please note that color coding of the infrastructural elements (subway lines and underground public car parks) refers to the HFB package color in Groundwater Vistas 8. Public car parks have been represented as triangles to differentiate them from wells.

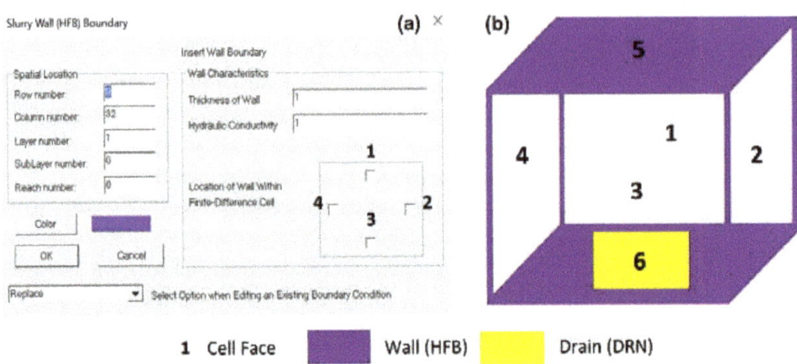

Figure 5. (**a**) Traditional application scheme for HFBs cell; insertion mask taken from Groundwater Vistas 8; (**b**) conceptual model of the adopted approach to model all the UIs.

At the exchange stations, lines were positioned at their real depth, thus properly separating the deepest and more recent lines (i.e., M3, M4, and M5) from the shallowest and older ones (M1 and M2).

Conductance, which can be defined as the ratio between hydraulic conductivity (K) and the wall thickness (d), is the single parameter that controls the ability of the wall to transmit water. The absence/presence of lining systems was represented through different conductance values. With regard to the wall thickness, a value of 1 m was considered representative of all the modeled UIs.

The drainage system was assumed to provide no resistance to GW flow, imposing a value of conductance higher than the wall conductance and the aquifer conductivity [25,35].

3.1.3. Further Modeling Aspects

The hydraulic conductivity parametrization was readapted from a previous project on the study area developed within the same research group [73], where the lithological information, stored within the Tangram database [74] in the form of stratigraphic data and pumping tests, was numerically coded and interpolated into GOCAD software using the kriging method [75]. Initial values to the continuous distribution of hydraulic conductivity were assigned from Tangram reference tables. A refined investigation was conducted which analyzed 3 cross-sections built along public-supply well fields from Airoldi and Casati [76], to infer the spatial distribution of fine materials (i.e., clay lenses).

With regard to calibration, sensitivity analysis using different multiplying factors (from 0.5 to 1.5) and a "trial and error" method were adopted to calibrate the steady-state model, focusing on GHB values and conductance, aquifer recharge (3 out of 5 zones), hydraulic conductivity, and well discharge. A total of 30 head targets, representing field water table measurements, were considered, showing an uneven distribution over the entire domain, with a limited amount of information for the western sector. The calibration process was conducted against the maximum groundwater condition of Mar15, the highest in the last 30 years [51], evaluating the goodness of the obtained results and analyzing the model statistics (i.e., residual sum of squares, scaled RMSE). In this way, the most critical situation for the UIs should be considered; this is also recommended for UIs currently under construction.

3.2. Decision Management Support

Different scenarios were analyzed to quantify GW infiltrations into UIs. Further engineering aspects, such as possible subsidence issues due to the drainage effect, or potential negative effects determined by buoyancy as a result of the aquifer pressure (i.e., uplift risks), were not considered within the aims of the project.

The conductance value for waterproofed subway lines (Table 1) was defined from the literature references [43,77,78]. Different conductance values (S1–S3) were tested for subway lines M1 and M2 due to a higher uncertainty; considering the absence of lining systems, the conductance was modified simulating possible deteriorations due to a prolonged interaction with the water table over time. In fact, infiltrations may be regarded as a gradual process, ranging from an unsaturated to a saturated flow induced by groundwater flow [79]. Since for public car parks' conductance no information was available, it was decided to attribute the lowest conductance value to all car parks.

Table 1. Conductance value for all the considered scenarios.

UI	Waterproofed	Initial Conductance (m^2/d) (S1-S2-S3)	Fractures Conductance (m^2/d) (S4-S5-S6)
M1	No	$1.16 \times 10^{-11}/10^{-10}/10^{-9}$	$1.16 \times 10^{-7}/10^{-6}/10^{-5}$
M2	No	$1.16 \times 10^{-11}/10^{-10}/10^{-9}$	$1.16 \times 10^{-7}/10^{-6}/10^{-5}$
M3	Yes	1.16×10^{-13}	1.16×10^{-13}
M4	Yes	1.16×10^{-13}	1.16×10^{-13}
M5	Yes	1.16×10^{-13}	1.16×10^{-13}
Car Parks	—	1.16×10^{-13}	1.16×10^{-9}

The most impacted locations of S1–S3 were then analyzed, locally increasing the conductance value of the HFB cells to simulate possible wall fractures. A focus was provided only for subway lines M1, M2, and car parks, as historically they have shown the most revealed interference. To reproduce fractures, wall conductance was only modified close to the infiltration area, increasing the initial value of four order magnitudes, as considered in studies on fractured rocks [80]. The change in the conductance was applied to the minimum model dimension (i.e., one cell). In this way, it was possible to compare the amounts of infiltration of intact and leaky walls.

The identified infiltrations were then analyzed to discuss some management proposals with regard to the design of dewatering systems in the most critical locations of the subsurface network, and also proposing the implementation of monitoring systems to manage possible infiltration issues in advance.

4. Results

4.1. Model Calibration and Statistics

The final values of GHBs were 127 m a.s.l. for the northern GHB and 102.2 m a.s.l. in the south, while the western and the eastern boundaries varied from 126 to 103 m a.s.l. and from 124 to 103 m a.s.l. from north to south, respectively. Calibrated values of hydraulic conductivity ranged from 235 to 1.15×10^{-3} m/d, as visible in Figure 6.

Final recharge values, and their spatial distribution, are represented in Figure 7. Finally, well discharge was reduced by 25% for the GWHPs and private wells, while for well fields, the reduction, when applied, ranged from 25% up to 50% (for the southernmost well field) of the initial value.

With regard to the calibration, the calibrated model generally provided good statistics (Table 2, Figures 8 and 9) for most of the 30 head targets considered. The most critical targets were located in the western and southernmost portions of the domain, quite far from the subsurface network that was the main focus of the study. Although these values could represent some modeling issues for some local areas of the domain, the scaled RMSE (4.6%) respects the international criteria that indicate the goodness of a solution in a scaled RMSE to be less than 8% [81,82].

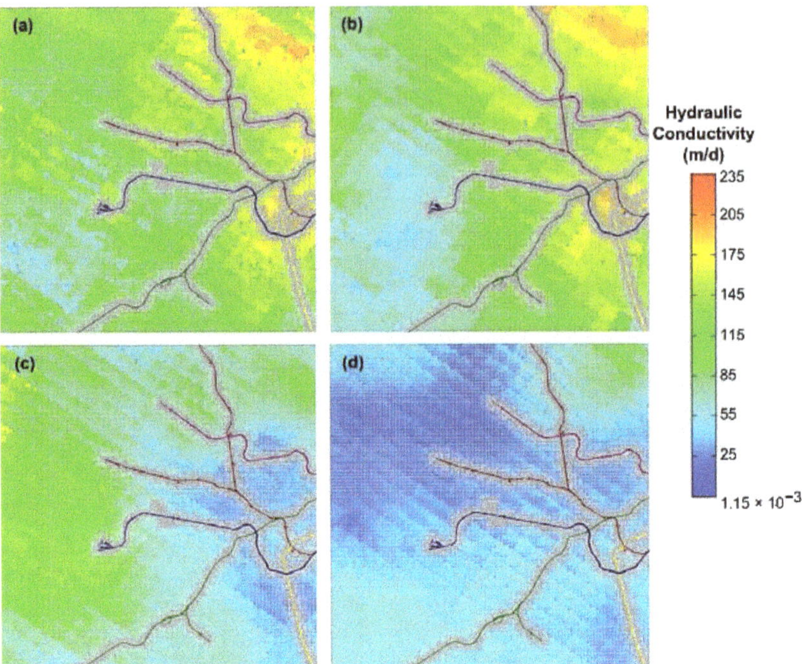

Figure 6. Hydraulic conductivity values for layers (**a**) 1 (**b**) 4 (**c**) 11 (**d**) 14. Please note that subway line tracks are plotted inside all layers to provide refence points, since the grid is not rotated in these images.

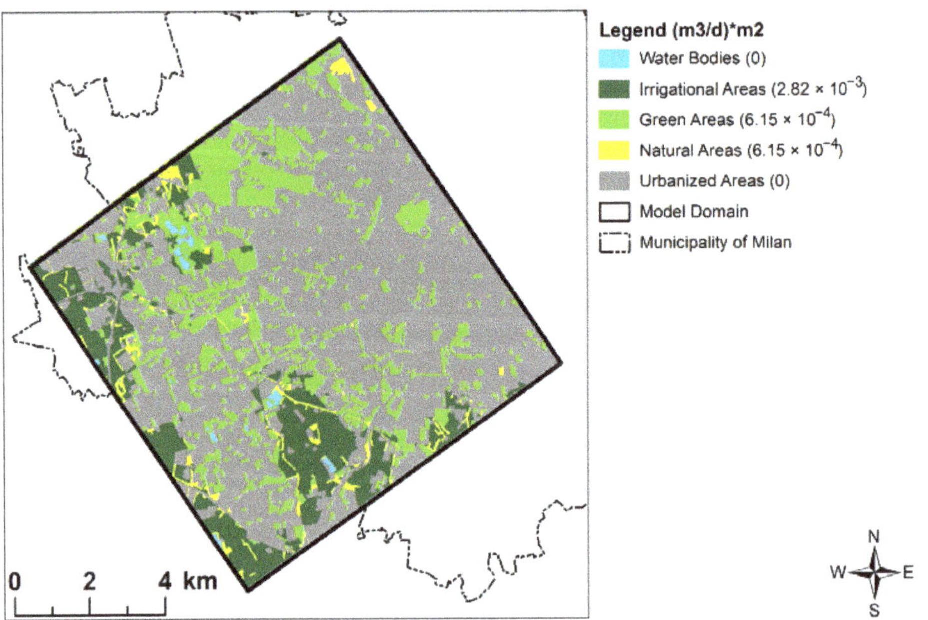

Figure 7. Areal distribution of the 5 recharge zones; final recharge values are provided in legend.

Table 2. Model statistics for the considered head targets. Statistics refer to S1.

Statistical Parameter	Target Value
Absolute Residual Mean	0.32
Residual Sum of Squares (RSS)	21.8
RMSE	0.85
Minimum Residual	−1.15
Maximum Residual	2.16
Range of Observations	18.39
Scaled RMSE (nRMSE)	0.046

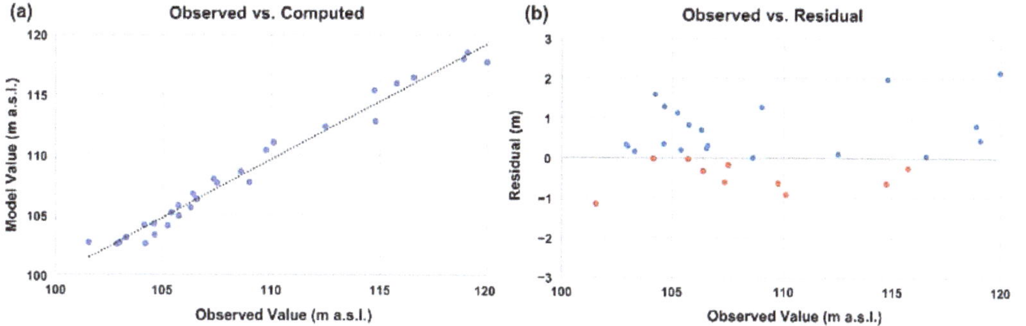

Figure 8. Comparison of (**a**) observed (m a.s.l.) vs. computed (m a.s.l.) values and (**b**) observed values (m a.s.l.) vs. residuals (m).

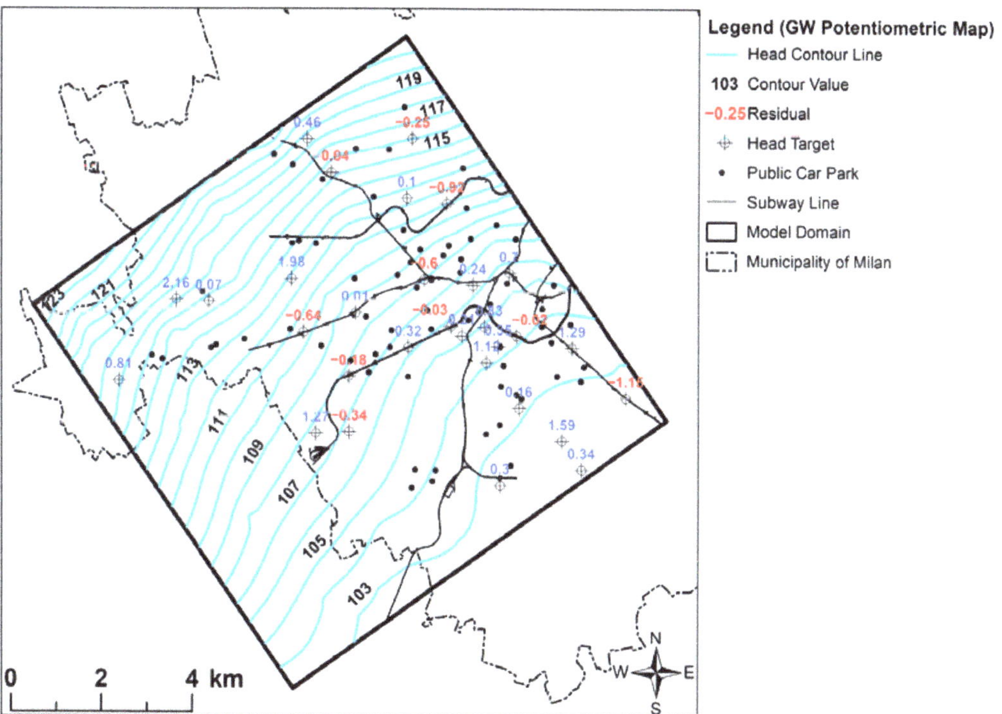

Figure 9. GW potentiometric map of the study area.

The potentiometric map for the shallow aquifer is represented in Figure 9. From the visualization of the head targets, it is visible that the water table map is generally well represented close to the subsurface network, thus allowing for proper assessment of GW/UI interactions and the consequent infiltrations. In the eastern part of the models, located close to Milan's downtown area, the contour lines' behavior is influenced by the pumping effect of both public well fields and GWHPs (Figure 4).

Model mass balance (Table 3) evidences the importance of well discharge inside the domain, both for the outflows and the inflows; the latter are exclusively due to the injection wells of GWHP systems. The water amounts withdrawn by the drains indicate the GW infiltration into the UIs; despite being a limited amount of water, quantification of the water amounts is important to compare them with the results of the other scenarios in the framework of urban underground management. Model percentage discrepancy is considered to be low (4.59×10^{-3}). A good coherence was detected between the drain outflows and the mass balance with neighboring zones (i.e., surrounding aquifer and the UIs), thus validating the obtained results.

Table 3. Model mass balance.

Mass Balance	Inflow (m^3/d)	Outflow (m^3/d)	% Error
GHB	419,633.07	72,039.06	
Wells	115,743.33	515,438.32	
Drain	—	2.93×10^{-5}	
Recharge	52,101.25	—	
Total	587,477.65	587,477.38	4.59×10^{-5}

4.2. Modeling Scenarios

Groundwater inflow for all the UIs was calculated, and results are summarized in Figure 10. As can be seen in Figure 10, an absence of inflow was detected for some subway line branches, as the water table level was lower than the bottom of the UIs [29]. Particularly, these inflow gaps were visible in the north, along subway line M1, and in the central portion of the domain, close to Cadorna exchange station (subway lines M1 and M2). The tunnel sections more exposed to GW inflows are the westmost stretch of M1, towards Bisceglie Station (M1-a), and the stretches close to Uruguay station (M1-b) and between QT8 and Lotto (M1-c) for line M1; and the sections from Porta Genova to Sant'Agostino station (M2-a) and from Lanza to Moscova (M2-b) for subway line M2. Due to their major depth, subway lines M3 and M4 were completely submerged by the water table, which also occurred for subway line M5 (Table 4). With regard to public car parks, 34 out of 67 resulted in infiltration; in the central area, Washington/Piemonte (P-a) and Carducci (P-b) turned out to be among the most impacted infrastructures, while, for example, Betulle Est was impacted in the west (P-c). Critical sections for M1 and M2 were already identified as areas where a historical interaction (i.e., submersion) with the water table was evidenced [46,83]. In particular, Sant'Agostino (M2-a) was impacted for both GW minimum and maximum conditions.

As summarized in Table 4, groundwater inflows for S1-S3 are limited, with low orders of magnitude. The highest values of inflows ($10^{-5}/10^{-3}$ order of magnitude) were detected for the oldest subway lines M1 and M2, modeled with higher conductance values to simulate the absence of waterproofing systems and a progressive saturation of the walls over time. As for the deepest lines, such small values are attributable to the low conductance representing lining systems. The spatial distribution of these infiltrations is different, as for shallow lines the infiltrations are detected only at certain spots, as visible in Figure 10, thus evidencing local but more critical situations to manage.

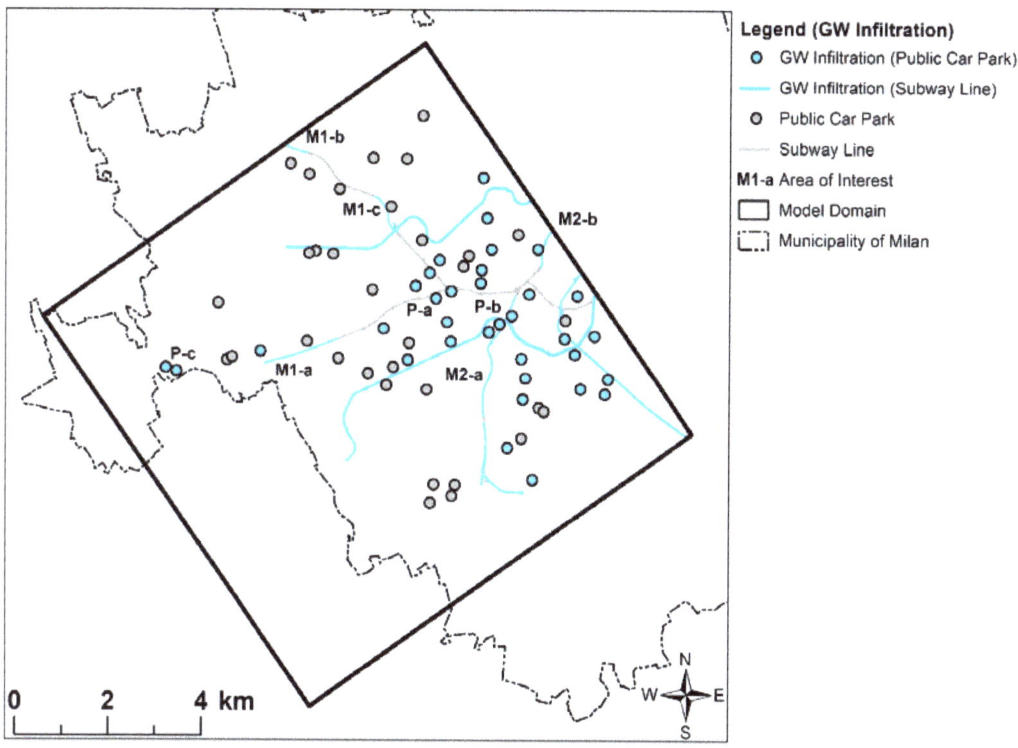

Figure 10. Areas showing GW infiltrations into UIs.

Table 4. GW inflows into UIs (m^3/d) for S1–S3. Please remember that for M3, M4, M5, and parks, K was always set equal to 1.16×10^{-3} m/d. Percentage below the water table is intended as the sections of UIs where the bottom of the infrastructure is lower than the hydraulic head.

UI Category	Amount of Infiltration (m^3/d)			% Below the Water Table
	S1 (K = 1.16×10^{-11} m/d)	S2 (K = 1.16×10^{-10} m/d)	S3 (K = 1.16×10^{-9} m/d)	
M1	3.70×10^{-6}	5.83×10^{-5}	4.23×10^{-4}	8.37
M2	2.00×10^{-5}	2.34×10^{-4}	2.27×10^{-3}	71.38
		S1–S3 (K = 1.16×10^{-13} m/d)		
M3		6.24×10^{-7}		100
M4		1.94×10^{-6}		100
M5		2.70×10^{-6}		100
Car Parks		3.00×10^{-7}		50.75

At the most critical points highlighted in S1–S3 (Figure 10) for subway lines M1 and M2, and for some public car parks, locally punctual wall fractures have been simulated to quantify the variation in GW infiltrations. The results of these spots are summarized in Table 5. As is visible, the most critical effects, also considering the features of the UIs (i.e., depth, volume), have been identified for M2-a, around Sant'Agostino station. The infiltration for these points generally increased linearly to one or two orders of magnitude.

As for the two public car parks under construction (Figure 1b), an absence of infiltration was detected in both cases, with respect to the considered groundwater maximum condition, due to a lack of interaction with the water table (Figure 11) which was contrastingly evidenced for the close branches of subway line M4.

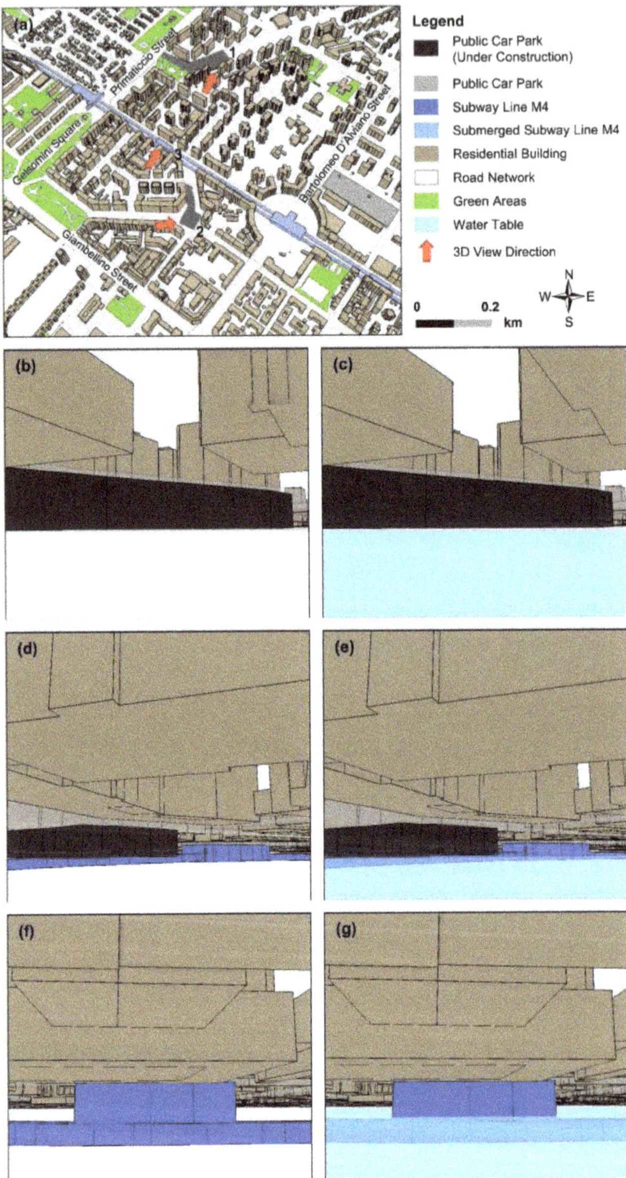

Figure 11. (**a**) 3D geographical setting of the area close to the car parks currently under construction. The car parks and the subway line M4 are visible below the road network; the names of some roads are indicated to provide more geographic details. Three-dimensional underground reconstruction of (**b**) Brasilia car park, (**d**) Scalabrini car park, (**f**) Lorenteggio 124 intervention point. GW/UIs interaction for (**c**) Brasilia car park, (**e**) Scalabrini car park, (**g**) Lorenteggio 124 intervention point. (**b**,**c**) refer to point 1 in (**a**); (**d**,**e**) refer to point 2 in (**a**); (**f**,**g**) refer to point 3 in (**a**). Transparency has been adopted to represent the volumes submerged by the water table; as visible in (**c**,**e**,**g**) this occurs only for subway line M4, and not for public car parks. The red arrows indicate the viewpoints and the view directions adopted in the 3D visualization of the subsurface elements. Images were realized using ArcGIS Pro.

Table 5. Comparison of GW inflows into UIs (m^3/d) for the initial scenario (S1–S3, intact walls) and their corresponding final scenario (S4–S6, leaky walls). Please remember that, for car parks, K was always set equal to 1.16×10^{-13} m/d for S1–S3 and to 1.16×10^{-9} m/d for wall fractures in S4–S6. S means station, T means tunnel, P means park. Depth (m) has been provided for subway stations and parks, as they are designed from the ground field; as for tunnels, since they are not designed from the ground field, thickness was provided rather than depth.

Type	Name	Thickness/ Depth (m)	Volume × 10 (m^3)	Amount of Infiltration (m^3/d) (S1–S4)	Amount of Infiltration (m^3/d) (S2–S5)	Amount of Infiltration (m^3/d) (S3–S6)
S	Bisceglie (M1-a)	11.93	33.49	1.13×10^{-7}/ 4.35×10^{-6}	1.12×10^{-6}/ 4.35×10^{-5}	2.04×10^{-4}/ 4.35×10^{-4}
T	Bisceglie—Inganni (M1-a)	6.5	42.88	2.04×10^{-6}/ 2.71×10^{-5}	2.04×10^{-5}/ 2.71×10^{-4}	4.30×10^{-5}/ 2.71×10^{-3}
S	Inganni (M1-a)	10.92	26.77	3.98×10^{-7}/ 1.14×10^{-5}	3.98×10^{-6}/ 1.15×10^{-4}	1.20×10^{-4}/ 1.15×10^{-3}
T	Bonola—Uruguay (M1-b)	6.5	42.81	1.75×10^{-7}/ 4.05×10^{-6}	2.43×10^{-6}/ 4.50×10^{-5}	1.81×10^{-5}/ 3.94×10^{-4}
T	QT8—Lotto (M1-c)	6.5	71.89	9.92×10^{-7}/ 1.06×10^{-6}	9.34×10^{-6}/ 1.77×10^{-5}	9.34×10^{-5}/ 1.17×10^{-4}
T	Romolo—Porta Genova (M2-a)	7	55.77	5.67×10^{-6}/ 2.37×10^{-5}	5.67×10^{-5}/ 2.37×10^{-4}	5.33×10^{-4}/ 2.71×10^{-3}
T	Porta Genova—Sant'Agostino (M2-a)	7	37.05	5.34×10^{-6}/ 5.86×10^{-5}	5.34×10^{-5}/ 5.86×10^{-4}	8.04×10^{-5}/ 5.86×10^{-3}
S	Sant'Agostino (M2-a)	17.35	23.77	8.24×10^{-7}/ 5.29×10^{-5}	8.24×10^{-6}/ 5.29×10^{-4}	2.64×10^{-4}/ 5.29×10^{-3}
T	Lanza—Moscova (M2-b)	7	36.41	1.03×10^{-6}/ 6.03×10^{-6}	1.03×10^{-5}/ 6.03×10^{-5}	2.84×10^{-5}/ 6.02×10^{-4}
P	Washington/ Piemonte	20	60.38	$1.72 \times 10^{-8}/1.49 \times 10^{-6}$		
P	Carducci Olona	17	58.14	$1.32 \times 10^{-8}/4.82 \times 10^{-7}$		
P	Betulle Est	5	23.02	$2.56 \times 10^{-9}/3.04 \times 10^{-7}$		

5. Discussion

Managing GW/UIs interaction in urban areas is a challenging issue. Different problems can arise regarding GW quality, quantity, and thermal issues [5,6,15], but stability, erosion, and infiltration for UIs are some further topics to consider. With regard to GW infiltration into UIs, the scientific literature deals both with water inrush calculation during the construction of tunnels [26,27] and problems regarding already-operating underground tunnels [31,33]; in this study, a local scale numerical model was developed for the western sector of Milan city, applying a methodology to quantify GW infiltrations into completed and operative UIs.

5.1. Modeling Scenarios

Model results in terms of calibration were generally acceptable (Table 2, Figures 8 and 9). However, some head targets did not show an optimal result. This happened for a couple of targets in the north-western portion of the domain, and for one target in the south. In the north-west, not far from the critical targets, the behavior of the water table is presumably influenced by a multitude of local situations. The proximity of a group of quarries and a public-supply well field with high discharge (Figure 4), and the presence of clay lenses determining the existence of perched aquifers with seasonal oscillations [51,52] make predictions more uncertain. This response could highlight the presence of local mechanisms, possibly uninformed by the targets, that have been neglected. In this sense, the model provides a guide for future data collection, that could allow the improvement of the appropriateness of the conceptual model [84]. In addition, acquiring further data could contribute to making more effective predictions, thus improving the model use in supporting management decisions [85].

Identifying the areas more exposed to infiltrations is important to predicting future risks due to a more severe water inrush; thus, adopting strategies to ensure these infrastructures are preserved is vital [86,87]. Some of the impacted areas (i.e., M1-a, M1-b, M1-c, M2-a) have already been identified as critical in previous works [46,83]. In this case, only a qualitative GW–UIs interaction was detected through a GIS methodology. This spatial coherence among the results could be considered as a validation of the numerical model. At the same time, model findings could represent a step forward in the definition of the urban conceptual model; through this approach, GW infiltrations resulting from GW/UI interactions could be estimated. As for M2-a, P-a, and P-b, the highest depth of downtown infrastructures (Table 5) plays a key role in influencing GW/UI interactions. This is due both to a high population density, thus requiring more space for subsurface infrastructures [88], and to the adoption of specific construction methods; as an example, Sant'Agostino station was built with two overlapping pipes [61]. As for the western sector, the complex geological situation explained above could be a possible driver of the infiltrations both for subway lines (M1-a) and underground car parks (P-c), despite their limited depth (Table 5). To counteract this situation, in the framework of creating a more sustainable and resilient city [5], some residential constructions have been designed with superficial car parks occupying the first floors of the buildings. With regard to public car parks, the new buildings currently under construction have been designed as two floors deep; at this time, this results in an absence of impact even considering a groundwater maximum condition (Figure 11). However, prolonged monitoring should be useful to cope with the evolution of GW/UI interactions. Finally, in the north, a reduced GW/UIs interaction is attributable to a wide unsaturated thickness of the shallow aquifer [51], with the water table located around 10–12 m from the ground field.

5.2. Considerations of the Adopted Modeling Approach

The applied modeling strategy aimed to quantitatively evaluate the interaction between the GW system and the subsurface structures. With regard to the calibration process, it is not tied to the prediction of interest; in fact, it is based on head targets whose hydraulic measures are not directly connected to the final goal (i.e., GW infiltrations into UIs). The information content on which the head targets are based is not informative about the degree of connection between the UIs and the water table. In technical terms, the K of the walls is completely in the null space and outside the solution space of the model. This does not mean that the calibration is useless, but it does mean that the model could not be so much a predictive tool as a way to understand a phenomenon (inflow across leaky walls) in general terms. The geometry of the UIs is realistic [46], but one can only make hypotheses about the permeability of the intact and leaky walls that are not in any way informed by the calibration. To limit this uncertainty, a literature analysis was conducted to choose the initial conductance values for subsurface impervious structures [78] and the conductance to simulate isolated fractures [80]. Moreover, an ensemble of scenarios [89] was defined to deal with non-lined systems, testing different conductance values. In this way, stakeholders are enabled to visualize a range of impacts and they could consider them to apply different management options [90,91]. As for S1–S3 (Table 4), GW infiltrations are very limited, especially for waterproofed subway lines; thus, the model allowed for the gaining of insights into the conductance values that are needed to simulate an almost impermeable element.

Anyway, obtaining good calibration results was crucial, since they allow GW/UI interactions to be well represented and, consequently, they allow the obtaining of a more reliable estimate of the infiltrations originated by the relationship between the aquifer and the subsurface infrastructures. As visible in Figures 8 and 9, this is mostly true for this specific case, especially for the targets located in the central part of the domain that lie in proximity of the main UIs' elements.

Using MODFLOW-USG as numerical code allowed the refining of the grid horizontally, therefore properly representing the UIs. Moreover, through the implementation of the unstructured grid, the key numerical computations could be limited within the required bounds [50], making the simulations less computationally intensive. Above all, the adoption of MODFLOW-USG was pivotal to model the UIs, as it allowed the WALL package to be used to represent not only the cells' lateral sides through HFB, but also the top and the bottom of the UIs. In this way, the subsurface elements could be modeled with their real depth and volumes, thus refining previous applications of the HFB package to simulate UI fully penetrating single-layered models [37,43,44]; hence, a precise estimate of further modeling aspects (i.e., evaluation of the barrier effect on groundwater flow paths) should also be guaranteed. In MODFLOW-USG, to reduce numerical instability, desaturated cells (i.e., dry cells) are not inactivated, so there could be a small amount of flow from one cell to another. The adoption of the DRN package helped to solve this possible issue, especially in the upper portion of the domain where unsaturated aquifer was present. As a drain is activated only when the hydraulic head is at least equal to the drain elevation, it was possible to unravel where an effective infiltration was present. The choice of the DRN package also came after its previous applications to quantifying flooding episodes during the construction of tunnels [34,36]. Through the developed methodology, modeling GW/UI interactions could be enhanced. In fact, combining the use of HFB, DRN, and mass balance zones to quantify infiltrations depending on different conductance values is possible, instead of deactivating cells of impervious structures. Thus, a step forward could be taken in the development of the urban conceptual model, supporting previous approaches conducted within the same domain [92,93], or in other areas [94] where different aspects of GW/UI interactions have been investigated but GW infiltrations into subsurface elements were not quantified.

The methodology has been tested on a steady-state numerical model. Future applications on transient numerical models would be possible depending on long-term data collections [95]; this could raise awareness about infiltration issues, supporting a deeper interpretation of GW/UI interactions and making the model a useful management tool to make long-term predictions [84].

5.3. Decision Management

The infiltration issue of UIs in Milan city is historical. Different episodes have been documented over time [46,60,83], leading both to economic and management problems for Metropolitana Milanese Spa, the subway managing company. For example, the section between Piola and Lambrate stations, along subway line M2 (outside the numerical model domain), was closed during summer 2019 to complete lining works because of GW infiltrations, thus forcing the use of surface public transport. Although the water inflow is small with respect to water inrush into subway tunnels during their construction [28,80,96], this situation could trigger further issues over a long time period (i.e., corrosion of foundations), resulting in a decline of the subway system efficiency; thus, this problem should not be underestimated.

To ensure sustainable development of GW/UI interactions, effective engagement of the stakeholders should be of great value [97–99]. Open communication is needed to raise awareness about the importance of data to describe the system and conceptualize and develop a model [91,100] with increased predictive capabilities. For this specific case, monitoring, estimation, and control are essential aspects for tunnel management [96]. Having access to existing infiltration measures, if available, or implementing monitoring of the punctual inflows along the tunnels or for car parks would also improve the calibration process; in this way, model uncertainty would be reduced, thus strengthening the usefulness of hydrogeologic models for decision-making bodies [84,85]. The collection of field data could focus on the most critical sectors highlighted (i.e., M1-a, M2-a) by the model results. Amongst these areas, dewatering solutions could be adopted to manage the issue, thus contributing to preserving the status of the subway network, avoiding the development of more serious issues as occurred for the surrounding areas of Piola and Lambrate stations. In particular, the historical issues of Sant'Agostino station, also due to the adoption of specific construction methods [61], impose an increased degree of attention for this limited branch of subway line M2.

However, applying these solutions would be a consequence of effective GW infiltrations into UIs. A move away from reacting and correcting measures, focusing on preventive actions [6] to secure the UIs, should be evaluated. In a previous work by Sartirana et al. [51], underground car parks were classified as possibly critical for different GW conditions if the difference between the reference plan (i.e., bottom) of the UI and the water table was less than one meter. To avoid infiltration issues, activating localized pumping when a certain threshold is locally exceeded would be a possible measure [101]. To do so, early warning monitoring solutions, such as integrating GIS, BIM, and GPS techniques [102,103], with continuous online data measurements should be implemented in proximity of the most critical UIs.

Moreover, groundwater is not only an annoyance for its side effects, but it is also a heritage [6] in urban frameworks; therefore, further management strategies could be proposed. For example, as GW is a valuable energy reservoir [15,93], increasing the adoption of GWHP systems, possibly only due to extraction wells, could keep the water table levels controlled close to the UIs, thus not only limiting the infiltration issues but also exploiting the thermal potential of these subsurface elements [104].

Finally, in the framework of the goals of the Plan of Government for the Territory, this local-scale urban model could help the decision makers to understand and manage the relationship between new UIs and water table levels, testing possible urban underground development scenarios.

6. Conclusions

This work aimed to adopt a methodology to quantify GW infiltrations into UIs (subway lines and public car parks) with the view of assisting urban underground management. In this sense, the realization of a local-scale, urban numerical model allowed the following:

- Verification of the usefulness of the applied methodology to model the UIs, quantifying GW infiltrations through the combination of HFB and DRN packages. In particular, the adoption of MODFLOW-USG allowed the use of the HFB package to model the top and the bottom of the UIs, thus considering the interaction with the water table along the vertical direction as well. The existence of a 3D GDB of the UIs for the city of Milan helped to accurately model the UIs' depth.
- Identification of the UI sectors more exposed to GW infiltrations under different conductance scenarios (from intact to leaky walls), providing a qualitative and quantitative overview intended for both the municipality decision makers and the subway managing company. The westmost stretch of subway line M1 and the sector around Sant'Agostino station for line M2 were among the most critical areas. Moreover, for the first time, public car parks have been deeply considered in a 3D groundwater flow numerical model for the city of Milan. Groundwater infiltrations were detected both for deep car parks in the central portion of the domain and shallow car parks in the western sectors. This resulted in an improvement of the already-existing urban conceptual model of the area.
- Support for the decision makers in designing possible dewatering systems, also proposing early warning monitoring systems and proactive solutions to secure the UIs from potential groundwater infiltration damages.

The overall findings of this study could provide a useful tool to the stakeholders to properly design new UIs in the framework of the planned underground development of the city. In this sense, the numerical model could be used to realize different GW scenarios, testing their effects on the designed UIs. Furthermore, modeling their tops and bottoms through the HFB package could improve the evaluation of their barrier effect on groundwater flow paths. For future applications, reasoning the combination of the

HFB package with different third-type boundary conditions (i.e., River, GHB) to model other subsurface elements (i.e., sewer systems, buried channels, etc., to evaluate their leakance) could represent a challenging task. The methodology has been tested for the city of Milan—nonetheless it should be worth considering its application to other urban realities to enhance the analysis of GW/UI interactions.

Author Contributions: Conceptualization, D.S.; methodology, D.S. and T.B.; validation, M.R., M.D.A., L.F. and T.B; formal analysis, D.S. and C.Z.; data curation, D.S. and M.D.A.; writing—original draft preparation, D.S.; writing—review and editing, D.S., C.Z., M.R., M.D.A., M.C., A.R., L.F. and T.B.; visualization, D.S., C.Z., M.C. and A.R.; supervision, M.R., M.D.A., L.F. and T.B.; project administration, T.B. All authors have read and agreed to the published version of the manuscript.

Funding: This research did not receive any external funding.

Acknowledgments: The authors are grateful to Metropolitana Milanese S.p.a for providing both the altimetric profiles of the subway lines and the piezometric data that were used as information for model calibration. Moreover, the authors would like to warmly thank Daniel T. Feinstein of USGS and Gennaro Alberto Stefania for their support and suggestions during the development of the work. The authors would also like to thank the three anonymous reviewers for their comments, which helped to improve this article.

Conflicts of Interest: The authors declare no conflict of interest.

References

1. Vázquez-Suñé, E.; Sánchez-Vila, X.; Carrera, J. Introductory review of specific factors influencing urban groundwater, an emerging branch of hydrogeology, with reference to Barcelona, Spain. *Hydrogeol. J.* **2005**, *13*, 522–533. [CrossRef]
2. Epting, J.; Huggenberger, P.; Rauber, M. Integrated methods and scenario development for urban groundwater management and protection during tunnel road construction: A case study of urban hydrogeology in the city of Basel, Switzerland. *Hydrogeol. J.* **2008**, *16*, 575–591. [CrossRef]
3. Arshad, I.; Umar, R. Status of urban hydrogeology research with emphasis on India. *Hydrogeol. J.* **2020**, *28*, 477–490. [CrossRef]
4. Un-Habitat. *State of the World's Cities 2008/9: Harmonious Cities*; Routledge: Oxford, UK, 2012; ISBN 1136556729.
5. La Vigna, F. Review: Urban groundwater issues and resource management, and their roles in the resilience of cities. *Hydrogeol. J.* **2022**, *30*, 1657–1683. [CrossRef]
6. Schirmer, M.; Leschik, S.; Musolff, A. Current research in urban hydrogeology—A review. *Adv. Water Resour.* **2013**, *51*, 280–291. [CrossRef]
7. Calderhead, A.I.; Martel, R.; Garfias, J.; Rivera, A.; Therrien, R. Pumping dry: An increasing groundwater budget deficit induced by urbanization, industrialization, and climate change in an over-exploited volcanic aquifer. *Environ. Earth Sci.* **2012**, *66*, 1753–1767. [CrossRef]
8. Parriaux, A.; Tacher, L.; Kaufmann, V.; Blunier, P. *Underground Resources and Sustainable Development in Urban Areas*; The Geological Society of London: London, UK, 2006.
9. Li, H.; Li, X.; Parriaux, A.; Thalmann, P. An integrated planning concept for the emerging underground urbanism: Deep City Method Part 2 case study for resource supply and project valuation. *Tunn. Undergr. Sp. Technol.* **2013**, *38*, 569–580. [CrossRef]
10. Li, H.-Q.; Parriaux, A.; Thalmann, P.; Li, X.-Z. An integrated planning concept for the emerging underground urbanism: Deep City Method Part 1 concept, process and application. *Tunn. Undergr. Sp. Technol.* **2013**, *38*, 559–568. [CrossRef]
11. Vähäaho, I. An introduction to the development for urban underground space in Helsinki. *Tunn. Undergr. Sp. Technol.* **2016**, *55*, 324–328. [CrossRef]
12. Koziatek, O.; Dragićević, S. iCity 3D: A geosimualtion method and tool for three-dimensional modeling of vertical urban development. *Landsc. Urban Plan.* **2017**, *167*, 356–367. [CrossRef]
13. Bobylev, N. Mainstreaming sustainable development into a city's Master plan: A case of Urban Underground Space use. *Land Use Policy* **2009**, *26*, 1128–1137. [CrossRef]
14. Attard, G.; Winiarski, T.; Rossier, Y.; Eisenlohr, L. Review: Impact of underground structures on the flow of urban groundwater-Revue: Impact des structures du sous-sol sur les écoulements des eaux souterraines en milieu urbainRevisión: Impacto de las estructuras del subsuelo en el flujo del agua subterránea. *Hydrogeol. J.* **2015**, *24*, 5–19. [CrossRef]
15. Noethen, M.; Hemmerle, H.; Bayer, P. Sources, intensities, and implications of subsurface warming in times of climate change. *Crit. Rev. Environ. Sci. Technol.* **2022**, 1–23. [CrossRef]
16. Wilkinson, W. Rising groundwater levels in London and possible effects on engineering structures. *IAHS-AISH Publ.* **1985**, *154*, 145–157.
17. Hernández, M.A.; González, N.; Chilton, J. Impact of Rising Piezometric Levels on Greater Buenos Aires Due to Partial Changing of Water Services Infrastructure. 1997. Available online: http://sedici.unlp.edu.ar/handle/10915/26650 (accessed on 10 November 2022).

18. Vazquez-sune, E.; Sanchez-vila, X. Groundwater modelling in urban areas as a tool for local authority management: Barcelona case study (Spain). In *Impacts of Urban Growth on Surface Water and Groundwater Quality: Proceedings of the International Symposium Held during IUGG 99, the XXII General Assembly of the International Union of Geodesy and Geophysics, Birmingham, UK, 18–30 July 1999*; IAHS Press: Wallingford, UK, 1999; Volume 259, pp. 65–72.
19. Hayashi, T.; Tokunaga, T.; Aichi, M.; Shimada, J.; Taniguchi, M. Effects of human activities and urbanization on groundwater environments: An example from the aquifer system of Tokyo and the surrounding area. *Sci. Total Environ.* **2009**, *407*, 3165–3172. [CrossRef]
20. Lamé, A. Modélisation Hydrogéologique des Aquifères de Paris et Impacts des Aménagements du Sous-Sol sur Les Écoulements Souterrains. Available online: https://theses.hal.science/pastel-00973861/2013 (accessed on 10 November 2022).
21. Ducci, D.; Sellerino, M. Groundwater Mass Balance in Urbanized Areas Estimated by a Groundwater Flow Model Based on a 3D Hydrostratigraphical Model: The Case Study of the Eastern Plain of Naples (Italy). *Water Resour. Manag.* **2015**, *29*, 4319–4333. [CrossRef]
22. Colombo, L.; Gattinoni, P.; Scesi, L. Influence of underground structures and infrastructures on the groundwater level in the urban area of Milan, Italy. *Int. J. Sustain. Dev. Plan.* **2017**, *12*, 176–184. [CrossRef]
23. Allocca, V.; Coda, S.; Calcaterra, D.; De Vita, P. Groundwater Rebound and Flooding in the Naples' Periurban Area (Italy). *J. Flood Risk Manag.* **2021**, *15*, e12775. [CrossRef]
24. El Tani, M. Circular tunnel in a semi-infinite aquifer. *Tunn. Undergr. Sp. Technol.* **2003**, *18*, 49–55. [CrossRef]
25. Butscher, C. Steady-state groundwater inflow into a circular tunnel. *Tunn. Undergr. Sp. Technol.* **2012**, *32*, 158–167. [CrossRef]
26. Hassani, A.N.; Katibeh, H.; Farhadian, H. Numerical analysis of steady-state groundwater inflow into Tabriz line 2 metro tunnel, northwestern Iran, with special consideration of model dimensions. *Bull. Eng. Geol. Environ.* **2016**, *75*, 1617–1627. [CrossRef]
27. Farhadian, H.; Hassani, A.N.; Katibeh, H. Groundwater inflow assessment to Karaj Water Conveyance tunnel, northern Iran. *KSCE J. Civ. Eng.* **2017**, *21*, 2429–2438. [CrossRef]
28. Xia, Q.; Xu, M.; Zhang, H.; Zhang, Q.; Xiao, X. xuan A dynamic modeling approach to simulate groundwater discharges into a tunnel from typical heterogenous geological media during continuing excavation. *KSCE J. Civ. Eng.* **2018**, *22*, 341–350. [CrossRef]
29. Hassani, A.N.; Farhadian, H.; Katibeh, H. A comparative study on evaluation of steady-state groundwater inflow into a circular shallow tunnel. *Tunn. Undergr. Sp. Technol.* **2018**, *73*, 15–25. [CrossRef]
30. Li, X.; Zhang, W.; Li, D.; Wang, Q. Influence of underground water seepage flow on surrounding rock deformation of multi-arch tunnel. *J. Cent. South. Univ. Technol.* **2008**, *15*, 69–74. [CrossRef]
31. Guo, Y.; Wang, H.; Jiang, M. Efficient Iterative Analytical Model for Underground Seepage around Multiple Tunnels in Semi-Infinite Saturated Media. *J. Eng. Mech.* **2021**, *147*, 04021101. [CrossRef]
32. Gao, C.L.; Zhou, Z.Q.; Yang, W.M.; Lin, C.J.; Li, L.P.; Wang, J. Model test and numerical simulation research of water leakage in operating tunnels passing through intersecting faults. *Tunn. Undergr. Sp. Technol.* **2019**, *94*, 103134. [CrossRef]
33. Ai, Q.; Yuan, Y.; Jiang, X.; Wang, H.; Han, C.; Huang, X.; Wang, K. Pathological diagnosis of the seepage of a mountain tunnel. *Tunn. Undergr. Sp. Technol.* **2022**, *128*, 104657. [CrossRef]
34. Golian, M.; Teshnizi, E.S.; Nakhaei, M. Prediction of water inflow to mechanized tunnels during tunnel-boring-machine advance using numerical simulation. *Hydrogeol. J.* **2018**, *26*, 2827–2851. [CrossRef]
35. Zaidel, J.; Markham, B.; Bleiker, D. Simulating seepage into mine shafts and tunnels with MODFLOW. *Ground Water* **2010**, *48*, 390–400. [CrossRef]
36. Lagudu, S.; Rao, V.V.S.G.; Nandan, M.J.; Khokhar, C. Application of MODFLOW for groundwater Seepage Problems in the Subsurface Tunnels. *J. Ind. Geophys. Union* **2015**, *19*, 422–432.
37. Golian, M.; Abolghasemi, M.; Hosseini, A.; Abbasi, M. Restoring groundwater levels after tunneling: A numerical simulation approach to tunnel sealing decision-making. *Hydrogeol. J.* **2021**, *29*, 1611–1628. [CrossRef]
38. Abd-Elaty, I.; Pugliese, L.; Straface, S. Inclined Physical Subsurface Barriers for Saltwater Intrusion Management in Coastal Aquifers. *Water Resour. Manag.* **2022**, *36*, 2973–2987. [CrossRef]
39. Abd-Elaty, I.; Zelenakova, M. Saltwater intrusion management in shallow and deep coastal aquifers for high aridity regions. *J. Hydrol. Reg. Stud.* **2022**, *40*, 101026. [CrossRef]
40. Chaussard, E.; Bürgmann, R.; Shirzaei, M.; Fielding, E.J.; Baker, B. Predictability of hydraulic head changes and characterization of aquifer-system and fault properties from InSAR-derived ground deformation. *J. Geophys. Res. Solid Earth* **2014**, *119*, 6572–6590. [CrossRef]
41. Medici, G.; Smeraglia, L.; Torabi, A.; Botter, C. Review of Modeling Approaches to Groundwater Flow in Deformed Carbonate Aquifers. *Groundwater* **2021**, *59*, 334–351. [CrossRef]
42. Bonomi, T.; Sartirana, D.; Toscani, L.; Stefania, G.A.; Zanotti, C.; Rotiroti, M.; Redaelli, A.; Fumagalli, L. Modeling groundwater/surface-water interactions and their effects on hydraulic barriers, the case of the industrial area of Mantua (Italy). *Acque Sotter.-Ital. J. Groundw.* **2022**, *11*, 43–55. [CrossRef]
43. Bonomi, T.; Bellini, R. The tunnel impact on the groundwater level in an urban area: A modelling approach to forecast it. *RMZ-Mater. Geoenviron.* **2003**, *50*, 45–48.
44. Boukhemacha, M.A.; Gogu, C.R.; Serpescu, I.; Gaitanaru, D.; Bica, I. A hydrogeological conceptual approach to study urban groundwater flow in Bucharest city, Romania. *Hydrogeol. J.* **2015**, *23*, 437–450. [CrossRef]

45. Di Salvo, C.; Mancini, M.; Cavinato, G.P.; Moscatelli, M.; Simionato, M.; Stigliano, F.; Rea, R.; Rodi, A. A 3d geological model as a base for the development of a conceptual groundwater scheme in the area of the colosseum (Rome, Italy). *Geosciences* **2020**, *10*, 266. [CrossRef]
46. Sartirana, D.; Rotiroti, M.; Zanotti, C.; Bonomi, T.; Fumagalli, L.; De Amicis, M. A 3D geodatabase for urban underground infrastructures: Implementation and application to groundwater management in Milan metropolitan area. *ISPRS Int. J. Geo-Inf.* **2020**, *9*, 609. [CrossRef]
47. Parriaux, A.; Blunier, P.; Maire, P.; Tacher, L. The DEEP CITY Project: A Global Concept for a Sustainable Urban Underground Management. In Proceedings of the 11th ACUUS International Conference, Underground Space: Expanding the Frontiers, Athens, Greece, 10–13 September 2007; pp. 255–260.
48. Delmastro, C.; Lavagno, E.; Schranz, L. Underground urbanism: Master Plans and Sectorial Plans. *Tunn. Undergr. Sp. Technol.* **2016**, *55*, 103–111. [CrossRef]
49. Moghadam, S.T.; Delmastro, C.; Lombardi, P.; Corgnati, S.P. Towards a New Integrated Spatial Decision Support System in Urban Context. *Procedia-Soc. Behav. Sci.* **2016**, *223*, 974–981. [CrossRef]
50. Panday, S.; Langevin, C.D.; Niswonger, R.G.; Ibaraki, M.; Hughes, J.D. *MODFLOW–USG Version 1: An Unstructured Grid Version of MODFLOW for Simulating Groundwater Flow and Tightly Coupled Processes Using a Control Volume Finite-Difference Formulation*; US Geological Survey: Reston, VA, USA, 2013; p. 66.
51. Sartirana, D.; Rotiroti, M.; Bonomi, T.; De Amicis, M.; Nava, V.; Fumagalli, L.; Zanotti, C. Data-driven decision management of urban underground infrastructure through groundwater-level time-series cluster analysis: The case of Milan (Italy). *Hydrogeol. J.* **2022**, *30*, 1157–1177. [CrossRef]
52. Bonomi, T.; Fumagalli, L.; Dotti, N. Fenomeno di inquinamento da solventi in acque sotterranee sfruttate ad uso potabile nel nord-ovest della provincia di Milano. *Eng. Hydro Environ. Geol.* **2009**, *12*, 43–59.
53. Pulighe, G.; Lupia, F. Multitemporal geospatial evaluation of urban agriculture and (non)-sustainable food self-provisioning in Milan, Italy. *Sustainability* **2019**, *11*, 1846. [CrossRef]
54. Istat. *L'italia Del Censimento. Struttura Demografica e Processo di Rilevazione, Lombardia*; Istat: Rome, Italy, 2011.
55. Boscacci, F.; Camagni, R.; Caragliu, A.; Maltese, I.; Mariotti, I. Collective benefits of an urban transformation: Restoring the Navigli in Milan. *Cities* **2017**, *71*, 11–18. [CrossRef]
56. Regione Lombardia & ENI Divisione AGIP. *Geologia Degli Acquiferi Padani Della Regione Lombardia*; Cipriano Carcano, A.P., Ed.; S.EL.CA: Florence, Italy, 2002.
57. Regione Lombardia. *Regione Lombardia Piano di Tutela ed Uso delle Acque (PTUA) 2016*; Regione Lombardia: Milano, Italy, 2016.
58. Bonomi, T.; Cavallin, A.; De Amicis, M.; Rizzi, S.; Tizzone, R.; Trefiletti, P. Evoluzione della dinamica piezometrica nell'area milanese in funzione di alcuni aspetti socio-economici. In Proceedings of the Atti Della Giornata Mondiale dell'Acqua Acque Sotterranee: Risorsa Invisibile, Rome, Italy, 23 March 1998; pp. 9–17.
59. Bonomi, T. *Groundwater Level Evolution in the Milan Area: Natural and Human Issues*; IAHS-AISH Publication: Brunswick, ME, USA, 1999; pp. 195–202.
60. Gattinoni, P.; Scesi, L. The groundwater rise in the urban area of Milan (Italy) and its interactions with underground structures and infrastructures. *Tunn. Undergr. Sp. Technol.* **2017**, *62*, 103–114. [CrossRef]
61. De Caro, M.; Crosta, G.B.; Previati, A. Modelling the interference of underground structures with groundwater flow and remedial solutions in Milan. *Eng. Geol.* **2020**, *272*, 105652. [CrossRef]
62. García-Gil, A.; Epting, J.; Ayora, C.; Garrido, E.; Vázquez-Suñé, E.; Huggenberger, P.; Gimenez, A.C. A reactive transport model for the quantification of risks induced by groundwater heat pump systems in urban aquifers. *J. Hydrol.* **2016**, *542*, 719–730. [CrossRef]
63. Milan Metropolitan City. *Documento di Piano Milano 2030 Visione, Costruzione, Strategie, Spazi*; Milan Metropolitan City: Milano, Italy, 2019.
64. Regione Lombardia Open Data Regione Lombardia. 2021. Available online: https://dati.lombardia.it/ (accessed on 15 December 2021).
65. Rumbaugh, J.; Rumbaugh, O. *Groundwater Vistas Version 7.24, Build 211*; Environment Simulations Inc.: Reinholds, PA, USA, 2020.
66. Beretta, G.P.; Avanzini, M.; Pagotto, A. Managing groundwater rise: Experimental results and modelling of water pumping from a quarry lake in Milan urban area (Italy). *Environ. Geol.* **2004**, *45*, 600–608. [CrossRef]
67. Regione Lombardia Geoportal of the Lombardy Region, Italy. 2021. Available online: http://www.geoportale.regione.lombardia.it/ (accessed on 1 December 2021).
68. ARPA Lombardia Agenzia Regionale per la Protezione dell'Ambiente [Regional Environmental Monitoring Agency]. 2021. Available online: https://www.arpalombardia.it/ (accessed on 13 November 2021).
69. Hsieh, P.A.; Freckleton, J.R. *Documentation of a Computer Program to Simulate Horizontal-Flow Barriers Using the U.S. Geological Survey's Modular Three-Dimensional Finite-Difference Ground-Water Flow Model*; US Geological Survey: Reston, VA, USA, 1993.
70. Harbaugh, A.W. *MODFLOW-2005, the US Geological Survey Modular Ground-Water Model: The Ground-Water Flow Process*; US Department of the Interior, US Geological Survey: Reston, VA, USA, 2005.
71. Harbaugh, A.W. A Computer Program for Calculating Subregional Water Budgets Using Results from the U.S. Geological Survey Modular Three-Dimensional Finite-Difference Ground-Water Flow Model. 1990. Available online: https://pubs.er.usgs.gov/publication/ofr90392 (accessed on 10 November 2022).

72. Toscani, L.; Stefania, G.A.; Masut, E.; Prieto, M.; Legnani, A.; Gigliuto, A.; Ferioli, L.; Battaglia, A. Groundwater flow numerical model to evaluate the water mass balance and flow patterns in Groundwater Circulation Wells (GCW) with varying aquifer parameters. *Acque Sotter.-Ital. J. Groundw.* **2022**. [CrossRef]
73. Bonomi, T.; Del Rosso, F.; Fumagalli, L.; Canepa, P. Assessment of groundwater availability in the Milan Province aquifers. *Mem. Descr. Della Cart. Geol. D'italia* **2010**, *90*, 31–40.
74. Bonomi, T.; Fumagalli, L.; Rotiroti, M.; Bellani, A.; Cavallin, A. The hydrogeological well database TANGRAM©: A tool for data processing to support groundwater assessment. *Acque Sotter.-Ital. J. Groundw.* **2014**, *3*, 98. [CrossRef]
75. *Paradigm Paradigm GOCAD 2009.1 User Guide*; Paradigm: Houston, TX, USA, 2009. Available online: https://www.scribd.com/document/475499865/01-Getting-Started# (accessed on 10 November 2022).
76. Airoldi, R.; Casati, P. *Le Falde Idriche del Sottosuolo di Milano*; Comune di Milano: Rome, Italy, 1989.
77. Dassargues, A. Groundwater modelling to predict the impact of tunnel on the behavior of water table aquiefer in urban condition. In Proceedings of the XXVII IAH Congress: Groundwater in the Urban Environment, Balkema, Nottingham, UK, 21–27 September 1997; pp. 225–230.
78. Attard, G.; Cuvillier, L.; Eisenlohr, L.; Rossier, Y.; Winiarski, T. Deterministic modelling of the cumulative impacts of underground structures on urban groundwater flow and the definition of a potential state of urban groundwater flow: Example of Lyon, FranceModélisation déterministe des impacts cumulés des structures so. *Hydrogeol. J.* **2016**, *24*, 1213–1229. [CrossRef]
79. Wang, X.; Lei, Q.; Lonergan, L.; Jourde, H.; Gosselin, O.; Cosgrove, J. Heterogeneous fluid flow in fractured layered carbonates and its implication for generation of incipient karst. *Adv. Water Resour.* **2017**, *107*, 502–516. [CrossRef]
80. Huang, Z.; Zhao, K.; Li, X.; Zhong, W.; Wu, Y. Numerical characterization of groundwater flow and fracture-induced water inrush in tunnels. *Tunn. Undergr. Sp. Technol.* **2021**, *116*, 104119. [CrossRef]
81. Middlemis, H.; Merrick, N.; Ross, J.B. *Groundwater Flow Modelling Guideline*; Prepared for Murray-Darling Basin; Aquaterra Consulting Pty Ltd.: West Palm Beach, FL, USA, 2000; Project No. 125.
82. Feinstein, D.T.; Hunt, R.J.; Reeves, H.W. *Regional Groundwater-Flow Model of the Lake Michigan Basin in Support of Great Lakes Basin Water Availability and Use Studies*; U. S. Geological Survey: Reston, VA, USA, 2010.
83. Colombo, A. Milano e l'innalzamento della falda. *Cave e Cantieri* **1999**, *2*, 26–36.
84. Bredehoeft, J. The conceptualization model problem-Surprise. *Hydrogeol. J.* **2005**, *13*, 37–46. [CrossRef]
85. Lotti, F.; Borsi, I.; Guastaldi, E.; Barbagli, A.; Basile, P.; Favaro, L.; Mallia, A.; Xuereb, R.; Schembri, M.; Mamo, J.A.; et al. Numerically enhanced conceptual modelling (NECoM) applied to the Malta Mean Sea Level Aquifer. *Hydrogeol. J.* **2021**, *29*, 1517–1537. [CrossRef]
86. Shi, S.; Xie, X.; Bu, L.; Li, L.; Zhou, Z. Hazard-based evaluation model of water inrush disaster sources in karst tunnels and its engineering application. *Environ. Earth Sci.* **2018**, *77*, 1–13. [CrossRef]
87. Wang, X.; Li, S.; Xu, Z.; Li, X.; Lin, P.; Lin, C. An interval risk assessment method and management of water inflow and inrush in course of karst tunnel excavation. *Tunn. Undergr. Sp. Technol.* **2019**, *92*, 103033. [CrossRef]
88. Bobylev, N. Transitions to a High Density Urban Underground Space. *Procedia Eng.* **2016**, *165*, 184–192. [CrossRef]
89. Ferré, T.P.A. Revisiting the Relationship Between Data, Models, and Decision-Making. *Groundwater* **2017**, *55*, 604–614. [CrossRef] [PubMed]
90. Wu, J.S.; Lee, J.J. Climate change games as tools for education and engagement. *Nat. Clim. Chang.* **2015**, *5*, 413–418. [CrossRef]
91. Castilla-Rho, J.C. Groundwater Modeling with Stakeholders: Finding the Complexity that Matters First Things First: We Are Dealing. *Groundwater* **2017**, *55*, 620–625. [CrossRef] [PubMed]
92. Colombo, L.; Gattinoni, P.; Scesi, L. Stochastic modelling of groundwater flow for hazard assessment along the underground infrastructures in Milan (northern Italy). *Tunn. Undergr. Sp. Technol.* **2018**, *79*, 110–120. [CrossRef]
93. Previati, A.; Epting, J.; Crosta, G.B. The subsurface urban heat island in Milan (Italy)-A modeling approach covering present and future thermal effects on groundwater regimes. *Sci. Total Environ.* **2022**, *810*, 152119. [CrossRef]
94. Attard, G.; Rossier, Y.; Winiarski, T.; Eisenlohr, L. Urban underground development confronted by the challenges of groundwater resources: Guidelines dedicated to the construction of underground structures in urban aquifers. *Land Use Policy* **2017**, *64*, 461–469. [CrossRef]
95. Naranjo, R.C. Knowing Requires Data. *Groundwater* **2017**, *55*, 674–677. [CrossRef]
96. Liu, J.Q.; Sun, Y.K.; Li, C.J.; Yuan, H.L.; Chen, W.Z.; Liu, X.Y.; Zhou, X.S. Field monitoring and numerical analysis of tunnel water inrush and the environmental changes. *Tunn. Undergr. Sp. Technol.* **2022**, *122*, 104360. [CrossRef]
97. Blunier, P.; Tacher, L.; Parriaux, A. Systemic approach of urban underground resources exploitation. In Proceedings of the 11th ACUUS Conference, Athens, Greece, 10–13 September 2007; pp. 43–48.
98. Admiraal, H.; Cornaro, A. Engaging decision makers for an urban underground future. *Tunn. Undergr. Sp. Technol.* **2016**, *55*, 221–223. [CrossRef]
99. Di Salvo, C.; Ciotoli, G.; Pennica, F.; Cavinato, G.P. Pluvial flood hazard in the city of Rome (Italy). *J. Maps* **2017**, *13*, 545–553. [CrossRef]
100. Peeters, L.J.M. Assumption Hunting in Groundwater Modeling: Find Assumptions Before They Find You. *Groundwater* **2017**, *55*, 665–669. [CrossRef] [PubMed]
101. Carneiro, J.; Carvalho, J.M. Groundwater modelling as an urban planning tool: Issues raised by a small-scale model. *Q. J. Eng. Geol. Hydrogeol.* **2010**, *43*, 157–170. [CrossRef]

102. Lyu, H.M.; Shen, S.L.; Zhou, A.; Yang, J. Perspectives for flood risk assessment and management for mega-city metro system. *Tunn. Undergr. Sp. Technol.* **2019**, *84*, 31–44. [CrossRef]
103. Du, H.; Du, J.; Huang, S. GIS, GPS, and BIM-based risk control of subway station construction. In Proceedings of the ICTE 2015, the Fifth International Conference on Transportation Engineering, Dalian, China, 26–27 September 2015; pp. 1478–1485.
104. Bayer, P.; Attard, G.; Blum, P.; Menberg, K. The geothermal potential of cities. *Renew. Sustain. Energy Rev.* **2019**, *106*, 17–30. [CrossRef]

Article

Differentiating Nitrate Origins and Fate in a Semi-Arid Basin (Tunisia) via Geostatistical Analyses and Groundwater Modelling

Kaouther Ncibi [1], Micòl Mastrocicco [2], Nicolò Colombani [3,*], Gianluigi Busico [2], Riheb Hadji [4], Younes Hamed [5] and Khan Shuhab [5]

1 Laboratory for the Application of Materials to the Environment, Water and Energy (LAMEEE), Faculty of Sciences of Gafsa, University of Gafsa, Gafsa 2112, Tunisia
2 Department of Environmental, Biological and Pharmaceutical Sciences and Technologies (DiSTABiF), University of Campania "Luigi Vanvitelli", Via Vivaldi 43, 81100 Caserta, Italy
3 Department of Materials, Environmental Sciences and Urban Planning (SIMAU), Marche Polytechnic University, Via Brecce Bianche 12, 60131 Ancona, Italy
4 Institute of Architecture and Earth Sciences (IAST), Department of Earth Sciences, University of Sétif 1, El Bez, Sétif 19000, Algeria
5 Department of Earth and Atmospheric Sciences, Science and Research Building 1, University of Houston, 3507 Cullen Blvd, Room 312, Houston, TX 77204, USA
* Correspondence: n.colombani@univpm.it; Tel.: +39-0712204743

Citation: Ncibi, K.; Mastrocicco, M.; Colombani, N.; Busico, G.; Hadji, R.; Hamed, Y.; Shuhab, K. Differentiating Nitrate Origins and Fate in a Semi-Arid Basin (Tunisia) via Geostatistical Analyses and Groundwater Modelling. *Water* **2022**, *14*, 4124. https://doi.org/10.3390/w14244124

Academic Editors: Elias Dimitriou and Cristina Di Salvo

Received: 20 September 2022
Accepted: 15 December 2022
Published: 18 December 2022

Abstract: Despite efforts to protect the hydrosystems from increasing pollution, nitrate (NO_3^-) remains a major groundwater pollutant worldwide, and determining its origin is still crucial and challenging. To disentangle the origins and fate of high NO_3^- (>900 mg/L) in the Sidi Bouzid North basin (Tunisia), a numerical groundwater flow model (MODFLOW-2005) and an advective particle tracking (MODPATH) have been combined with geostatistical analyses on groundwater quality and hydrogeological characterization. Correlations between chemical elements and Principal Component Analysis (PCA) suggested that groundwater quality was primarily controlled by evaporite dissolution and subsequently driven by processes like dedolomitization and ion exchange. PCA indicated that NO_3^- origin is linked to anthropic (unconfined aquifer) and geogenic (semi-confined aquifer) sources. To suggest the geogenic origin of NO_3^- in the semi-confined aquifer, the multi-aquifer groundwater flow system and the forward and backward particle tracking was simulated. The observed and calculated hydraulic heads displayed a good correlation (R^2 of 0.93). The residence time of groundwater with high NO_3^- concentrations was more significant than the timespan during which chemical fertilizers were used, and urban settlements expansion began. This confirmed the natural origin of NO_3^- associated with pre-Triassic embankment landscapes and located on domed geomorphic surfaces with a gypsum, phosphate, or clay cover.

Keywords: groundwater hydrochemistry; principal component analysis; multi-aquifer system; flow model; contaminant sources

1. Introduction

Groundwater is the main source of water supply worldwide, especially in arid and semi-arid regions, where it also plays a key role for their proper economic and social development [1,2]. Groundwater suitable for human consumption or crop irrigation must contain mineral salts in a well-balanced quantity. However, groundwater is most often subject to natural and/or anthropogenic constraints affecting its quality degradation. Globally, the main issue related to groundwater quality degradation is nutrients and/or chemical enrichment from a chemical such as nitrate (NO_3^-), which is recognized as the main water pollutant [3,4]. There are several sources of NO_3^- in groundwater, and anthropogenic activities are indeed the main ones. Diffuse agricultural pollution due to the development of intensive practices, new methods of cultivation and breeding characterized by a massive spreading of effluents and fertilizers, and urban and industrial discharges contribute to groundwater pollution and quality degradation [5–8]. Bioavailable nitrogen (N) deposition at the land surface constitutes a natural origin of NO_3^- [9,10]. The main source of N resides in the atmosphere in the molecular form (N_2), representing approximately 78% of the atmospheric composition [11]. Another natural source of NO_3^-, is represented by the accumulations of NO_3^- rich salts which have also been found in regions with an arid climate and/or deserts, such as the Death Valley of the Mojave Desert, southern California [12]. Regardless of the NO_3^-'s origin, this ion has harmfully affected groundwater quality, biodiversity, and ecosystem functioning worldwide [13,14].

The Sidi Bouzid basin (Central Tunisia, North Africa) is an example of a semi-arid area that is characterized by an aquifer system with high concentrations of NO_3^-, which often exceed the standard of the World Health Organization for drinking water of 10 mg-N/L [15,16]. Previous research on this aquifer system has already focused on groundwater quality [13], groundwater vulnerability to NO_3^- pollution [17,18], and on the health risk assessment of NO_3^- in groundwater [9]. Additionally, some studies have used stable isotopes to assess groundwater recharge [19]. Nevertheless, none of the aforementioned studies focused on the origin of NO_3^- in groundwater. Numerous methods have been proposed to identify the NO_3^- sources around the world, including: (i) geophysical approaches [20], (ii) statistical techniques [21], (iii) and via stable isotopes [22–25]. Petelet-Giraud et al. [20] used detailed geological and geophysical profiles, such as electric tomography, to improve a local structural model. The authors in this case studied the heterogeneity of the hydrogeological system, where some compartments were disconnected from the general groundwater flow and explained the presence of young and old groundwater via NO_3^- concentrations and environmental tracers. Kendall and Aravena [26] defined the use of the stable N and O isotopes of NO_3^- molecules as tracers to evaluate the sources and processes that affect NO_3^- in groundwater. Widory et al. [27] investigated the viability of an isotopic multi-tracer approach ($\delta^{15}N$, $\delta^{11}B$, $^{87}Sr/^{86}Sr$) to determine the source(s) of NO_3^- pollution in groundwater in the Arguenon watershed (France). Xuan et al. [25] in southern China also applied N isotope analyses to identify the source and transformations of NO_3^- in groundwater in a mixed land use watershed, but the studies of the isotopes are limited to knowing an area punctually. However, most of the previous studies did not explicitly model the retention time within the aquifer and possible different sources of NO_3^- except for small field sites [28–30]. One of the few exceptions is the study of Koh et al. [31] that modelled an aquifer system characterized by complex hydrogeology and mixing of groundwater with different ages via environmental tracers, but a single source of NO_3^- from fertilizers was employed. To the best of authors' knowledge, no environmental studies have evaluated NO_3^- origins (anthropogenic versus geogenic) in an aquifer system in a semi-arid region using a combined approach of geostatistical analysis on hydrochemical data and numerical flow and advective modeling. Here, two popular codes, MODFLOW-2005 v.11 [32] and MODPATH v.7 [33], have been used for the Sidi Bouzid North basin in central Tunisia. The objectives of this study are: (1) to evaluate the spatial distribution of NO_3^- in the aquifer system of the Sidi Bouzid basin; and (2) to identify possible NO_3^- sources in groundwater.

2. Study Area

2.1. Geography and Climate

The study area, situated between longitudes 9°10′00″ E to 9°45′00″ E and latitudes 34°55′00″ N to 35°20′00″ N, constitutes the Sidi Bouzid North basin located in central Tunisia. It covers an area of approximately 1508 Km^2 (Figure 1) and extends from Jebel Rakhmet, Jebel Hamra, and Jebel Mghila in the West to the North-South axis (NOSA) in the East, and from the Zawiya-Roua chain in the North to Jebel Kebar in the South. To the Southwest, it is limited by the Jebel Al Hfay. The study area is made of three sub-basins: (i) the Southern part: Sidi Bouzid, (ii) the Western part: Awled Asker, and (iii) the Eastern part: Oued El Hjal. The overall elevation of the area ranges between 238 m and 640 m above sea level (a.s.l.). This area belongs to a semi-arid Mediterranean climate with a mean annual precipitation of 223 mm and an average yearly temperature of 19 °C [34]. The average annual evapotranspiration is 180 mm.

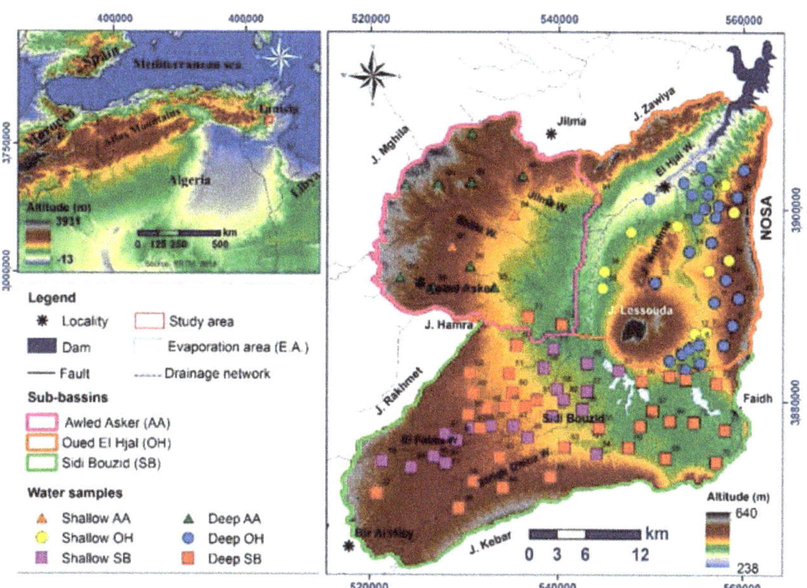

Figure 1. Location of the study area: Sidi Bouzid basin.

2.2. Geology

The exposed geological units in the study area include: (i) Mesozoic (Triassic, Jurassic, and Cretaceous) and (ii) Cenozoic (Paleogene, Neogene, and Quaternary) aged rocks. The oldest rocks are Triassic and Jurassic in age and outcrop on raised structures bordering the study area. The Triassic strata consist of diapiric intrusions of a complex combination of gypsum, clays, and dolomites. Jurassic aged rocks are calcareous-dolomitic deposits of the Nara Formation. The Cretaceous aged rocks begin with clay and sandstone deposits and continue with a series of limestones, dolomites, clays, and gypsums (Figure 2). The Paleogene rocks in the study area include a succession of gypsum, marl, phosphate, and limestone [35,36]. The Neogene is characterized by a variety of continental and lagoon facies with red clayey silts, small calcareous concretions, and gypsum. Finally, the Quaternary strata constituting diversified fluvial deposits of sandy clays, silts, gypsum crust, sandy silts, and sands are distributed throughout the study area.

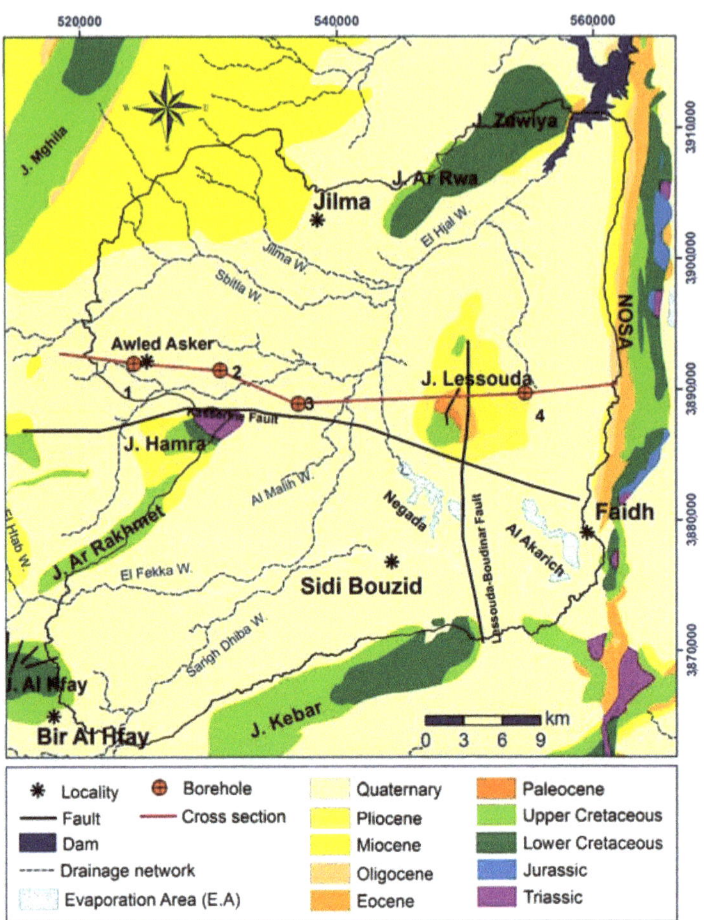

Figure 2. Geological map of the study area (modified from the geological map of Tunisia at 1/500,000) [37].

2.3. Hydrogeology

The conceptual model of the Sidi Bouzid basin has been schematized and presented in Figure 3. The Sidi Bouzid basin is a detrital Mio-Plio-Quaternary complex hosting two aquifers. The borehole cross-section represents the superposition of three layers. The first layer, widely distributed throughout the study area, constitutes the shallow reservoir (with an average thickness of 40 m) made of sand with gravel and clayey intercalations. The second layer is an impermeable and sometimes semi-permeable aquitard with an average thickness of 45 m, while the third layer is the deeper aquifer with an average thickness of 25 m. Groundwater is recharged through atmospheric precipitation, supplemented by lateral runoff and irrigation return flow. Groundwater discharge occurs by lateral outflow, evapotranspiration, evaporation areas, and artificial extraction (for domestic and agricultural use). The groundwater flows from the mountainous boundary area to the northeast of the study area. Finally, it discharges to the evaporation area (Negada and Al Akarich) and El Hjal Wadi, for both aquifers shallow and deep [38]. Lastly, the communication between the aquifers is only downstream of the basin. The hydrodynamics of the water are influenced by the aquifer geometry and the tectonic structures. The groundwater flow converges from Miocene outcrops in two directions: (i) the main direction is from West to East, while (ii) the secondary flow is from NW to SE. The groundwater overexploitation is intense in the central and downstream parts, and it has experienced a constant increase during the last years [36].

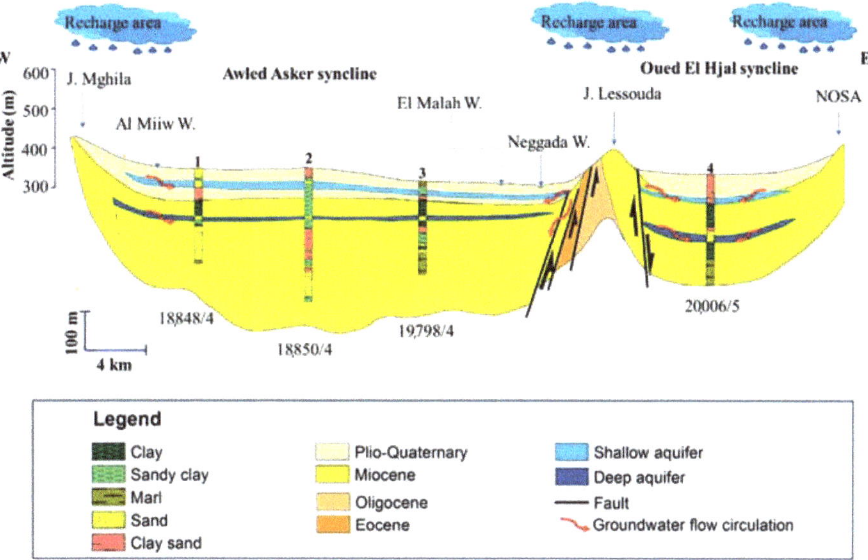

Figure 3. Simplified cross section elaborated from lithostratigraphic sections of boreholes.

3. Materials and Methods

3.1. Groundwater Flow Simulation

Governing Equations and Groundwater Model Selection

The worldwide popular groundwater flow numerical model MODFLOW-2005 v.11, based on Darcy's law and mass conservation concept has been used in this study. MODFLOW-2005 employs a three-dimensional simulation of groundwater flow circulation in porous media, for both aquifers, shallow and deep, which is represented by the mathematically following equation [32]:

$$\frac{\partial}{\partial x}\left[K_x\frac{\partial h}{\partial x}\right]+\frac{\partial}{\partial y}\left[K_y\frac{\partial h}{\partial y}\right]+\frac{\partial}{\partial y}\left[K_z\frac{\partial h}{\partial z}\right]-w=S_s\frac{\partial h}{\partial t} \quad (1)$$

where Kx, Ky, and Kz are the hydraulic conductivity values along the x, y, and z coordinate axes, which are assumed to be parallel to the major hydraulic conductivity axes, w is the volumetric flow per unit volume and represents the sources (negative values) and/or the sinks (positive values) of water per unit time, h is the hydraulic head, Ss is the specific storage of the porous material if the aquifer is confined or specific yield if the aquifer is unconfined, and t is the time.

Following the groundwater flow field calculated by MODFLOW-2005, an advective particle tracking numerical code MODPATH v.7 was employed to define the direction of solute particles' migration and their retention time within the aquifers system.

3.2. Data Collection and Processing

In this study, a two-step approach was employed: (i) the first step aimed at identifying the relationship between NO_3^- and other chemical elements in groundwater, and (ii) the second step aimed at identifying the different NO_3^- origins developing a numerical model to simulate the groundwater flow and particle circulation (Figure 4). This approach was used to test two hypotheses to explain NO_3^- accumulation in groundwater. The first hypothesis describes a "top-down" mechanism where NO_3^- is thought to be of anthropogenic origin and infiltrated from the surface through stratified layers of sand and sandstone. In this hypothesis, the particles (nitrates in our case) reach the groundwater and migrate inside it according to the existing flow in a transient state. It is determined by the circulation time of the particles from surface to groundwater. The second hypothesis is a "bottom-up-laterally" mechanism where NO_3^- is deposited onto a stable soil surface through aerosol deposition and has concentrated in the subsoil over time. In this hypothesis, it is considered that the nitrates are of natural origin (linked to sedimentation) and are found in the subsurface. This mechanism has been used to explain NO_3^- reservoirs found in arid and semi-arid areas [39].

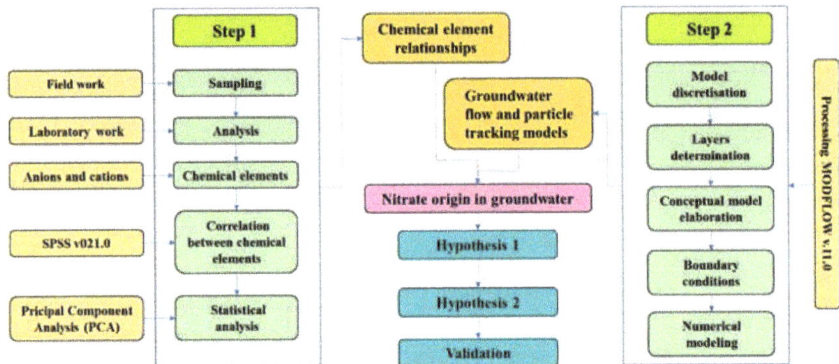

Figure 4. Flowchart showing the methodology adopted for the determination of the NO_3^- origin in groundwater.

Temperature, pH, and electrical conductivity (EC) were measured in the field using a HI 99301 multiparameter analyzer. A total of 103 water samples were collected from 38 shallow wells and 65 deep wells to describe the physicochemical characteristics of groundwater in 2019. The samples were collected in sterilized bottles after purging at least 3 volumes from the well casing. Water samples were delivered to the Laboratory of Physico-chemical Analyses of Soil and Water of the Regional Commissariat of Agricultural Development of Sidi Bouzid for major ions analysis. NO_3^- concentrations were also measured directly in the field using a portable NO_3^- meter (LAQUAtwin B-743) after two-point calibration. The reliability of the results of the chemical analyzes was determined by the calculation of the ionic balance ($IB\% = (\sum cations - \sum anions)/(\sum cations + \sum anions)$). The analysis is declared acceptable if $-6 \leq BI \leq 6\%$.

The data used to develop the conceptual and numerical models of the groundwater flow circulation (piezometric and exploitation histories since 1990 and the hydrodynamic data of the aquifer) were collected from the Regional Commissariat for Agricultural Development (CRDA). Groundwater was exploited by 6970 wells for the shallow aquifer and 195 boreholes for the deep aquifer in 2020. The observation wells used are 53 and 39 wells for the phreatic and deep aquifer, respectively.

ArcGIS v.10.5 was used to prepare input data maps. The DEM to define the vadose zone thickness and ground elevation was obtained from STRM, while the digital geologic map and borehole cross-section were used to determine the distribution of rock and vadose zone types.

The statistical analysis was carried out by using SPSS v21.0. The Principal Component Analysis (PCA) with varimax rotation was conducted to assess the strength of relationships between variables (NO_3^- and other major ions) in the study area. Processing MODFLOW-2005 v11.0 [40] and MODPATH v.7 was used to simulate groundwater flow modeling and particle tracking, respectively.

4. Results and Discussion

4.1. Hydrochemical Characteristics

4.1.1. NO_3^- Concentrations in Groundwater

Figure 5, generated using distribution of NO_3^- concentrations in groundwater through Kriging interpolation technique, shows that the lowest NO_3^- concentrations are in the western area of the Sidi Bouzid North basin and in the Awled Asker sub-basin. While towards the East of the study area, NO_3^- concentrations become very high, often exceeding 300 mg/L in both the shallow (Figure 6a) and deep (Figure 6b) aquifers. In the south, NO_3^- concentrations vary from 30 mg/L to 120 mg/L. The highest levels of NO_3^- are around 930 mg/L and are recorded in the Oued El Hjal sub-basin. This differentiation of spatial concentration distributions of nitrate is due to the hydrodynamic functioning of the aquifer system which is also influenced by lithostratigraphic variations.

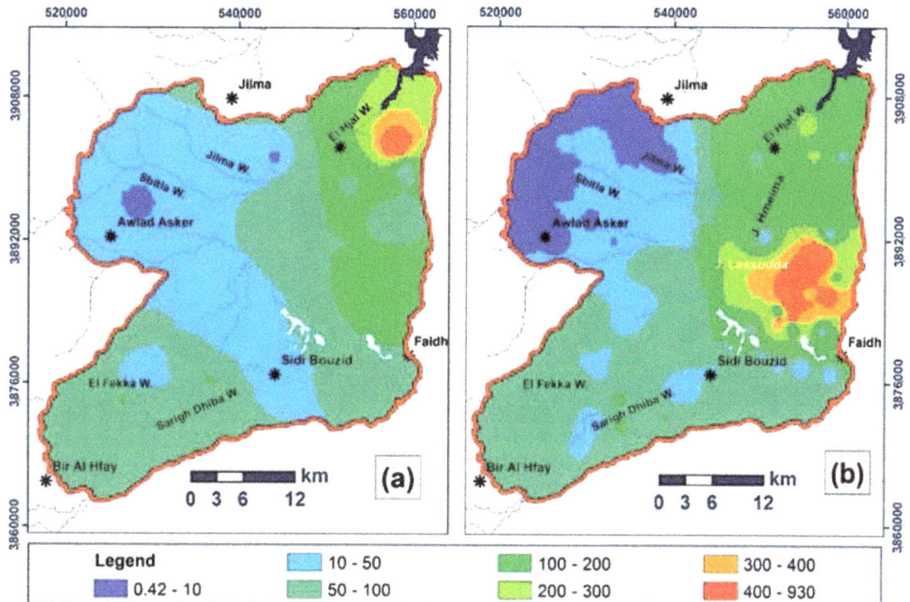

Figure 5. Spatial distribution of NO_3^- concentrations in groundwater: (**a**) shallow aquifer and (**b**) deep aquifer.

4.1.2. Comparison Nitrate with Other Ions

The relationship between the nitrate concentration and chemical elements was investigated for geochemical characterization of groundwater and to trace the origin of NO_3^-. Cl^- and Na^+ in groundwater are often linked to halite dissolution (NaCl). The evolution of Na^+ has been studied as a function of Cl^-, which is considered a stable and conservative tracer of evaporites [41]. The graph in Figure 6a shows that several samples line up on the slope line 1:1, indicating coexistence of the two ions and possible NaCl dissolution. Other samples, especially from the deep aquifer of the Oued El Hjal sub-basin, have an excess of Cl^- compared to Na^+, this can also be explained by the dissolution of halite, but with subsequent sorption of Na^+ via cation exchange. The role of carbonate and evaporite dissolution on groundwater composition was investigated through the scatter plots of (Ca^{2+} + Mg^{2+}) versus (HCO_3^- + SO_4^{2-}) as shown in Figure 6b. Most water samples are located near and below the slope line 1:1 on the side of HCO_3^- + SO_4^{2-}, indicating no preferential dissolution of evaporitic or carbonate rocks as major hydrochemical process of the entire aquifer system of the study area. One exception is the deep aquifer of Oued El

Hjal sub-basin, in which the excess of Ca^{2+} and Mg^{2+} could be due to cation exchange with Na^{+}, as previously postulated.

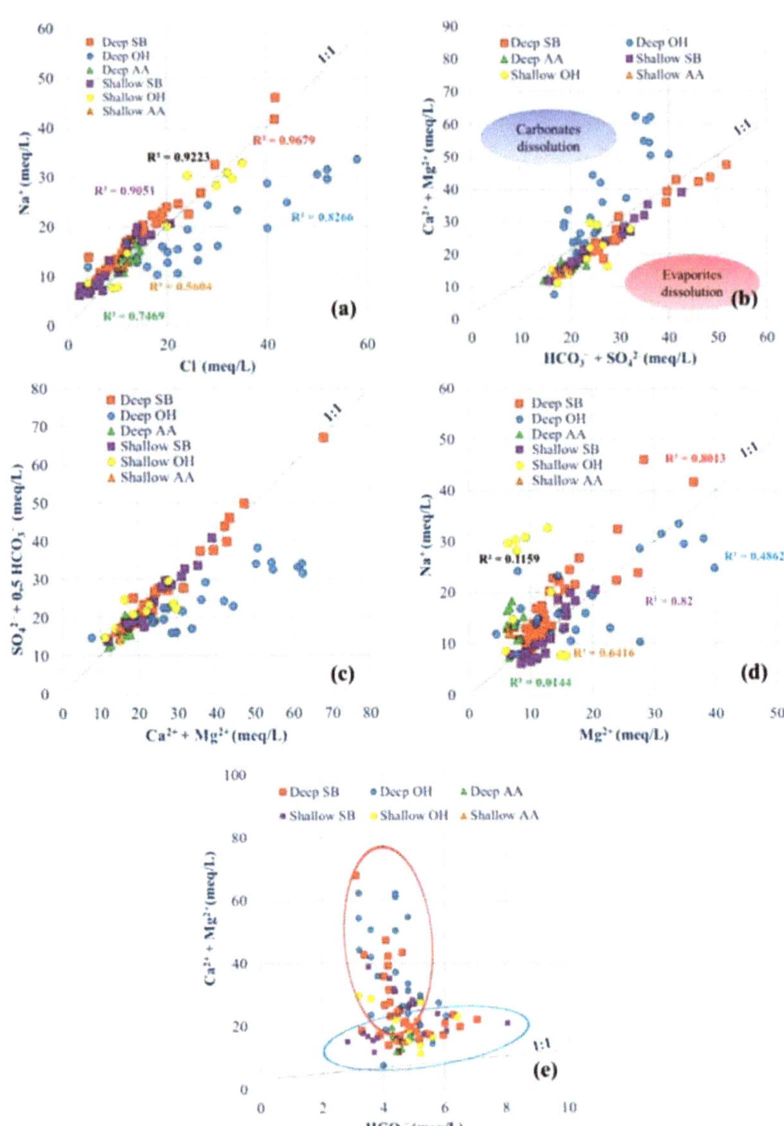

Figure 6. Scatter plots of major ions in the deep and shallow aquifers.

The predominance of SO_4^{2-} over HCO_3^- and the lack of a strong link between the species Ca^{2+}, HCO_3^-, and SO_4^{2-} indicate that other processes control the water chemistry, such as dedolomitization, which involves dissolution reactions with carbonate minerals and gypsum. Dedolomitization is often caused by gypsum-to-anhydrite conversion [42] accompanied by the dissolution of dolomite and the precipitation of calcite ($CaMg(CO_3)_2 + Ca^{2+} \Leftrightarrow 2CaCO_3 + Mg^{2+}$). The dissolution of gypsum consequently increased the concentration of Ca^{2+} by the same Ca^{2+}/Mg^{2+} ratio. This ratio once greater than 0.5 thermodynamically causes dedolomitization [43]. All samples from the study area show a Ca^{2+}/Mg^{2+} ratio greater than 0.5, thus explaining that the dedolomitization process seems to mark the water chemistry. The dissolution of dolomite consequently led to an increase in the concentration of Mg^{2+} in groundwater. This process, which includes the dissolution of gypsum, is highlighted through the correlation (Ca^{2+} + Mg^{2+}) vs. (SO_4^{2-} + $0.5HCO_3^-$), where most of the samples are organized along the straight line 1:1 (Figure 6c), except the ones pertaining to the deep aquifer of the Oued El Hjal sub-basin, in which the excess of Ca^{2+} and Mg^{2+} could be due to cation exchange with Na^+ which takes place within and on colloid particles. The overall reaction of the dedolomitization process can be written as follows:

$$CaMg(CO_3)_{2(s)} + CaSO_4x2H_2O_{(s)} + H^+ = CaCO_{3(s)} + Ca^{2+} + Mg^{2+} + SO_4^{2-} + HCO_3^- + 2H_2O \quad (2)$$

To better discriminate all the cation exchange phenomena within the aquifers system, a plot of Na^+ vs. Mg^{2+} is shown (Figure 6d). Cation exchange takes place with the colloids of organic matter and rich clay minerals present in the aquifer matrix, which release Na^+ and adsorb Ca^{2+} and Mg^{2+}, leading to an increase in Na^+ concentrations in groundwater. Here, all samples show a slight abundance of Na^+ with respect to Mg^{2+}, except for the samples from the deep aquifers of Oued El Hjal, in which excess of Mg^{2+} is released in groundwater via dedolomitization which triggers Na^+ adsorption. Finally, Figure 6e shows that the sum of Ca^{2+} and Mg^{2+} correlates very poorly with HCO_3^-, excluding simple dolomite dissolution mechanism as the main driver of the observed patterns.

4.1.3. Principal Component Analysis

The PCA was applied to the chemical elements (the variables) of groundwater in the study area for the three sub-basins, reducing the dimensions of the data to two main components (F1 and F2) (Table 1), which are visualized graphically in Figure 7. The correlations between the variables and the main axes show that the first two axes F1 and F2 express 69.6%, 69.6%, and 82.0% of the total variance for the sub-basins of Awled Asker, Oued El Hjal, and Sidi Bouzid, respectively.

The NO_3^- content in the Awled Asker and Oued El Hjal basins is found to be associated with other dissolved species, which explains a common origin between them, and thus a geogenic origin (Figure 7). While in the Sidi Bouzid, NO_3^- content is associated with HCO_3^- that is often related to an increase in inorganic carbon due to heterotrophic denitrification, thus suggesting fertilizer application is the source.

Table 1. The contribution of the factorial axes in the total values and the attributed eigenvalues of groundwater in the three sub-basins of the study area.

	F1	F2
Oued El Hjal		
Eigenvalue	5.227	1.737
Variance (%)	52.268	17.374
Cumulated variance (%)	52.268	69.642
Sidi Bouzid		
Eigenvalue	7.151	1.051
Variance (%)	71.508	10.510
Cumulated variance (%)	71.508	82.018
Awled Asker		
Eigenvalue	4.637	2.326
Variance (%)	46.369	23.263
Cumulated variance (%)	46.369	69.633

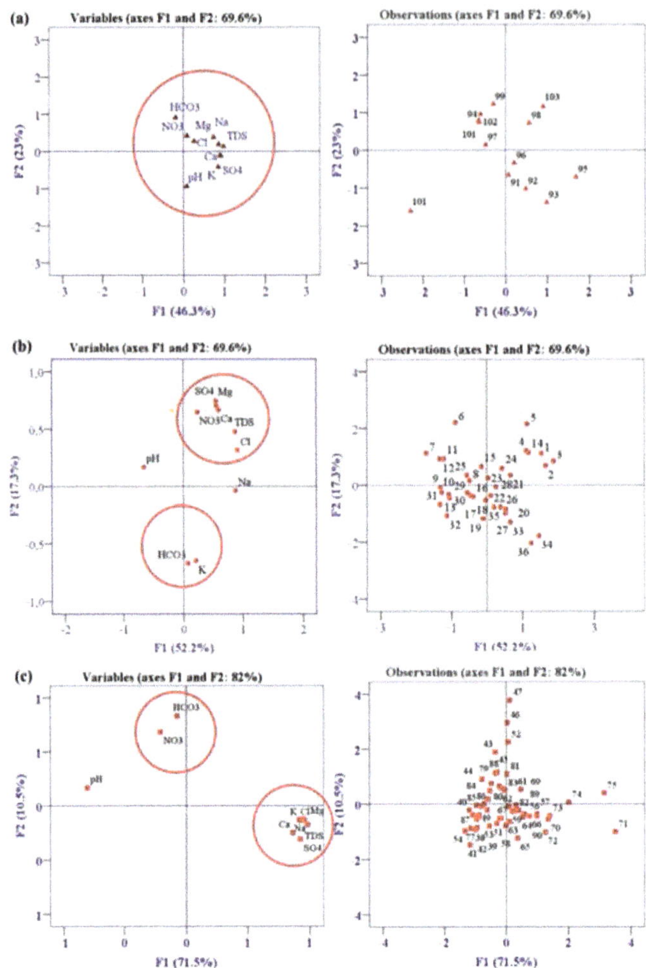

Figure 7. Projection of observations and variables into the factorial plane (F1 × F2): (**a**) sub-basin of Awled Asker (13 samples), (**b**) sub-basin of Oued El Hjal (36 samples), and (**c**) sub-basin of Sidi Bouzid (57 samples).

4.2. Model Discretization and Calibration

The numerical model domain covers an area of 1508 km^2 (43 km × 41 km). The UTM global coordinate system has been used to create the model and the database. The modeling grid consists of a cell dimension of 500 m × 500 m and 4 layers (Figure 8), the shallow aquifer was subdivided into 2 layers to better represent the surface features, while layer 3 represented the confining unit and layer 4 the deep aquifer. The grid cells are designated as inactive outside the model domain and in the impermeable areas, and as active inside the model domain. The regional Shuttle Radar Topography Mission (STRM) Digital Elevation Model (DEM) with a spatial resolution of 20 m × 20 m cells was used and interpolated over the model grid to reproduce the basin topography. The hydraulic conductivity (K) of lithological units was obtained from pumping tests: 41 pumping tests were performed in the shallow aquifer and 30 in the deep aquifer. The resulting mean K values and the respective standard deviation were $5.35 \times 10^{-4} \pm 3.54 \times 10^{-4}$ m/s for the shallow aquifer and $1.44 \times 10^{-3} \pm 7.73 \times 10^{-4}$ m/s for the deep aquifer. Thus, the deep aquifer is more permeable than the shallow one, and both are characterized by relatively homogeneous K distributions. The confining unit was simulated with a K of 1.0×10^{-9} m/s to ensure sealing among the unconfined and confined aquifers to maintain the observed heads differences among the aquifers. The vertical anisotropy was set equal to 1:10 in all layers as suggested by the PM11 manual [40].

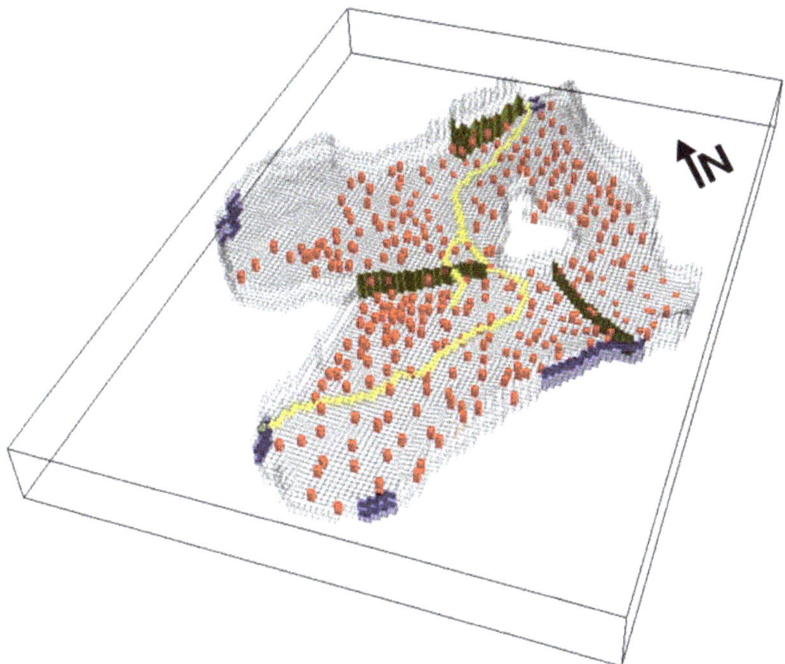

Figure 8. 3D discretization and boundary conditions of the Mio-Plio-Quaternary aquifer system: drain cells representing the Wadis (yellow), pumping wells (red), General Head Boundary representing the inflow and outflow from the basin (blue), and HFB representing the major faults (olive green). Vertical exaggeration is 1:20.

The boundary conditions are presented by assumed or known supplies and/or flows [33]. The lateral contribution from the nearby basins were determined based on the hydraulic potentials and were simulated via the General Head Boundary package (Figure 8), as well as the contribution of the Sidi Saad dam to the supply of the downstream part of the basin. To ensure a good connection with the aquifers, a very high value (0.1 m^2/s) of water supply from Sidi Saad dam was set up. The Well package was employed to distribute an average pumping rate in more than 300 wells scattered throughout the model domain in both aquifers. Groundwater drainage from the beds of wadis (El Fekka, Sarigh Dhiba, Sbitla, and Jilma) was simulated with the Drain package (Figure 8), using a conductance of 0.001 m^2/s to allow a good drainage from the nearby cells and setting the elevation of the drain 2 m below the topographic surface. Recharge was initially set to 43 mm/y (approximately 20% of precipitation) and multiplied by an altitude factor of 1.44 over 100 m. This factor was calculated comparing 12 meteorological stations ranging from 297 to 413 m a.s.l. to include the higher recharge from the mountain ranges that border the basin. The Evapotranspiration package was used to simulate the evapotranspiration from groundwater, using the mean value for the area from 1990 to 2010 [44] as maximum uptake rate (630 mm/y) and an average extinction depth of 3 m. The horizontal flow barrier (HFB) package was used to simulate the compressive faults, using a thickness of 1 m and an equivalent K of 1.0×10^{-9} m/s (Figure 8).

The model calibration was initially obtained manually by a trial-and-error approach, then employing PEST for an automated calibration and sensitivity analysis [45]. The vertical K were tied to the horizontal K parameters that were log transformed, and their minimum and maximum values were set using the observed ones (see Supplementary Information of K values in Table S3), while the recharge and evapotranspiration rates were calibrated as relative parameters. The observations used in the model calibration include the observed hydraulic heads in the 53 observation wells for the shallow aquifer and 38 observation wells for the deep aquifer. Quantitatively, the model calibration performance was evaluated by the criterion of the mean error (ME), the mean square error (MSE), and the determination coefficient (R^2).

The model calibration was carried out by automatically tuning the K values and other parameters like maximum evapotranspiration rate, recharge rate, and drain conductance, reported in Table 2 with their composite sensitivities. The comparison between observed and simulated head for both the shallow and the deep aquifers is shown in Figure 9. The points are scattered along the 1:1 line, with no apparent pattern. However, there is a slight underestimation in the calculated heads as suggested by the mean error (ME) that denotes approximately −0.26 m of error, which is acceptable compared to the whole piezometric range simulated here (more than 150 m) with the R^2 being higher than 0.9 (Figure 9). The relatively large MSE of calculated heads could have been due to local variability of hydraulic conductivity here not considered to maintain the simplicity of the model as much as possible.

Table 2. Calibrated parameters values and their composite sensitivity via PEST.

Parameter	Value	Composite Sensitivity
K layer 1 Shallow aquifer (m/s)	8.50×10^{-5}	4705
K layer 2 Shallow aquifer (m/s)	1.00×10^{-3}	364
K layer 4 Deep aquifer (m/s)	1.37×10^{-3}	707
Recharge rate (mm/y)	38.7	134
Evapotranspiration rate (mm/y)	625	13
GHB Conductance (m^2/s)	0.1	22
Drain Conductance (m^2/s)	0.24	122

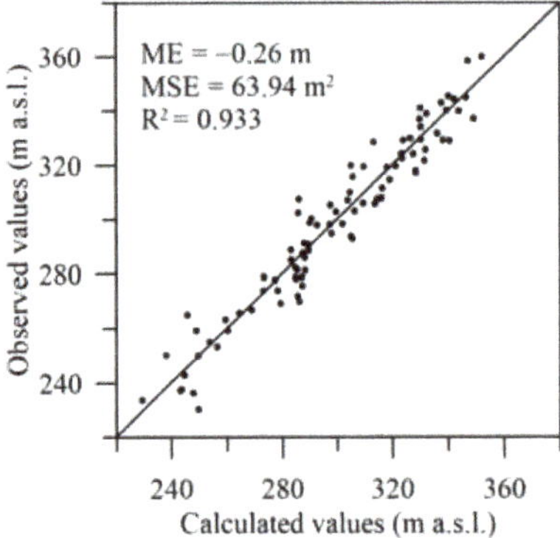

Figure 9. Scatter diagram of the observed versus calculated head values (dots) for the simulated aquifers system.

The most sensitive parameter was the K of the first layer which was mostly unsaturated, followed by the K value of the deep aquifer indicating that the most uncertain parameters are the hydraulic conductivities of the aquifers.

For the shallow aquifer, the hydraulic head ranged between 350 and 260 m, and high values were observed in the western part of the Awled Asker sub-basin (Figure 10a). The main flow direction goes from the Southwest and the West draining Mghila and Al Rakhmet mountains to the discharge area in the evaporation areas and the El Hjal Wadi in the north and the Centre-East of the study area.

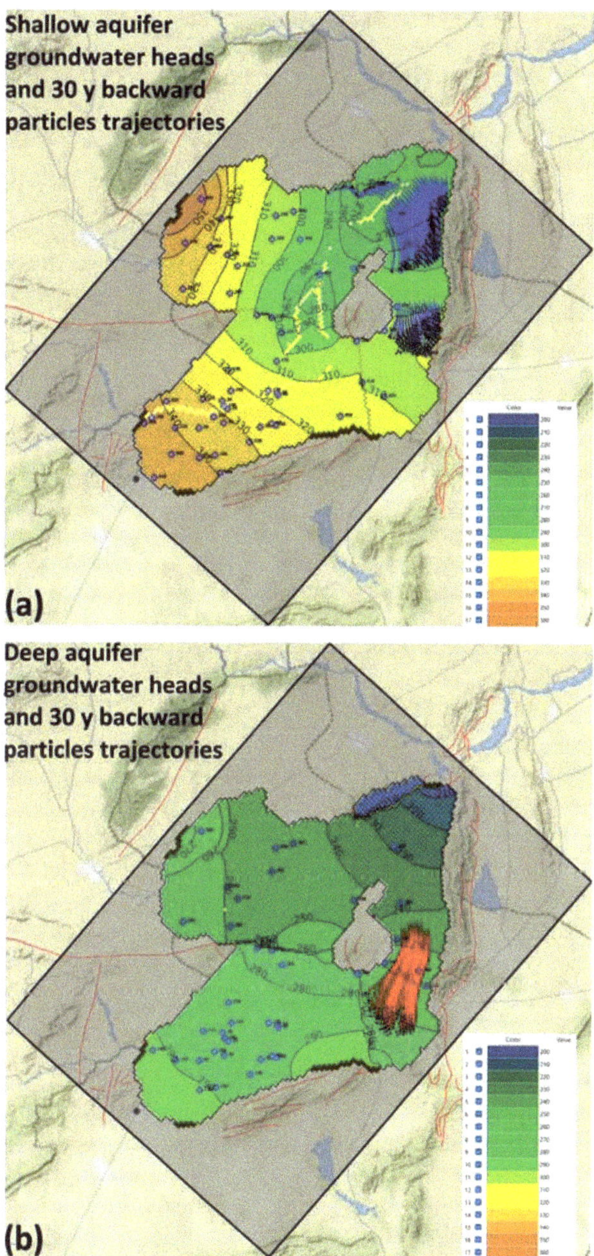

Figure 10. Contours of the calculated groundwater heads for the shallow (**a**) and deep aquifers (**b**). The backward particle trajectories from the zones with high NO_3^- concentration for the period 1990–2020, in blue for the shallow aquifer and red for the deep aquifer; black points delineate the NO_3^- source zones in 1990.

For the deep aquifer, the hydraulic head map (Figure 10b) shows values between 270 m and 210 m, thus potentially draining the shallow aquifer. The flow direction is mostly from the West and Southwest towards Northeast. From the model, the flow converges in the south of Lessouda mountain (here simulated as no flow boundary and the center of the model). Then it is divided into two directions: (1) to the East of the study area where discharge is in the evaporation area of Al Akarich and Negada, and (2) to the North towards the El Hjal Wadi where the drainage takes place. This division of the flow in the two directions imposes the assumption of the existence of the deep faults of Lessouda Boudinar and Kassrine that act as horizontal flow barriers. Without using the HFB package, it was not possible to reach an acceptable calibration due to groundwater heads that were too low or too high with respect to the measured ones near faults.

In the Sidi Bouzid North basin, the groundwater budget, in the steady state condition, shows a good balance between input and output flows with a percent error equal to 0.16% and −0.35% for the shallow and deep aquifers, respectively (Table 3). The main input to the shallow aquifer is the recharge through porous deposits, which is estimated at 2.78 m^3/s and constitutes 71.5% of the total inflow of the shallow aquifer and the 36% of the whole aquifers system. The contribution of the GHB estimated by the model are: (1) for the inputs, 1.106 m^3/s and 3.59 m^3/s for the shallow and deep aquifer, respectively; and (2) for the outputs, 8.55×10^{-2} m^3/s towards the Sidi Saad dam for the shallow aquifer and 0.80 m^3/s for the deep aquifer towards the basin located in the north of the study area. For the vertical groundwater leakage between the aquifers, the input from the shallow to the deep aquifer is estimated at 0.300 m^3/s. The main outflow of the aquifers system is divided into exploitation by wells (1.01 m^3/s from the shallow aquifer and 3.1 m^3/s from the deep aquifer), drainage towards the wadis, and evapotranspiration which is found only in the shallow aquifer with estimated values of 2.34 m^3/s and 0.157 m^3/s, respectively. It can be noticed that the actual overexploitation from wells is not sustainable by the recharge occurring in the study area; in fact, a water table drawdown (approximately 20 m) has been experienced by both the shallow and the deep aquifer in the last decades.

Table 3. Groundwater balance of the shallow and deep aquifers: GHB (Head Dependent flux Boundaries).

	Shallow Aquifer		Deep Aquifer	
Flow in	**m^3/s**	**%**	**m^3/s**	**%**
GHB	1.106	28.5	3.59	92.0
Recharge	2.78	71.5	0	0
From Shallow aquifer	0	0	0.300	8.0
Flow out	m^3/s	%	m^3/s	%
Wells	1.01	25.7	3.10	79.4
Drains	2.34	60.3	0	0
Evapotranspiration	0.157	4.0	0	0
GHB	8.55×10^{-2}	2.2	0.80	20.5
To Deep aquifer	0.300	8.7	0	0
Total	3.88	100	3.90	100
IN-OUT	6.39×10^{-3}		-1.35×10^{-2}	
Percent error	0.16		−0.35	

4.3. Particle Tracking Results

The results of MODPATH via forward particle transport highlighted that it was not possible to bypass the thick aquitard between the shallow and deep aquifer since the beginning of urban settlement growth and use of synthetic fertilizers since the early 1970s. Moreover, it must be stressed that the choice of a steady state model is a conservative option, since the overexploitation of the aquifer experienced a dramatic increase in the last 30 years.

The simulation shown in Figure 10 backtracked particles from the NO_3^- enriched zones (see Figure 5 for location) towards their possible origin before the overexploitation of the aquifer system for a period of 30 years (from 1990 to 2020). This approach allows the interpretation of flow directions and travel times and possibly reconstructs the location of the source zones. The flow paths of advective transport in the shallow aquifer differ from those obtained in the deep aquifer. While the shallow aquifer has the NO_3^- source zones located near the contaminated zones and within agricultural areas, the source zone for the deep aquifer is located approximately 10 to 15 m upgradient and southward to Oued El Hjal sub-basin (between Jebel Lessouda and Jebel Faidh), where no agricultural fields are present, since the small valley is a former Sabkha.

In the Sidi Bouzid basin, the NO_3^- accumulations are associated with pre-Triassic embankment landscapes and occur on domed geomorphic surfaces with a gypsum, phosphate or clay cover. The rough surfaces of surficial materials trap fine textured aeolian sediments rich in organic matter, which are then washed under or over the sides of the syncline during intense episodic rainfall events. For thousands of years, the eolian deposits accumulated and raised a fine mosaic of an alluvial embankment to form a desert stone pavement. These deposits stabilized over time, decreasing infiltration, increasing surface runoff, and allowing soluble salts to accumulate below the surface. Heaton [40] suggested that the high NO_3^- concentrations in groundwater result from N fixation by cyanobacteria and subsequent mineralization and nitrification of the organic matter over time. However, the first hypothesis could not be confirmed with the available data. The hydrochemical data and Figure 10 indicate that the NO_3^- in the deep aquifer of Oued El Hjal is likely to have come from natural sources and could not have been derived from anthropogenic sources.

As reported by Kaplan et al. [46], according to the logic of nature, the accumulation of contaminants that comes from the surface takes place in the unconfined aquifer, which is not the case of this study where the highest NO_3^- concentrations were found in the deep aquifer. Also, the examination of the land cover map shows that the areas with the highest NO_3^- concentrations coincide with bare land or olive trees that do not require excessive use of N fertilizers. Thus, the above-mentioned features suggest that the first hypothesis is far from being accepted. Nevertheless, new data on groundwater ages and environmental tracers should be collected in future studies to independently confirm or reject this conceptual model.

5. Conclusions

In this study, NO_3^- origin in groundwater of the Mio-Plio-Quaternary aquifer of Sidi Bouzid North basin was assessed. To evaluate the origin of extremely high NO_3^- concentrations in deep groundwater samples, hydrochemical investigations and groundwater flow modeling were employed. Geostatistical analyses were used for hydrochemistry assessment, and the correlation among solute species showed that groundwater is affected by evaporite dissolution and water quality changes in the whole studied area. PCA showed that NO_3 in most samples have origins associated with other chemical elements related to evaporitic salts dissolution. Groundwater flow modeling highlighted that recharge was the most important groundwater inflow into the aquifer system, while exploitation by wells is the most important outflow. Moreover, the particle tracking simulation showed that leakage of NO_3^- through the aquitard between the shallow unconfined aquifer and the deep aquifer was negligible, which further suggested that the NO_3^- origin in the deep aquifer is geogenic. On the contrary, the NO_3^- origin in the shallow aquifer is anthropogenic and mainly due to fertilizers leaching.

The approach developed in this study can be a valuable decision support tool for groundwater resource managers in the Sidi Bouzid North basin, and the approach can be replicated in similar environmental settings. However, to make the implemented model a more robust tool for integrated water resources management in the study area, future simulations must be applied using transient state models based on several scenarios and new data on groundwater age and origin must confirm the proposed conceptual model.

Supplementary Materials: The following supporting information can be downloaded at: https://www.mdpi.com/article/10.3390/w14244124/s1, Table S1: Factor Loadings, Table S2: Hydrochemical data, Table S3: Transmissivity thickness and K.

Author Contributions: N.C. and M.M. conceived and designed the study; K.N. performed field activities; K.N. and G.B. elaborated the data in GIS; K.N. elaborated the hydrochemical data with the supervision of N.C., R.H. and Y.H.; N.C. developed the numerical model; K.N. wrote the draft of the paper. K.S. improved the conceptual model and revised the manuscript along with all the other authors. All authors have read and agreed to the published version of the manuscript.

Funding: This research received no external funding.

Data Availability Statement: Data will be made available by the authors upon request.

Conflicts of Interest: The authors declare no conflict of interest.

References

1. Guermazi, E.; Milano, M.; Reynard, E.; Zairi, M. Impact of climate change and anthropogenic pressure on the groundwater resources in arid environment. *Mitig. Adapt. Strateg. Glob. Chang.* **2019**, *24*, 73–92. [CrossRef]
2. Nadiri, A.A.; Moghaddam, A.A.; Tsai, F.T.; Fijani, E. Hydrogeochemical analysis for Tasuj plain aquifer, Iran. *J. Earth Syst. Sci.* **2013**, *122*, 1091–1105. [CrossRef]
3. Hamed, Y.; Awad, S.; Ben Sâad, A. Nitrate contamination in groundwater in the Sidi Aïch–Gafsa oases region, Southern Tunisia. *Environ. Earth Sci.* **2013**, *70*, 2335–2348. [CrossRef]
4. Nadiri, A.A.; Aghdam, F.S.; Khatibi, R.; Moghaddam, A.A. The problem of identifying arsenic anomalies in the basin of Sahand dam through risk-based 'soft modelling'. *Sci. Total Environ.* **2018**, *613*, 693–706. [CrossRef] [PubMed]
5. Busico, G.; Alessandrino, L.; Mastrocicco, M. Denitrification in intrinsic and specific groundwater vulnerability assessment: A review. *Appl. Sci.* **2021**, *11*, 10657. [CrossRef]
6. Nadiri, A.A.; Aghdam, F.S.; Razzagh, S.; Barzegar, R.; Jabraili-Andaryan, N.; Senapathi, V. Using a soft computing OSPRC risk framework to analyze multiple contaminants from multiple sources; a case study from Khoy Plain, NW Iran. *Chemosphere* **2022**, *308*, 136527. [CrossRef]
7. Nadiri, A.A.; Sadeghfam, S.; Gharekhani, M.; Khatibi, R.; Akbari, E. Introducing the risk aggregation problem to aquifers exposed to impacts of anthropogenic and geogenic origins on a modular basis using 'risk cells'. *J. Environ. Manag.* **2018**, *217*, 654–667. [CrossRef]
8. Hamed, Y.; Hadji, R.; Ncibi, K.; Hamad, A.; Ben Saad, A.; Melki, A.; Khelifi, F.; Mokadem, N.; Mustafa, E. Modelling of potential groundwater artificial recharge in the transboundary Algero-Tunisian Basin (Tebessa-Gafsa): The application of stable isotopes and hydroinformatics tools. *Irrig. Drain.* **2022**, *71*, 137–156. [CrossRef]
9. Ascott, M.J.; Gooddy, D.C.; Wang, L.; Stuart, M.E.; Lewis, M.A.; Ward, R.S.; Binley, A.M. Global patterns of nitrate storage in the vadose zone. *Nat. Commun.* **2017**, *8*, 1416. [CrossRef]
10. Nadiri, A.A.; Sedghi, Z.; Khatibi, R. Qualitative risk aggregation problems for the safety of multiple aquifers exposed to nitrate, fluoride and arsenic contaminants by a 'Total Information Management' framework. *J. Hydrol.* **2021**, *595*, 126011. [CrossRef]
11. Cox, A.N. *Allen's Astrophysical Quantities*, 4th ed.; AIP Press: New York, NY, USA, 2000; ISBN 0-387-98746-0.
12. Walvoord, M.A.; Phillips, F.M.; Stonestrom, D.A.; Evans, R.D.; Hartsough, P.C.; Newman, B.D.; Striegl, R.G. A reservoir of nitrate beneath desert soils. *Science* **2003**, *302*, 1021–1024. [CrossRef] [PubMed]
13. Ncibi, K.; Hadji, R.; Hamdi, M.; Mokadem, N.; Abbes, M.; Khelifi, F.; Zighmi, K.; Hamed, Y. Application of the analytic hierarchy process to weight the criteria used to determine the Water Quality Index of groundwater in the northeastern basin of the Sidi Bouzid region, Central Tunisia. *Euro-Mediterr. J. Environ. Integr.* **2020**, *5*, 19. [CrossRef]
14. Valiente, N.; Gil-Márquez, J.M.; Gómez-Alday, J.J.; Andreo, B. Unraveling groundwater functioning and nitrate attenuation in evaporitic karst systems from southern Spain: An isotopic approach. *Appl. Geochem.* **2020**, *123*, 104820. [CrossRef]
15. World Health Organization's (WHO). *Guidelines for Drinking-Water Quality: Fourth Edition Incorporating the First and second Addenda*; WHO: Geneva, Switzerland, 2022; p. 614. ISBN 978-92-4-004506-4.
16. Razzagh, S.; Nadiri, A.A.; Khatibi, R.; Sadeghfam, S.; Senapathi, V.; Sekar, S. An investigation to human health risks from multiple contaminants and multiple origins by introducing 'Total Information Management'. *Environ. Sci. Pollut. Res.* **2021**, *28*, 18702–18724. [CrossRef] [PubMed]
17. Ncibi, K.; Mosbahi, M.; Gaaloul, N. Assessment of groundwater risk to Plio-quaternary aquifer's contamination: Semi-arid climate case (central Tunisia). *Desalin. Water Treat.* **2018**, *124*, 211–222. [CrossRef]
18. Ncibi, K.; Hadji, R.; Hajji, S.; Besser, H.; Hajlaoui, H.; Hamad, A.; Mokadem, N.; Ben Saad, A.; Hamdi, M.; Hamed, Y. Spatial variation of groundwater vulnerability to nitrate pollution under excessive fertilization using Index Overlay Method in Central Tunisia. *Irrig. Drain.* **2021**, *70*, 1209–1226. [CrossRef]
19. Yangui, H.; Zouari, K.; Trabelsi, R. Recharge mode and mineralization of groundwater in a semi-arid region: Sidi Bouzid plain (central Tunisia). *Environ. Earth Sci.* **2011**, *63*, 969–979. [CrossRef]

20. Petelet-Giraud, E.; Baran, N.; Vergnaud-Ayraud, V.; Portal, A.; Michel, C.; Joulian, C.; Lucassou, F. Elucidating heterogeneous nitrate contamination in a small basement aquifer. A multidisciplinary approach: NO_3 isotopes, CFCs-SF6, microbiological activity, geophysics and hydrogeology. *J. Contam. Hydrol.* **2021**, *241*, 103813. [CrossRef]
21. Busico, G.; Cuoco, E.; Kazakis, N.; Colombani, N.; Mastrocicco, M.; Tedesco, D.; Voudouris, K. Multivariate statistical analysis to characterize/discriminate between anthropogenic and geogenic trace elements occurrence in the Campania Plain, Southern Italy. *Environ. Pollut.* **2018**, *234*, 260–269. [CrossRef]
22. Hosono, T.; Tokunaga, T.; Kagabu, M.; Nakata, H.; Orishikida, T.; Lin, I.T.; Shimada, J. The use of $\delta^{15}N$ and $\delta^{18}O$ tracers with an understanding of groundwater flow dynamics for evaluating the origins and attenuation mechanisms of nitrate pollution. *Water Res.* **2013**, *47*, 2661–2675. [CrossRef]
23. Xue, Y.; Song, J.; Zhang, Y.; Kong, F.; Wen, M.; Zhang, G. Nitrate pollution and preliminary source identification of surface water in a semi-arid river basin, using isotopic and hydrochemical approaches. *Water* **2016**, *8*, 328. [CrossRef]
24. Lorette, G.; Sebilo, M.; Buquet, D.; Lastennet, R.; Denis, A.; Peyraube, N.; Charriere, V.; Studer, J.C. Tracing sources and fate of nitrate in multilayered karstic hydrogeological catchments using natural stable isotopic composition ($\delta^{15}N$-NO_3^- and $\delta^{18}O$-NO_3^-). Application to the Toulon karst system (Dordogne, France). *J. Hydrol.* **2022**, *610*, 127972. [CrossRef]
25. Xuan, Y.; Liu, G.; Zhang, Y.; Cao, Y. Factor affecting nitrate in a mixed land-use watershed of southern China based on dual nitrate isotopes, sources or transformations? *J. Hydrol.* **2022**, *4*, 127220. [CrossRef]
26. Kendall, C.; Aravena, R. Nitrate Isotopes in Groundwater Systems. In *Environmental Tracers in Subsurface Hydrology*; Cook, P.G., Herczeg, A.L., Eds.; Springer: Boston, MA, USA, 2000. [CrossRef]
27. Widory, D.; Kloppmann, W.; Chery, L.; Bonnin, J.; Rochdi, H.; Guinamant, J.L. Nitrate in groundwater: An isotopic multi-tracer approach. *J. Contam. Hydrol.* **2004**, *72*, 165–188. [CrossRef] [PubMed]
28. Levy, Y.; Shapira, R.H.; Chefetz, B.; Kurtzman, D. Modeling nitrate from land surface to wells' perforations under agricultural land: Success, failure, and future scenarios in a Mediterranean case study. *Hydrol. Earth Syst. Sci.* **2017**, *21*, 3811–3825. [CrossRef]
29. Mastrocicco, M.; Colombani, N.; Castaldelli, G.; Jovanovic, N. Monitoring and modeling nitrate persistence in a shallow aquifer. *Water Air Soil Pollut.* **2011**, *217*, 83–93. [CrossRef]
30. Pulido-Velazquez, M.; Peña-Haro, S.; García-Prats, A.; Mocholi-Almudever, A.F.; Henríquez-Dole, L.; Macian-Sorribes, H.; Lopez-Nicolas, A. Integrated assessment of the impact of climate and land use changes on groundwater quantity and quality in the Mancha Oriental system (Spain). *Hydrol. Earth Syst. Sci.* **2015**, *19*, 1677–1693. [CrossRef]
31. Koh, E.H.; Lee, E.; Kaown, D.; Green, C.T.; Koh, D.C.; Lee, K.K.; Lee, S.H. Comparison of groundwater age models for assessing nitrate loading, transport pathways, and management options in a complex aquifer system. *Hydrol. Process.* **2018**, *32*, 923–938. [CrossRef]
32. Harbaugh, A.W.; Langevin, C.D.; Hughes, J.D.; Niswonger, R.N.; Konikow, L.F. *MODFLOW-2005, Version 1.12.00*; The U.S. Geological Survey Modular Groundwater Model: U.S. Geological Survey Software Release, 3 February 2017; USGS: Reston, VA, USA, 2017. [CrossRef]
33. Pollock, D.W. *MODPATH, v7.2.01*; A Particle-Tracking Model for MODFLOW: U.S. Geological Survey Software Release, 15 December 2017; USGS: Reston, VA, USA, 2017. [CrossRef]
34. Missaoui, R.; Abdelkarim, B.; Ncibi, K.; Hamed, Y.; Choura, A.; Essalami, L. Assessment of Groundwater Vulnerability to Nitrate Contamination Using an Improved Model in the Regueb Basin, Central Tunisia. *Water Air Soil Pollut.* **2022**, *233*, 320. [CrossRef]
35. Hamdi, M.; Louati, D.; Rajouene, M.; Abida, H. Impact of spate irrigation of floodwaters on agricultural drought and groundwater recharge: Case of Sidi Bouzid plain, Central Tunisia. *Arab. J. Geosci.* **2016**, *9*, 653. [CrossRef]
36. Kocsis, L.; Ounis, A., Baumgartner, C.; Pirkenseer, C.; Harding, I.C.; Adatte, T.; Chaabani, F.; Neili, S.M. Paleocene-Eocene palaeoenvironmental conditions of the main phosphorite deposits (Chouabine Formation) in the Gafsa Basin, Tunisia. *J. African Earth Sci.* **2014**, *100*, 586–597. [CrossRef]
37. National Mining Office. *Geological Map of Tunisie*; National Mining Office: Tunis, Tunisia, 1987.
38. Gaiolini, M.; Colombani, N.; Busico, G.; Rama, F.; Mastrocicco, M. Impact of Boundary Conditions Dynamics on Groundwater Budget in the Campania Region (Italy). *Water* **2022**, *14*, 2462. [CrossRef]
39. Graham, R.C.; Hirmas, D.R.; Wood, Y.A.; Amrhein, C. Large near-surface nitrate pools in soils capped by desert pavement in the Mojave Desert, California. *Geology* **2008**, *36*, 259–262. [CrossRef]
40. Chiang, E. *Processing Modflow*, Version 11.0.3; A Graphical User Interface for MODFLOW, GSFLOW, MODPATH, MT3D, PEST, SEAWAT, and ZoneBudget, Simcore Software, Irvine, California; 4 July 2022. Available online: https://www.simcore.com/wp/processing-modflow-11/ (accessed on 1 December 2022).
41. Hallenberger, M.; Reuning, L.; Schoenherr, J. Dedolomitization Potential of Fluids from Gypsum-to-Anhydrite Conversion: Mass Balance Constraints from the Late Permian Zechstein-2-Carbonates in NW Germany. *Geofluids* **2018**, *2018*, 1784821. [CrossRef]
42. Besser, H.; Dhaouadi, L.; Hadji, R.; Hamed, Y.; Jemmali, H. Ecologic and economic perspectives for sustainable irrigated agriculture under arid climate conditions: An analysis based on environmental indicators for southern Tunisia. *J. Afr. Earth Sci.* **2021**, *177*, 104134. [CrossRef]
43. Schoenherr, J.; Reuning, L.; Hallenberger, M.; Lüders, V.; Lemmens, L.; Biehl, B.C.; Lewin, A.; Leupold, M.; Wimmers, K.; Strohmenger, C.J. Dedolomitization: Review and case study of uncommon mesogenetic formation conditions. *Earth-Sci. Rev.* **2018**, *185*, 780–805. [CrossRef]

44. Aschonitis, V.G.; Papamichail, D.; Demertzi, K.; Colombani, N.; Mastrocicco, M.; Ghirardini, A.; Castaldelli, G.; Fano, E.A. High-resolution global grids of revised Priestley–Taylor and Hargreaves–Samani coefficients for assessing ASCE-standardized reference crop evapotranspiration and solar radiation. *Earth Syst. Sci. Data* **2017**, *9*, 615–638. [CrossRef]
45. Doherty, J. *PEST-Model-Independent Parameter Estimation*, Version 12; Watermark Computing: Corinda, Australia, 2010. Available online: http://www.pesthomepage.org/ (accessed on 1 November 2022).
46. Kaplan, D.I.; Bertsch, P.M.; Adriano, D.C.; Miller, W.P. Soil-borne mobile colloids as influenced by water flow and organic carbon. *Environ. Sci. Technol.* **1993**, *27*, 1193–1200. [CrossRef]

MDPI

Article

Groundwater Modeling with Process-Based and Data-Driven Approaches in the Context of Climate Change

Matia Menichini [1], Linda Franceschi [1,2,*], Brunella Raco [1], Giulio Masetti [1], Andrea Scozzari [3] and Marco Doveri [1]

[1] Istituto di Geoscienze e Georisorse-Consiglio Nazionale delle Ricerche (IGG-CNR), Via G. Moruzzi 1, 56124 Pisa, Italy
[2] Dipartimento di Scienze della Terra, Università di Pisa, Via Santa Maria 53, 56126 Pisa, Italy
[3] Istituto di Scienza e Tecnologie dell'Informazione-Consiglio Nazionale delle Ricerche (ISTI-CNR), Via G. Moruzzi 1, 56124 Pisa, Italy
* Correspondence: linda.franceschi@igg.cnr.it

Abstract: In the context of climate change, the correct management of groundwater, which is strategic for meeting water needs, becomes essential. Groundwater modeling is particularly crucial for the sustainable and efficient management of groundwater. This manuscript provides different types of modeling according to data availability and features of three porous aquifer systems in Italy (Empoli, Magra, and Brenta systems). The models calibrated on robust time series enabled the performing of forecast simulations capable of representing the quantitative and qualitative response to expected climate regimes. For the Empoli aquifer, the process-based models highlighted the system's ability to mitigate the effects of dry climate conditions thanks to its storage capability. The data-driven models concerning the Brenta foothill aquifer pointed out the high sensitivity of the system to climate extremes, thus suggesting the need for specific water management actions. The integrated data-driven/process-based approach developed for the Magra Valley aquifer remarked that the water quantity and quality effects are tied to certain boundary conditions over dry climate periods. This work shows that, for groundwater modeling, the choice of the suitable approach is mandatory, and it mainly depends on the specific aquifer features that result in different ways to be sensitive to climate. This manuscript also provides a novel outcome involving the integrated approach wherein it is a very efficient tool for forecasting modeling when boundary conditions, which significantly affect the behavior of such systems, are subjected to evolve under expected climate scenarios.

Keywords: water management; groundwater forecasting; foothill aquifers; Brenta River plain; Empoli plain; Magra Valley

Citation: Menichini, M.; Franceschi, L.; Raco, B.; Masetti, G.; Scozzari, A.; Doveri, M. Groundwater Modeling with Process-Based and Data-Driven Approaches in the Context of Climate Change. *Water* **2022**, *14*, 3956. https://doi.org/10.3390/w14233956

Academic Editor: Cristina Di Salvo

Received: 5 October 2022
Accepted: 25 November 2022
Published: 5 December 2022

1. Introduction

Most of the available freshwater sources on Earth are stored underground; therefore, groundwater represents the main source of water supply [1]. Worldwide, more than 2 billion people depend on groundwater for their daily water use [2]. In many areas, groundwater bodies represent the most important and safest source of drinking water [3,4]. In European countries, for example, groundwater exploitation provides water for human consumption for 70% of the population on average [5]. Groundwater withdrawals supply 40% of industrial water [6], and groundwater use for irrigation is also significant and increasing. Siebert et al. [7] estimated that, globally, 38% of the area equipped for irrigation is provided by groundwater.

The reliance on this resource is continuously growing [8], given the key role that groundwater plays in mitigating climate change/variability and in addressing the significant increase in the global water demand, which has been predicted as a consequence of the future economic expansion, population growth, and urbanization [1,9].

Both present observations and simulations of climate conditions point out that changes are occurring or will occur in several regions in terms of higher temperatures, lower precipitation, increases in climatic extremes, and climate types [10–14]. Although groundwater systems are more resilient to climate change than surface waters, they are affected both directly and indirectly [15,16], especially when referring to local groundwater flow with low travel time [17].

Despite this, and unlike surface waters, groundwater bodies have not been widely studied, and there is a general paucity of quantitative information, especially in relation to climate change. The estimation of the entity of these effects is mandatory for the reliable management of this crucial resource, which must be protected by suitable actions in order to guarantee a safe water supply for the next generations [18].

Groundwater modeling is particularly crucial for the sustainable and efficient management of groundwater resources, even more in the context of expected climate change. Process-based or empirical (data-driven) numerical models can be used to model groundwater.

For the implementation of process-based models, it is necessary to know the boundary conditions, hydrogeological variables, and structural complexities of the aquifers [19]; in other words, it is necessary to know the conceptual model that is the synthesis of what is known and quantified on the system [20]. These models use deterministic and spatially distributed data, which must be characterized by significant accuracy. Their application to heterogeneous aquifers and particular situations within the hydrogeological domain (e.g., groundwater–river interactions) can lead to significant uncertainty associated with the number and complexity of parameters, as well as with the type of modeling itself based on the head-oriented approach (HOA) [21,22]. In such cases, aside from the alternative of the MODFLOW-based velocity-oriented approach (VOA) [22], an additional possible approach is to use the stochastic approach in defining the aquifer parameters [23].

In contrast, data-driven models involve mathematical equations that are not derived from prior knowledge of the physical process. Rather, they are based on the analysis of input/output relationships in the process under observation [24–26], even if the knowledge of the general conceptual model of the systems can significantly steer the development of the data-driven approach. To perform such analysis, mathematical tools, often involving machine-learning techniques, are used to approximate the behavior of the physical process on the basis of available datasets that describe its input/output transfer functions (e.g., linear regressions, multilinear regressions, neural network models, etc.). Therefore, data-driven models can be efficiently used to describe particular and specific processes [27,28] that are hard to determine with a physically based approach and can eventually be incorporated into larger physically based models [29–31].

Generally, process-based models are preferred over data-driven models because the first can make acceptable forecasts when a large number of observations are not available and when future conditions lie outside the range of stresses in the historical record, such as responses to climate change [19]. Moreover, the data-driven approach has the limitation of referring to punctual situations or specific processes in the groundwater system (e.g., relationships between superficial water and groundwater) and not to the entire volume or significant portions of the aquifer, as in the case of process-based models. Additionally, for this reason, the latter is generally preferred by decision-makers to steer groundwater management actions.

This manuscript provides different experiences of groundwater modeling applied to porous aquifer systems, with the aim of emphasizing the criteria of the methodology choice, its advantages and disadvantages, and the potentiality of a combined approach.

All the presented models represent operational tools for the reliable management of the water resource, as well as diagnostic tools to understand the functioning of the aquifer system better and how the effects of climate change in the short and long term affect the water resource quantity and quality.

2. Materials and Methods

Three case studies are presented (Figure 1): the Empoli plain aquifer system, which was modeled using a process-based model; the foothill aquifer system of the Brenta River plain, where a mathematical regression model was applied; and the Magra Valley aquifer system, where a combined approach with process-based and data-driven modeling was developed.

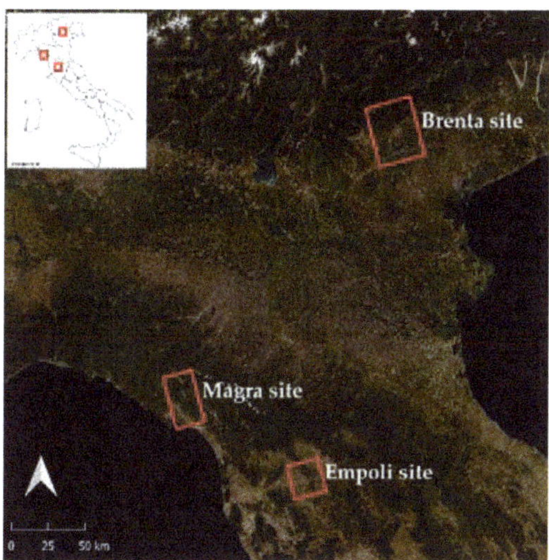

Figure 1. Location of the three case studies.

The considered case studies encompass different combinations of recharge mechanisms, thus providing cases with different implications of climate change for groundwater systems [32]. Particularly, we analyzed situations in which the diffuse recharge is dominant and situations in which the focused recharge (i.e., losing rivers) is from significant to principal and, therefore, very sensitive to changes in the weather and climate regime, both in terms of quantity availability and water quality [33].

The first step was to collect data on meteo-climatic parameters, geology, hydrogeology, geochemistry and isotope signatures, as well as on the main items entering or leaving the system (e.g., well withdrawal). All the data collected were processed and compared in order to define a conceptual model that summarizes and quantifies the main processes and constraints of the aquifer for the purpose of developing the mathematical model.

Most geological, hydrogeological, and geochemical information used to elaborate the conceptual model of the different areas derive from the numerous studies that the authors carried out in cooperation with local authorities (Tuscany Region, Water Authorities, Water Managers) and/or in the frame of doctoral theses.

As a consequence of the hydrogeological and hydrodynamic features of the systems and data availability, different types of mathematical modeling were chosen: process-based modeling, data-driven modeling, and a combined approach.

2.1. Process-Based Model

The process-based model uses processes and principles of physics to represent flow, and it consists of: (i) a governing equation that describes the physical process within the problem domain; (ii) boundary conditions that specify heads or flow along the boundaries; and (iii) for time-dependent problems, initial conditions that specify head at the beginning of the simulation [19]. Groundwater flow models can be solved analytically or numerically for the distribution of heads in space and in time for transient problems. Assumptions built into analytical solutions limit their application to relatively simple systems [34], and usually, they are inappropriate for real groundwater problems. Typically, numerical models based on finite differences are used to simulate groundwater flow.

In this work, the codes used to process the input data and solve the process-based equation that describes the groundwater system are MODFLOW [35] and related codes. Groundwater Vistas [36] and Groundwater Modeling Systems [37] were used as graphical user interfaces.

The model creation started with the implementation phase, where the domain in space and time was discretized, hydraulic parameters applied, and boundary conditions defined based on the conceptual model previously defined. Subsequently, the model was calibrated on the basis of a comparison between the simulated and collected data. The calibration was performed manually (trial-and-error method) and automatically using specific code, like the PEST code [38].

Once a sufficiently calibrated model was obtained, forecast simulations were then carried out with the aim of assessing how the expected extreme climatic regimes would affect the groundwater quantity and quality.

The development of this type of model requires a large amount of information and input data evenly distributed over the domain. In particular, it is necessary to know the geometries of the system under study, the hydraulic parameters of the various lithotypes that make it up and to identify and quantify the main components entering and leaving the system. Moreover, the representativeness of the model itself is inextricably linked to the availability and reliability of the data required in the calibration phase.

The process-based modeling was applied to a porous multilayer system (Empoli plain aquifer system) in the central part of the Arno River catchment (Central Italy, Tuscany) because previous knowledge and available data were deemed sufficient and suitable for the development of this type of model.

2.2. *Data-Driven Model*

Data-driven models use empirical or statistical equations derived from the available data to approximate the input/output relationships between physical variables characterizing a system without quantifying its process and physical properties [19].

Initially, a site-specific equation is developed by fitting parameters either empirically or statistically to reproduce the historical data (e.g., piezometric level) in response to other parameters (e.g., surface water level, precipitation). This equation is subsequently used to calculate the response to future stresses.

This type of modeling requires a large number of observations of heads that ideally encompass the range of all expected stresses to the system, but it is independent of the knowledge of numerous parameters distributed over the territory that is often difficult to find, as well as the hydrogeological processes taking place in the aquifer system. Data-driven models are very powerful, but they only estimate a given variable on one or more individual points and do not return information on a domain scale (e.g., water balance).

Considering the very long time series of numerous continuous monitoring stations of precipitation, hydrometric, and piezometric levels in the foothill aquifer system of the Brenta River (North-East Italy, Veneto), and pending a more detailed reconstruction of the hydro-structures, the first attempt of a mathematical regression model was developed for this system. In particular, a Multiple Linear Analysis (MLRA) was performed, considering dependent and independent variables; in this case, piezometric levels were estimated by means of hydrometric level and rainfall quantity.

Various regression models were carried out using transformation of the raw variable (e.g., moving average, shift) to maximize the correlation coefficient between the dependent and independent variables. The main steps of the workflow can be summarized as follows:

(1) Gap filling;
(2) Trend and outlier analysis;
(3) Autocorrelation and cross-correlation;
(4) MLRA;
(5) Residual analysis;
(6) Validation and forecasting.

All of these procedures were applied by selecting one piezometer and one hydrometric station that are representative of the main hydrological process occurring in the Brenta aquifer system, consisting of the groundwater–streamwater exchanges.

2.3. *Process-Based and Data-Driven Models Combined*

To develop a predictive flow and transport process-based model in order to evaluate the behavior of the aquifer in anticipated climate scenarios, it is often necessary to implement the value of expected precipitation plus expected values for additional constraints that are strongly related to the rainfall itself, such as river levels or heads of boundary conditions. This is particularly important in the systems where the groundwater flow response is deeply affected by such local constraints (e.g., losing a river whose feeding rate is relevant to the total water budget of the aquifer system) in comparison with the effects of diffuse rainwater infiltration. In order to address these issues, a possible solution is to realize a data-driven model able to reproduce over time boundary conditions (e.g., rivers and constant head or general head boundaries) of the physical model under hypothetic (or "synthetic") future climate scenarios.

The production of synthetic scenarios is a fundamental step in this type of approach, as it permits the analysis and evaluation of situations that are plausible but not necessarily recorded in historical data series. More in detail, the basic concept is to produce a synthetic time series inspired by historical series, to establish hypothetical but plausible trends of parameters that are insufficiently monitored or not monitored at all, and to thicken the statistical database used for training machine learning algorithms.

This type of combined approach, therefore, requires not only an appropriate knowledge of the aquifer system, as well as the availability of reliable data appropriately distributed throughout the territory for correct implementation and calibration of the process-based numerical model, but also the availability of robust data sets on strategic continuous monitoring points for the development of data-driven models.

The aquifer system of the Magra River plain (North-West Italy, Liguria) was taken as a case study for this type of approach due to the great number of measurement points and the availability of robust time series for significant boundary conditions. The first step was to create a suitably calibrated flow and transport process-based model. Subsequently, in order to create a useful dataset to implement the boundary conditions of the forecasting model, fully synthetic datasets were generated as a training set of the data-driven scheme, with input variables inspired by selected climate models and input/output relationships estimated by past observations (Figure 2). In particular, the rainfall time series was generated stochastically by means of the WeaGETS package [39], using as input the rainfall time series actually measured at a rain gauge located in the area of interest, whereas the synthetic time series of temperature was based on the MarkSim simulation system [40]. An experimental run of the flow-transport model for 30 years ahead was therefore performed based on such hypothetic scenarios.

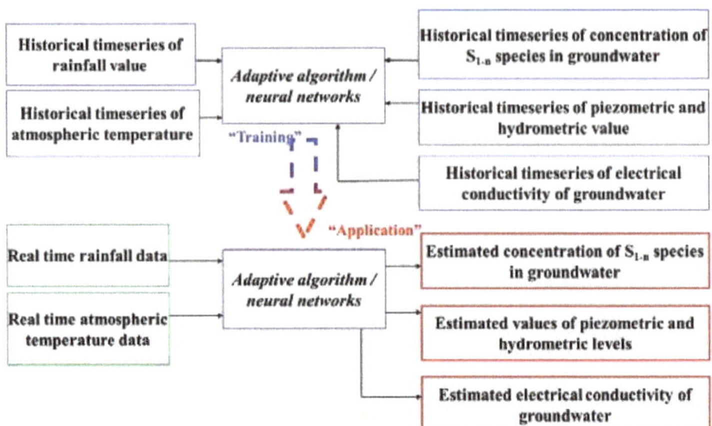

Figure 2. Block diagram describing the approach for developing a flow and transport model dedicated to specific parameters based on machine learning.

3. Results

3.1. Process-Based Models of the Empoli Aquifer System

As previously mentioned, given the knowledge degree of the Empoli aquifer system and considering the available data, it was decided to develop process-based models in order to provide the water manager with a tool for planning a sustainable use of the resource under possible future weather and climate conditions.

3.1.1. Conceptual Model

The Empoli aquifer system develops in the plain of the middle part of the River Arno catchment (Figure 3). From a geological point of view, the area corresponds to a wide depression filled by Neogene-Quaternary deposits and recent alluvium sediments that reaches a maximum thickness of around 40 m lying on a substratum of Pliocene marine deposits, mainly made up of clays. Overall, the aquifer system consists of two main permeable layers characterized by a significant lateral extension, to which lower permeability deposits are interbedded (Figure 3).

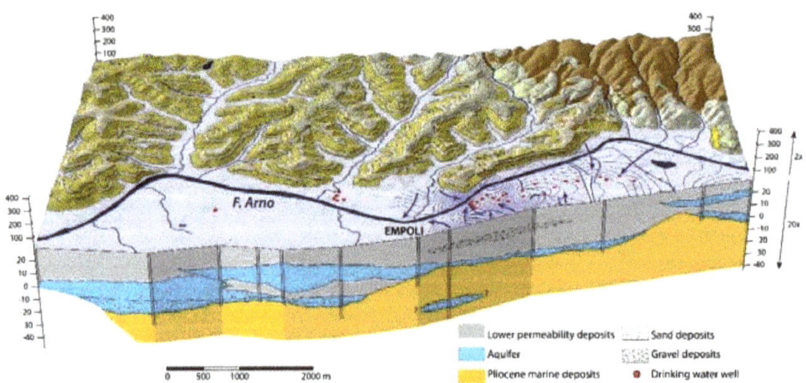

Figure 3. Geometrical reconstruction of the horizons constituting the Empoli aquifer system.

Based on the 3D reconstruction of the subsoil (Figure 3), it can be stated that:

- The chiefly confined lower aquifer layer consists of sandy gravels and gravels with an average thickness of approx. 7 m (maximum 14 m). The base of the aquifer is represented by the Pliocene substratum. This aquifer is separated from the upper aquifer by a clayey layer of variable thickness (average 7 m) and spatially discontinuous. Locally, where the clayey septum is not present, this aquifer level is hydraulically connected to the shallower one.
- The upper aquifer was reconstructed with continuity over the entire domain and represented the main aquifer in volumetric terms. It is composed of sand and gravel, often in mixed components; the grain size varies from predominantly gravelly sandy in the sectors facing the Arno River to a sandy loamy in the more distal areas. The average thickness of the aquifer is approx. 10 m, with varying values up to 20 m.
- Generally, a horizon defined in the reconstruction as an aquitard/aquiclude is above the most superficial aquifer, but its grain composition, typically consisting of clayey and sandy silts, is such that it does not impart a purely confined character to the most superficial aquifer.

Based on hydraulic tests performed on a lot of drinkable wells in the Empoli area, transmissivity values characterizing the aquifer horizons vary between about 2×10^{-3} and 2×10^{-2} m^2/s and hydraulic permeability values between about 2×10^{-4} and 2×10^{-3} m/s, in accordance with the gravelly sandy grain sizes.

Based on previous multidisciplinary studies [41–43] that involved geological, hydrogeological, and isotopic-hydrochemistry elaborations, the recharge of the system is mainly from direct infiltration. Furthermore, there are significant contributions from rivers (in particular from the Arno and Pesa rivers), as well as secondary underground transfers from the hill system along the foothill margins. As regards meteoric recharge, on the basis of historical data [44], it is possible to calculate an average annual value (data from 2000 to 2018) of 776 mm of rainfall and an average evapotranspiration value of 456 mm with an effective rainfall value of 320 mm [45]. The main outflows from the system are wells withdrawal, the drainage action of the Arno River in limited sectors, and natural underground outflowing of the modeled domain at the west. As far as outflows are concerned, the wells have been subdivided according to the use (domestic, agricultural, industrial, drinkable use) on the basis of available information and land use. For wells of drinking water supply, the consumptions were provided by Acque SpA (average annual flow rates for the period 2011–2016; average monthly flow rates for 2017), whereas for the wells used in agricultural and industrial activities, the pumping rates were estimated from a study of the Tuscan Region's Hydrological Service [46]. Table 1 shows the estimated consumption for each type of use in the model domain.

Table 1. Annual water consumption of each type of well.

Water Consumption (Mm^3)	Use
8	Drinkable
0.2	Domestic
2.2	Industrial
0.5	Agricultural

3.1.2. Models Implementation

The model domain is shown in Figure 4. The space has been discretized by 108 rows and 291 columns with cells of 50 × 50 m and four layers for a total of 125,712 cells, of which 81,856 are active. The thickness of the cells varies according to the geometry of the geological model (Figure 4). Specifically, the first layer is mainly representative of the horizon defined in the reconstruction as an aquitard/aquiclude; the second and fourth layers are representative of the shallower and deeper aquifers, respectively; the third layer was implemented to represent the impermeable interlayer separating the two main aquifers when present. The assigned hydraulic conductivity is also shown in Figure 4.

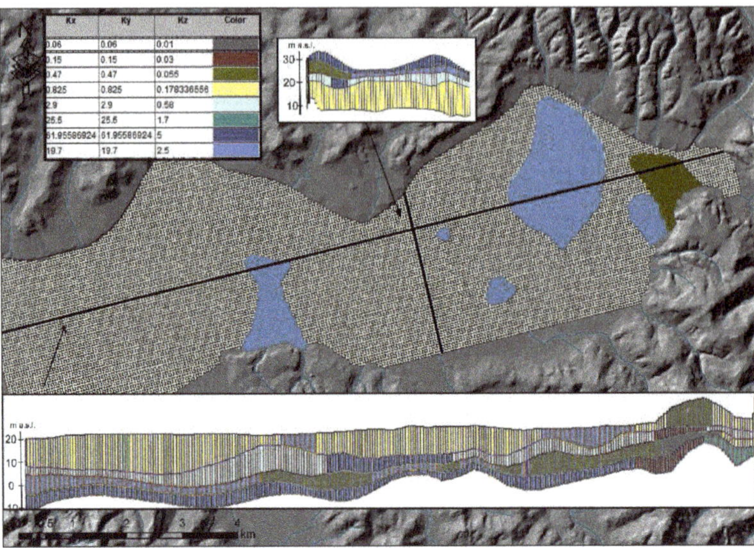

Figure 4. Spatial discretization in raw and column (cells of 50 × 50 m) and in 4 layers, and hydraulic properties (K in m/day) assigned.

From a temporal point of view, a steady-state model was initially implemented, thus taking into account the average values of all input data; subsequently, a transient-state model was implemented with monthly stress periods covering the period from January 2016 to December 2017.

The main components entering or leaving the system were mathematically represented through boundary conditions (Figure 5). Boundary conditions of the first type (Constant Head) were set in those sectors (boundaries to the south) where previous studies [42,43] indicate a significant feed component, as well as in the boundary to the west of the model to represent the natural subterranean outflow through that sector (downstream zone). The values assigned to these boundaries were defined on the basis of available monitoring data and subsequently refined during the calibration phase.

The main watercourses were represented using a third type of boundary condition (River). The implemented water level data are derived from those recorded by monitoring stations in the area (data available online at www.sir.toscana.it, accessed on 1 March 2018), the river bed elevation and width data from Lidar images (data available online at http://www502.regione.toscana.it/geoscopio/cartoteca.html, accessed on 1 March 2018). For the permeability of the river bed sediments, a value between 0.86 and 8.64 m/d was set, whereas their thickness was arbitrarily set at 1 m.

The effective rainfall value (precipitation-evapotranspiration) was used and multiplied by an infiltration coefficient related to land use to represent the effective infiltration (0.25 in an urban environment and 0.4 in a rural environment). For the steady-state model, the average effective rainfall value calculated for the period (2000–2018) was used, and for the transient-state model, the monthly values were calculated.

The initial conditions of hydraulic loading in the aquifer were set equal to the ground level elevation for the steady-state simulation, while for the transient-state model, they correspond to the piezometric surface returned by the steady-state model.

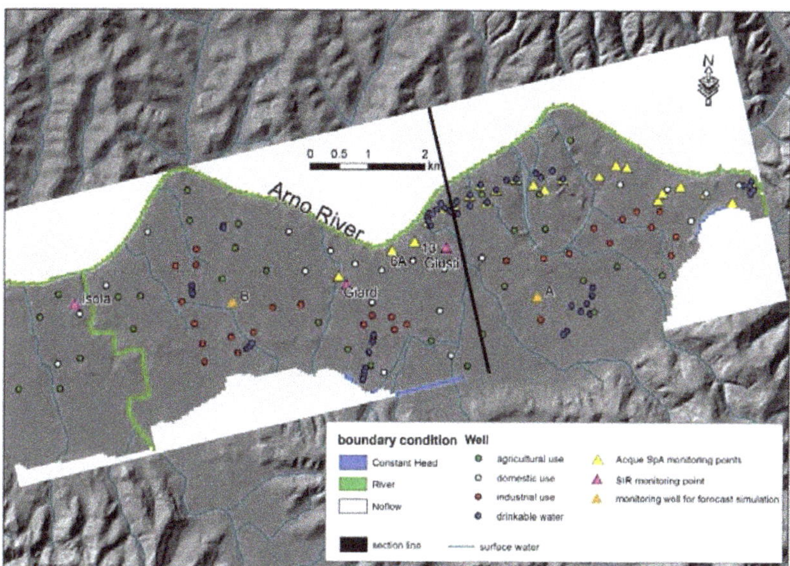

Figure 5. Boundary conditions and calibration target.

3.1.3. Models Calibration, Validation, and Output

For the calibration of the models, the data of three stations belonging to the continuous monitoring network of the SIR were used (Isola, Giardino Val Gardena, and Via Giusti, respectively named Isola, Giard, and Giusti in Figure 5), as well as the piezometric levels measured on some monitoring points provided by Acque SpA (Figure 5), which perform a monitoring activity on a network defined as the 'Empoli profile'.

For the steady-state model, the average values of all available data were used. For the transient-state model, on the other hand, monthly average values were calculated with the daily data of the SIR stations [44], while for the other points, the discrete measurement was attributed to the stress period in which it fell.

In conjunction with the calibration process, a sensitivity analysis was carried out in order to make the identification of those parameters whose changes have the greatest influence on the results of the simulations possible. Specifically, the most sensitive parameters were found to be the hydraulic conductivity of zone four and zone seven, representative of the surface semi-permeable horizon and the deep aquifer, respectively.

Following an initial coarser calibration using the trial-and-error method, the PEST code was subsequently used to better calibrate the model by varying the most sensitive parameters by no more than 10% compared to what was defined on the basis of the conceptual model.

The results of the calibrated steady-state model are shown in Figure 6a,b, whereas in the diagrams of Figure 6c,d, it is possible to observe the variations over time of the experimental piezometric levels in comparison to the level evolution simulated with the transient-state model in selected monitoring points during the calibration and validation periods.

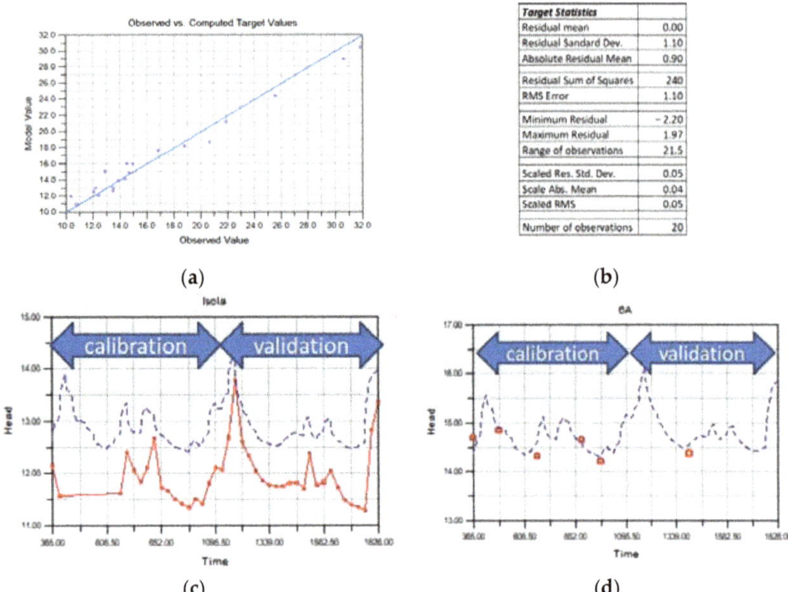

Figure 6. Calibration and validation of the model: (**a**) graph of observed vs. simulated level (in m a.s.l.) and (**b**) calibration statistics of the steady-state model; (**c**) piezometric level (in m a.s.l.) over time (day) measured (in red) and simulated (in blue) in the Isola target continuous monitoring point and (**d**) piezometric level (in m a.s.l.) over time (day) measured (in red) and simulated (in blue) in the 6A target point of the Empoli profile (only one experimental datum is available for the 6A piezometer over the validation period).

From the analysis of these diagrams (Figure 6), it can be observed that the model as a whole is sufficiently representative in terms of both absolute values and monthly trends, obviously in the areas where calibration targets are present. The modeled levels of the Isola target are very consistent in terms of evolution with respect to the experimental data. However, the residual values are significant (about 1 m), likely because of local conditions that do not allow knowing the available information.

The water balance of the steady-state model (Figure 7a) indicates that the main recharge component of the aquifer is the diffuse infiltration water (Rch). The rivers seepage (Riv in red) is also significant, and this component is particularly important in the area close to the Empoli wellfield (Figure 3), as shown in the section of Figure 7c, thus confirming the indications from water isotopes and hydro-chemical tracers analyzed in previous studies [42,43]. The main outflow component from the aquifer is the wells' withdrawal and, secondarily, with similar quantities, stream drainage (Riv in green) and natural subterranean outflow through the west sector (downstream zone).

By analyzing the water balance of the transient simulation (Figure 7b), the diffuse recharge occurring in the rainy periods is confirmed to be the most important input for the system. In the rainy seasons, almost 70% of such input contributes to water storage, which is then consumed in part during the dry season. However, in the period covered by the transient model (2016–2017 period, significantly rainy), there is a general increase in water storage with an accumulation of more than 2 million m^3 of water resources. The water pumping is the main output from the aquifer (well in green).

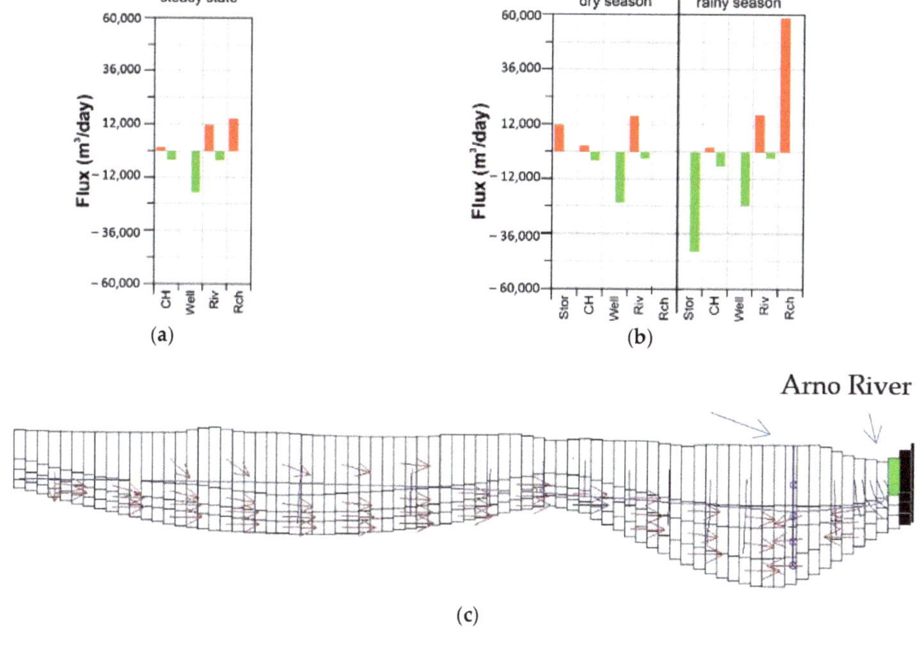

Figure 7. Water balance for (**a**) steady-state and (**b**) transient-state models in the dry and rainy seasons, and (**c**) section with flow line of the Empoli aquifer (for the section trace see Figure 5; CH: constant head, Riv: river, Rch: recharge).

3.1.4. Forecast Simulations

Based on the groundwater models of the Empoli aquifer system, forecast simulations were then carried out with the aim of assessing how the expected extreme climatic regimes would affect groundwater. In particular, it was assumed that a weather-climate regime similar to that of the current year and previous ones (i.e., 2003 and 2017), with several months of drought conditions, can repeat for five consecutive years.

A transient-state flow model was then created with 20 quarterly stress periods. Stress periods relating to the April–June and July–September periods were implemented with zero recharge, as occurred in 2022, whereas the stress periods in the January–March and October–December periods were implemented with low rainfall amounts similar to those of the 2003 and 2017 years. Considering that diffuse infiltration is the most important recharge component of the system, all other boundary conditions were kept constant with average values used for the steady-state model.

In Figure 8, the evolution of groundwater level at targets (wells 6A, Giardino, A and B; see Figure 5 for the location) and storage inflow in the domain during the simulation period are represented. Most observation wells (wells 6A, Giardino and B) present a more or less constant and low decrease rate that, over the entire period, results in about 1 m of drawdown. For well A, the forecasted drawdown appears more important, amounting to about 3 m in total over the period of simulation. The inflow of water storage in the groundwater flux also decreases over the entire period, reducing by about 50%, even if it presents a sub-cycle of increasing-decreasing correspondence to rainy and dry season alternations.

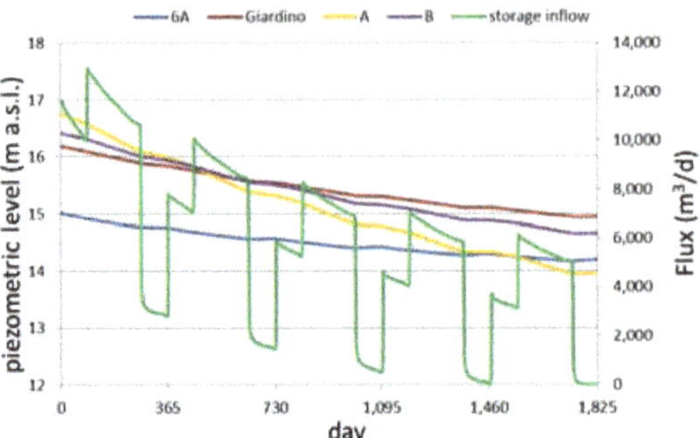

Figure 8. Simulated evolutions of storage inflow into the domain and piezometric level at observation wells A, B, and Giardino, 6A (wells location is in Figure 5).

3.2. Data-Driven Model of the Brenta Aquifer System

A consistent, continuous monitoring network made up of several meteoric, hydrometric, and piezometric stations is available for the foothill aquifer system of the River Brenta. The availability of a very long time series allowed for the development of an MLRA of the piezometric levels of groundwater in the middle-high plain. The regression model aims to make predictions on the development of the groundwater level in expected weather and climate conditions in order to provide useful information for steering the best water management practices in a zone where strategic groundwater exploitation systems are located.

3.2.1. Conceptual Model

The middle-high plain of the River Brenta is located in the Veneto region between the Pre-Alps, and a series of aligned springs called the "Linea delle Risorgive" (Figure 9a). The plain is mainly made up of glacio-fluvial and fluvial sediments (Late Pleistocene-Holocene) deposited by Astico, Brenta, and Piave rivers from the west to east [47].

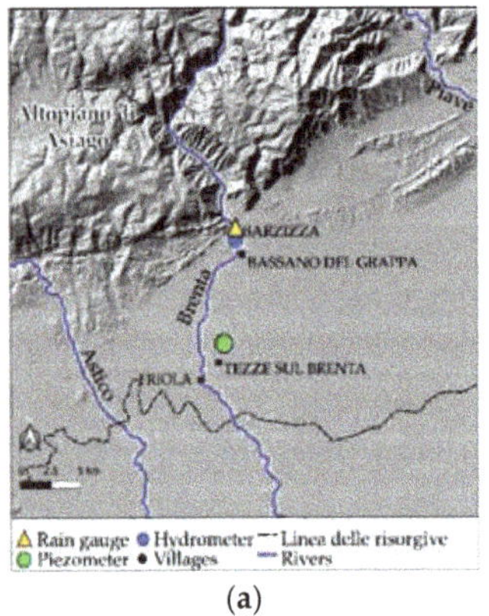

(a)

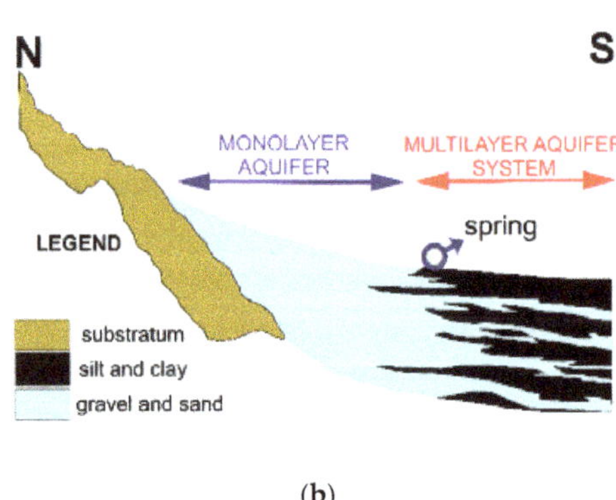

(b)

Figure 9. (**a**) Middle-high plain of the Brenta River and (**b**) schematic section of the aquifer system (modified after [48]).

In particular, the high plain is essentially made up of a very thick layer of unconsolidated conglomerates (gravels and pebbles), whereas the middle plain consists of gravelly deposits intercalated by levels of sporadic cemented sands and silt and clay [49–52]. Thanks to the general high permeability of deposits and to the high amount of annual rainfall, which varies between 1200 and 1500 mm/year, an important foothill aquifer system develop in this area. The northernmost portion of the aquifer system is substantially unconfined, being hosted into the thick single-layer of gravelly pebbly alluvial deposits (Figure 9b), whereas the southern portion of the aquifer becomes semi-confined and multilayer consisting of gravelly deposits intercalated by level with low permeability [49,51,53]. The inhomogeneous hydrogeological conditions are not only moving north–south, from downward the foothill zone, but also laterally, in the east–west direction, for the presence of other foothill alluvium fan systems belonging to different rivers. The recharge of groundwater resources depends on different mechanisms, whose prevalence varies according to the sectors of the aquifer system. The focused recharge is dominant in the zone crossed by the River Brenta over the high plain, where this watercourse loses important water quantity towards the aquifer [49,51,53]. Local rainfall infiltration becomes the main recharge mechanism away from the main river path.

3.2.2. Model Implementation and Validation

The MLRA of the piezometric level has been realized using data recorded by: one pluviometry station (PS), one hydrometric station (HS) and one piezometer (Figure 9a). These points were chosen based on the conceptual model discussed above in order to represent the relationship between surface water and groundwater that significantly affect the groundwater flow in the system. the selected pluviometry station is located at Bassano del Grappa (127 m a.s.l.), whereas the used hydrometric station is located in Barziza, near Bassano del Grappa, at an altitude of 106 m a.s.l. and immediately downstream of the mountain area. The selected piezometer is located on the hydrographic left of the River Brenta in the middle-high part of the plain, near Tezze sul Brenta, in a zone where the river feeds the aquifer and a few kilometers upstream with respect to a wells field of regional importance. The monthly data from 2010 to 2022 recorded from rainfall and hydrometric stations have been collected on the ARPAV website [54]. The piezometric levels derive from continuative monitoring performed by ETRA SpA, which has provided the hourly data from 2010 to 2020. The hourly piezometric level data has been transformed into average monthly data to compare them with rainfall and hydrometric level data.

To implement the MLRA of piezometric level, the first step of the data processing was the gaps filling by linear interpolation method, which mainly concerned the rainfall series. The second step was to obtain the maximum correlation between dependent and independent variables performing the transformation of raw data. In particular, rainfall (Rf) and the hydrometric level (HL) have been identified as independent variables, whereas the piezometric level (PL) is the dependent variable. The maximum correlation coefficient between PL and Rf has been obtained by shifting of 1 lag (i.e., 1 month) and by using the moving average at 12 months of the Rf series. On the other hand, the maximum correlation coefficient between PL and HL has been obtained only by shifting 1 lag (i.e., 1 month) in the HL series. Moreover, in order to obtain a more reliable model, the multicollinearity between the independent variables was taken into account. Multicollinearity exists whenever two or more of the independent variables in a regression model are moderately or highly correlated. For this reason, the VIF (Variation Inflation Factor [55,56]) has been calculated in Equation (1):

$$VIF_k = 1/(1 - R^2{}_k) \quad (1)$$

where R^2 represents the unadjusted coefficient of determination for regressing, and k, the predictor variable, on the remaining ones. In practice, VIF is calculated by performing a linear regression of predictor variables and then obtaining R^2 (coefficient of determination) from that regression.

Taking into account the correlation coefficient between Rf and HL, the VIF value is 1.8. Up to now, there is not an accepted cutoff value for VIF; nevertheless, values greater than 10 are generally considered indicators of multicollinearity problems.

The best MLRA model has been chosen by comparing the observed vs. predicted data (Figure 10). Data from January 2010 to December 2019 of the PL, HL and Rf have been used, and the mathematical expression is represented by the following equation:

$$PL = HL \times 3.225 + Rf \times 0.031 - 295.897 \quad (2)$$

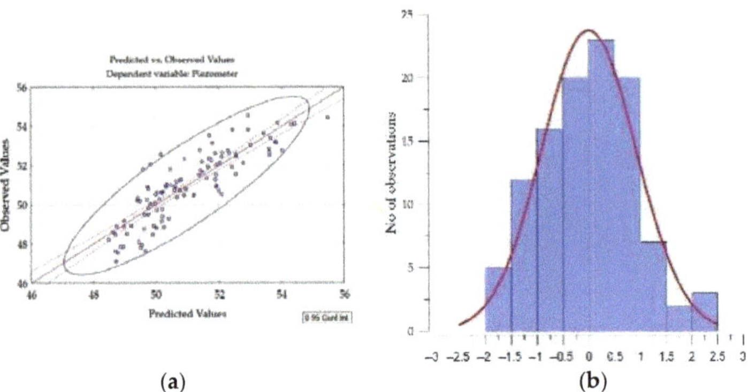

Figure 10. (**a**) Error values between the predicted and observed data along the regression line and (**b**) distribution of residuals.

Equation (2) shows that the River Brenta is more influential than the rainfall on the piezometric level variations at the selected point of monitoring.

The scatterplot in Figure 10a shows the regression line between predicted and observed values with R^2 = 0.69, and the regression bands and the ellipse with 95% confidence interval [57], whereas the Figure 10b shows that residuals have a Normal distribution (Lilliefors and Shapiro–Wilk tests) at 0.05 significance level. Moreover, no data exceed the threshold of $\pm 2.5\ \sigma$.

The model has been validated using rainfall and hydrometric level data from August 2019 to December 2020. Even for validation, a residual analysis has been performed, and it results in a Normal distribution (Lilliefors and Shapiro–Wilk tests) of residuals that have a 0.05 significance level and values that do not exceed the limit of $\pm 1.5\ \sigma$.

The MLR model replicates quite well the general trend and behavior of the piezometric level (Figure 11), aside from slight existing errors for the absolute value in a few sub-periods (e.g., 2016–2018).

3.2.3. Forecasting Simulation

The implemented MLR results are sufficiently reliable to perform a forecasting simulation of the piezometric level 6 months ahead. The forecast has been elaborated for the prospective of simulating a particularly dry period, such as the summer of 2003 and 2022. Data on hydrometric levels and rainfall similar to the dry periods of 2003 and 2022 have been used to achieve this goal. Figure 12 displays the model with the results of forecasting simulation (green line) and used data rainfall. The simulation shows that, in the case of a particularly dry period (monthly rainfall lower than 100 mm), the piezometric level abruptly decreases, reaching very low-level values in a few months. It is important to note that when rainfall occurs, even though in medium-low quantity, the groundwater level increases very quickly. However, such as recovering is not sufficient for return in the hydrodynamic state ante-dry period.

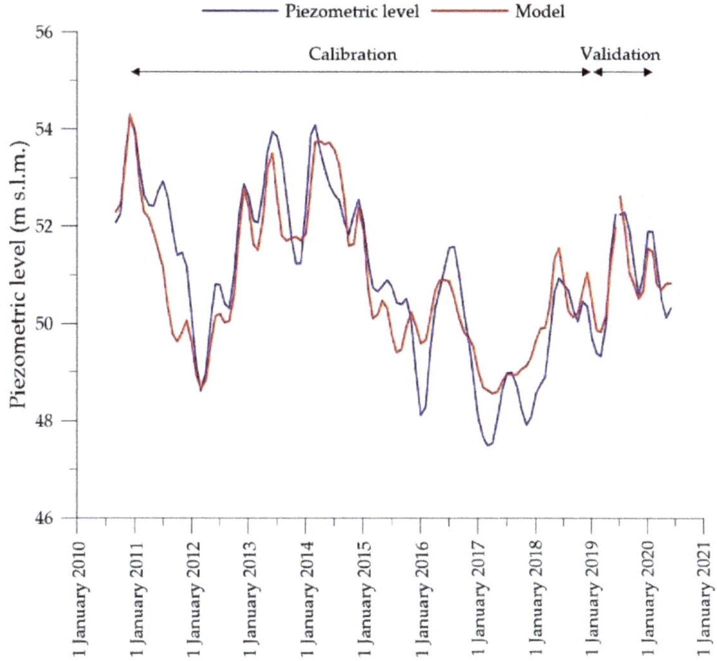

Figure 11. Calibration and validation of the piezometric level model.

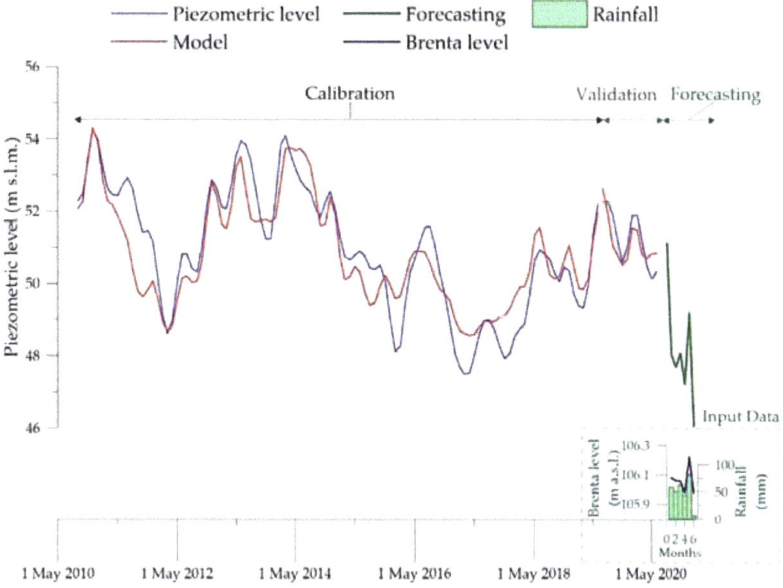

Figure 12. Forecasting simulation of 6 months using fictitious rainfall and hydrometric level data.

3.3. *Combined Approach of Data-Driven and Process-Based Models for the Magra Lower-Valley Aquifer System*

The aquifer of the Lower Magra Valley is the main source of water for drinking, industrial and agricultural purposes in the area of La Spezia (SE Liguria, Italy). Groundwater flow is, for a significant part, controlled by streamwater infiltration, which affects both groundwater level and quality [58]. So, the groundwater system is exposed to high vulnerability, both in terms of quality and quantity, not only in relation to human activities but also towards climate conditions.

In view of its importance, as well as vulnerability, the system has been the subject of numerous quantitative and qualitative monitoring activities appropriately distributed throughout the territory and with a robust time series of historical data. This availability, and the recharge mechanisms, made it possible and necessary to perform the data-driven and process-based combined approach, thus aspiring to a process-based model that, under anticipated climate conditions, predicts all groundwater systems response but also takes into account the climate-dependent evolutions of some boundary conditions of the same model.

This study is aimed to develop this kind of predictive flow and transport model in order to achieve information on the vulnerability "sensu lato" of the Magra Valley aquifer system and to evaluate its behavior in anticipated climate scenarios.

3.3.1. Conceptual Model

The aquifer of the Magra Lower Valley extends in a flat plain, within which two main rivers (Magra and Vara) flow. These rivers are characterized by a wide variation of water level and chemical composition due to the combination of rainfall regime and the presence of thermal springs in the inner part of the catchment area, which are characterized by a sulphate-dominant chemical composition [58,59].

The conceptual model was achieved by elaborating and comparing geology, stratigraphy, hydrogeology, geochemistry, and isotope data (Figure 13). The aquifer is mostly unconfined and made up of gravel and sand (K ranging from 10^{-5} to 10^{-3} m/s). The groundwater flow results show it is widely controlled by stream water infiltration, which affects water levels and water quality. In particular, the wide range of variation of some particular chemical species in the stream water influences groundwater chemistry on a seasonal basis (Figure 13d). Based on data elaboration, main recharge components and their mixing processes have been identified (Figure 14) and subsequently mathematically represented in the process-based model.

3.3.2. Process-Based and Data-Driven Models Implementation and Calibration

Based on the conceptual model, flow and transport process-based models were implemented and developed using MODFLOW and MT3DMS codes and have been calibrated in both steady-state and transient conditions over the 2004–2011 period (in which the experimental data are available). For more details on the implementation of flow and transport process-based models, see the work of El Mezouary et al. [60].

The calibration of the models showed a high congruence with the observed data for both piezometric levels and chemical concentrations (Figure 15), thus indicating very good representativeness of the models. The results of the process-based models confirmed the importance of the Magra river in the water balance of the aquifer (input of about 66%) and in the chemical composition of groundwater.

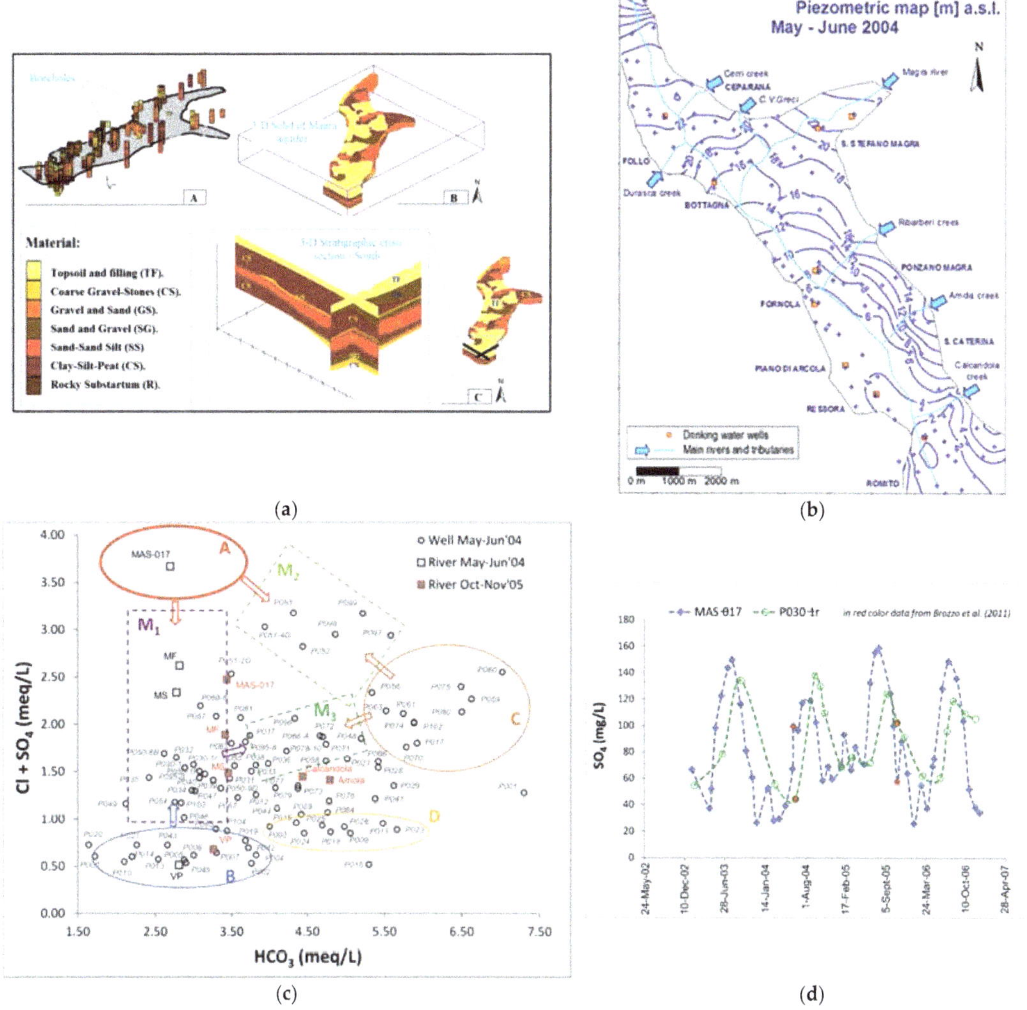

Figure 13. Multidisciplinary data elaboration to develop the conceptual model [58,59]: (**a**) 3D reconstruction by mean borehole information (where A is available boreholes used, B is 3-D solid of Magra Aquifer and C is an example of 3-D stratigraphic cross section); (**b**) Piezometric map (m) a.s.l. for the "2004 May–June" period; (**c**) Cl + SO_4 vs. HCO_3 diagram (A–D letters indicate the main recharge components of the groundwater system in Figure 14 and M1–3 indicate mixing processes); (**d**) SO_4 concentrations in the Magra River (station MAS-017) and in a drinking water well (P030-1r) located very close to the river.

Given the strong dependence of the aquifer on river waters, both quantitatively and qualitatively, the implementation of river-related boundary conditions is, therefore, essential for carrying out forecast simulations. This justifies the need to create plausible data sets of hydrometric level and concentration of chemical species in the river water, to investigate their dependence on the expected weather and climate data and implement such analysis in the forecasting simulations performed with the process-based model.

For this purpose, fully synthetic datasets have been generated as a training set of the data-driven scheme, with input variables inspired by selected climate models and input/output relationships estimated by past observations (Figure 2).

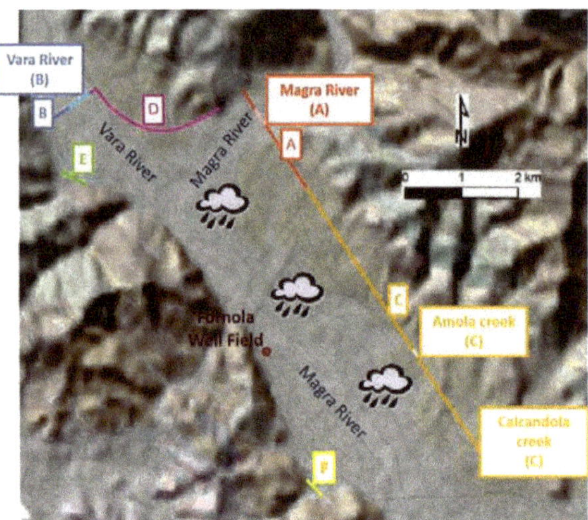

Figure 14. Simplified sketch map showing the main recharge components of the groundwater system [53]: A—feeding from River Magra and its alluvial fan; B—feeding from River Vara and its alluvial fan; C—groundwater transfer from western hills and secondary input from minor creeks; D—groundwater transfer from northern hills; E and F—groundwater recharge from secondary creeks flowing in the western sectors.

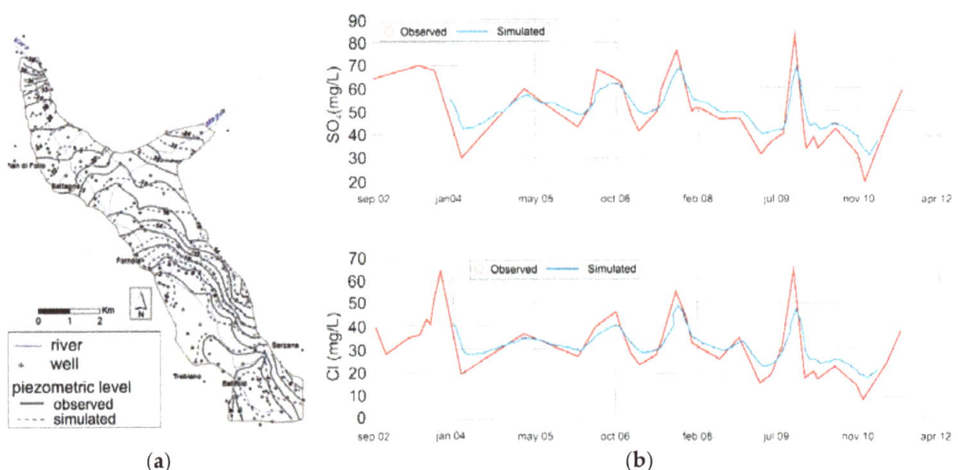

Figure 15. Calibration of flow and transport model: (**a**) Observed and simulated piezometric head [60] (**b**) SO_4 and Cl observed and simulated value at their respective time (2004–2011) [60].

In Figure 16, a good agreement between modeled and real data (in this case, groundwater SO_4 concentration in the River Magra) is shown, derived by the application of the data-driven model for the calibration period 2004–2011.

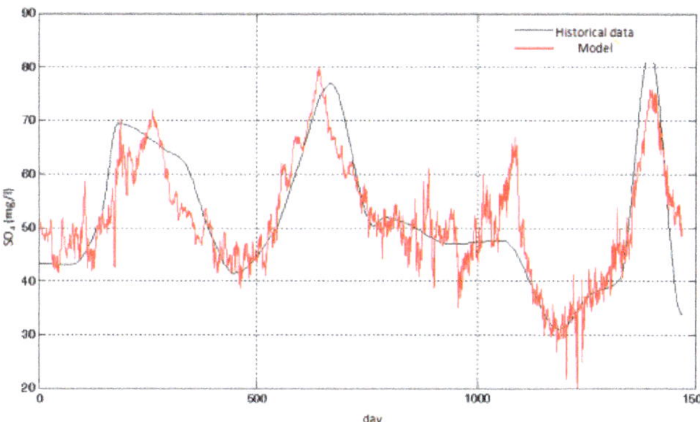

Figure 16. Observed SO_4 concentration (in grey) and calculated (in red) by the model for observation points of the Magra River.

3.3.3. Forecasting Simulation

As previously stated, to develop predictive flow and transport models in order to evaluate the aquifer behavior in anticipated climate scenarios, a process-based and data-driven modeling combined approach was applied. The data-driven models were built in order to determine boundary conditions (e.g., rivers and constant head or general head boundaries) of the process-based model under hypothetic future climate scenarios (Figure 17), characterized by some dry and low-rainy periods.

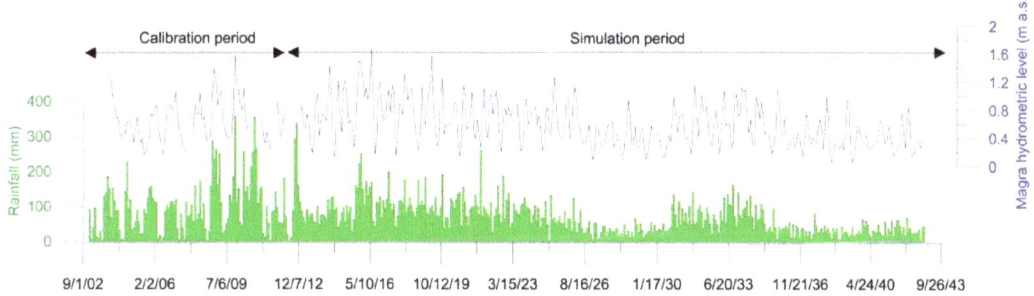

Figure 17. Rainfall and Magra hydrometric level for the calibration period (2004–2011) and forecasting simulation period (2012–2042).

Based on such hypothetic scenarios, an experimental and exemplified run of the flow-transport model for 30 years ahead was performed. In Figure 18, it is possible to observe an SO_4 concentration map of the aquifer and SO_4 time series of specific points. From the map, the strong chemical influence of the River Magra waters on groundwater quality is evident, especially close to Fornola wellfield, which attracts even more surface water. This influence can also be observed at the two representative points of the aquifer outside the wellfield (Figure 18), where a variation in the concentration of sulphates can be observed, mainly linked to the rainfall regime, especially at the point closest to the wellfield. It is interesting to observe how the sulphate value at the point located furthest from the wellfield remains higher and approximately constant over time following dry periods.

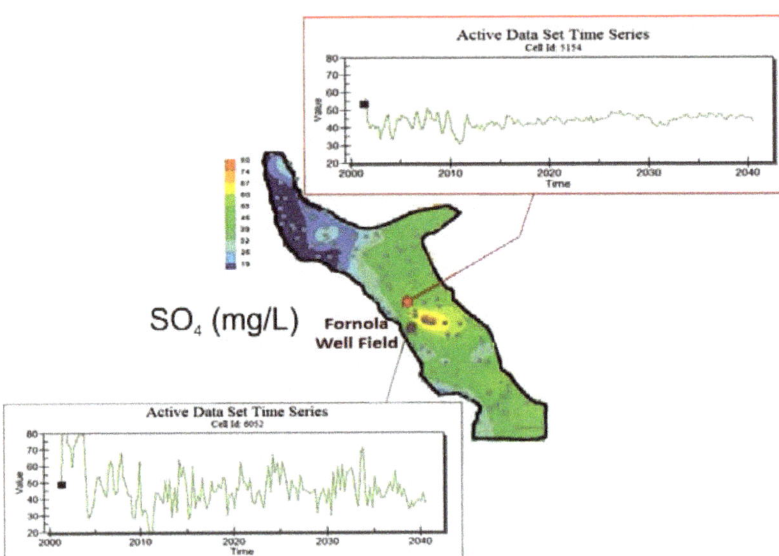

Figure 18. Map and time series of SO_4 concentration in the calibration and prediction periods.

Finally, by observing the piezometric level and sulphate concentration in a well for drinking water (Fornola 3) in the wellfield (Figure 19), it is possible to observe how, after a medium-long period characterized by a low value of rainfall (e.g., 2025–2030), the aquifer system reacts with significant decreasing of piezometric level and relative increasing of SO_4 solute concentration. After these periods, the sulphate concentration continues to vary greatly depending on the rainfall regime, although it generally seems to increase or at least fluctuate around a higher average value.

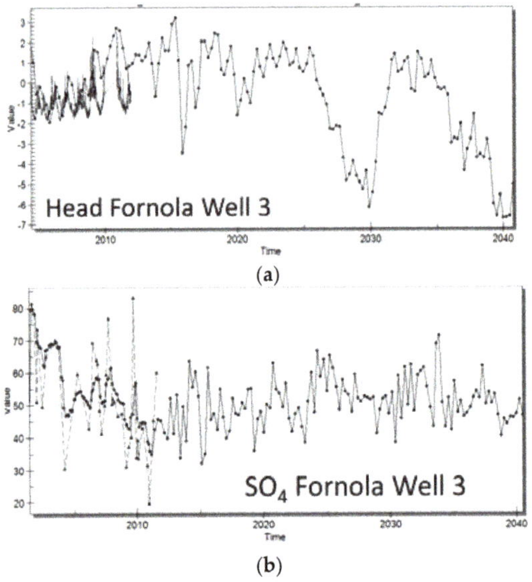

Figure 19. Time series of piezometric value (**a**) and SO_4 concentration (**b**) for Fornola Well 3 in the calibration and prediction periods.

4. Discussion and Conclusions

The performed groundwater modeling for the three different systems first provides information about the main hydrological processes and elements regulating the groundwater flow in the systems themselves. On the other hand, the different approaches adopted accordingly to the specificities of the sites enabled the forecasting of groundwater behavior under anticipated climate conditions.

The calibrated process-based model of the Empoli aquifer system as the diffuse recharge represents the main highlighted input in this case. Rivers also contribute to groundwater flow, but this component appears secondary with respect to diffuse infiltration. Furthermore, the system denotes a good capability for water storage, which is an important part of the total water budget (Figure 7). Such as hydrodynamic features are favorable for mitigating, at least in short periods (a few years), the effects of drought events, which have become more and more frequent in the last decades and also in the Mediterranean regions [61]. The forecasting provided by the model seems to confirm these characteristics and behaviors for the Empoli aquifer system. Under drought conditions (very similar to those that occurred in recent periods) repeated for some consecutive years, the model, in general, highlighted that the decreasing groundwater level is moderate. It is very likely that the storage capability of the system that is capitalizing the small amount of effective yearly rainfall attenuates and dilutes over time the groundwater drawdown (Figure 8).

The Brenta River's aquifer system is extensive, strategic for water supply, and characterized by inhomogeneity in terms of hydrogeology, recharge components, and water exploitation distribution as well. In these conditions, it becomes important to provide models able to describe the local groundwater behavior over time. This is the case of the data-driven model developed for a significant piezometric monitoring point sited near the losing River Brenta. The groundwater level evolution has been well-reproduced, starting from precipitation and hydrometric data as input. The equation describing the model confirms the close dependence of groundwater from the river in the analyzed zone and, therefore, the high sensitivity of the aquifer with respect to the meteo-climate regime, which directly and abruptly affects the regime of the River Brenta. In view of this high sensitivity, the forecast of groundwater level evolution under a relatively dry period of six months, similar to what occurred during the 2022 and 2003 years, was performed. Results point out a drawdown of groundwater level of more than 3 m (Figure 12) in a few months, thus remarking the very high sensitivity of the aquifer to climate extremes, as well as the need to plan actions for mitigating risks for water supply, given the regional importance of the wellfields present in this zone. Given the straight dependence of the local groundwater quantity from the hydrometric level, an advantage in making available a model that an equation estimates, such as a relationship, is the possibility of solving the same equation for a set of minimum progressive hydrometric levels with respect to when the piezometric levels do not decrease under safety threshold values. In this way, a water manager can provide actions over the catchment for maintaining the hydrometric levels over certain critical values once the territory dotes specific hydro-infrastructures able to regulate the dynamics of the watercourse.

For what concerns the aquifer of the River Magra Lower Valley, here a combined approach of modeling has been proposed in order to obtain more efficient information through the developed model. Indeed, the aquifer system is well-described by means of data from geology, hydrogeology and geochemistry, enough to develop a process-based model able to reproduce the overall behavior of the system for groundwater quantity and quality. At the same time, and as the same process-based calibrated models confirm, the groundwater flow and transport deeply depend on some boundary conditions very sensitive to the hydro-climate regime, such as the River Magra. In these situations, to provide hypothetic sets of data for such boundaries (e.g., streamwater level or streamwater chemical concentration) in case of forecasting by a process-based model can be a real risk of failure, given the probable incongruence between such input data and the input data concerning the diffuse recharge directly deducible from the anticipated climate condition. In this kind of context, it is therefore evident that the use of a data-driven model able to link the evolution of certain boundary conditions with the anticipated climate conditions represents a significant advantage in terms of the efficiency and reliability of the predictive model. The so-developed model for the River Magra aquifer demonstrates that, under relatively dry climate conditions, the river is responsible for a relative worsening of the water quality, mainly tied to an increase or a relatively high and constant value of sulphate in groundwater. The dry weather results in a sort of persistent base flow in the River Magra, which is mainly dependent on the mountain thermal springs with sulphate chemical facies. The losing character of the river and the general groundwater flow path network promote the diffusion of such sulphate components into the aquifer for an extensive sector (Figure 18).

All models developed in this work enable a better understanding of the aquifer systems' functioning and the sensitivity of groundwater flow to climate change. Generally speaking, the applicability, limitations and accuracy of the different modeling approaches are deeply dependent on the aquifer systems' features, functioning, and data availability as well.

The process-based model requires a large amount of information (e.g., the geometry of the main hydrogeological units) and data (e.g., hydraulic parameters value) well-distributed over the domain and time in order to achieve accurate results. At the same time, this type of approach makes it possible to model the behavior of the whole groundwater system under study.

The data-driven models require a number of observations that ideally encompass the values range of all expected stresses to the system, but it is independent of the knowledge of numerous parameters that are often difficult to find. Data-driven models are powerful, but they only estimate a given variable on one or more individual points and do not return information on a domain scale (e.g., water balance).

The integrated approach of data-driven/process-based modeling represents a novel outcome resulting in a very efficient tool for forecasting aquifer systems' evolution when also boundary conditions significantly affecting the behavior of such systems are subjected to evolve under expected climate scenarios. As these conditions are frequent in aquifer systems, the proposed integrated modeling represents a very powerful approach, even if it needs a deep knowledge of the system as a whole and large datasets.

As a general outcome, this manuscript remarks that, for modeling groundwater resources and forecasting its evolution with respect to climate change, the choice of the suitable numerical modeling methodology is mandatory, and it mainly depends on the specific aquifer features that result in different ways to be sensitive to climate. Only through this approach is it possible to provide efficient groundwater forecasts that are able to steer water management plans and actions aimed at mitigating the effects of climate change.

Author Contributions: Conceptualization, M.M. and M.D.; Methodology, M.M., B.R. and A.S.; Validation, M.M., L.F. and A.S.; Data curation, M.M., L.F., A.S. and M.D.; Writing—original draft preparation, M.M., M.D. and L.F.; Writing—review and editing, M.M., M.D. and L.F.; Visualization, M.M. and G.M.; Supervision, M.D. and M.M.; Project coordination, M.D. and M.M. All authors have read and agreed to the published version of the manuscript.

Funding: This research was co-funded in part by Tuscany Water Authorities, Brenta Basin Council and the Italian national project ACQUASENSE (Industria2015- Project number: MI01_00223).

Data Availability Statement: Data are available upon request to the corresponding author.

Acknowledgments: The authors would like to thank the Tuscan Water Authority (AIT) and the Brenta Basin Council for co-funding this study and Ingegnerie Toscane and ETRA water service for sharing their data. The authors would also like to thank the five referees and the editors who contributed to improving the manuscript with their advice.

Conflicts of Interest: The authors declare no conflict of interest.

References

1. United Nations. *The United Nations World Water Development Report 2022: Groundwater: Making the Invisible Visible*; UNESCO: Paris, France, 2022; p. 225, ISBN 978-92-3-100507-7.
2. Hiscock, K.M. Groundwater in the 21st century—meeting the challenges. In *Sustaining Groundwater Resources: A Critical Element in the Global Water Crisis, in International Year of Planet Earth*; Anthony, J., Jones, A., Eds.; Springer: Dordrecht, Netherlands, 2011; pp. 207–225. [CrossRef]
3. Zhu, Y.; Balke, K.D. Groundwater protection: What can we learn from Germany? *J. Zhejiang Univ. Sci.* **2008**, *9*, 227–231. [CrossRef] [PubMed]
4. Baoxiang, Z.; Fanhai, M. Delineation methods and application of groundwater source protection zone. In Proceedings of the Water Resource and Environmental Protection (ISWREP), 2011 International Symposium, (IEEE Conference Publications), Xi'an, China, 20–22 May 2011; Volume 1, pp. 66–69. [CrossRef]
5. Martınez Navarrete, C.; Grima Olmedo, J.; Duran Valsero, J.J.; Gomez, J.D.; Luque Espinar, J.A.; de la Orden, G.J.A. Groundwater protection in Mediterranean countries after the European water framework directive. *Environ. Geol.* **2008**, *54*, 537–549. [CrossRef]
6. WBCSD Facts and Trends–Water. World Business Council for Sustainable Development. Available online: https://www.wbcsd.org/Programs/Food-and-Nature/Water/Resources/Water-Facts-and-trends (accessed on 1 September 2022).
7. Siebert, S.; Burke, J.; Faures, J.M.; Frenken, K.; Hoogeveen, J.; Döll, P.; Portmann, F.T. Groundwater use for irrigation—A global inventory. *Hydrol. Earth Syst. Sci.* **2010**, *14*, 1863–1880. [CrossRef]
8. Wada, Y.; Van Beek, L.P.H.; Van Kempen, C.M.; Reckman, J.W.T.M.; Vasak, S.; Bierkens, M.F.P. Global depletion of groundwater resources. *Geophys. Res. Lett.* **2010**, *37*, L20402. [CrossRef]
9. Rosegrant, M.W.; Cai, X.; Cline, S.A. *World Water and Food to 2025: Dealing with Scarcity*; International Food Policy Research Institute: Washington, DC, USA, 2002; p. 322.
10. Milly, P.C.D.; Dunne, K.A.; Vecchia, A.V. Global pattern of trends in streamflow and water availability in a changing climate. *Nature* **2005**, *438*, 347–350. [CrossRef]
11. Hirabayashi, Y.; Mahendran, R.; Koirala, S.; Konoshima, L.; Yamazaki, D.; Watanabe, S.; Kim, H.; Kanae, S. Global flood risk under climate change. *Nat. Clim. Chang.* **2013**, *3*, 816–821. [CrossRef]
12. Chan, D.; Wu, Q. Significant anthropogenic-induced changes of climate classes since 1950. *Sci. Rep.* **2015**, *5*, 13487. [CrossRef]
13. Turco, M.; Palazzi, E.; von Hardenberg, J.; Provenzale, A. Observed climate change hotspots. *Geophys. Res. Lett.* **2015**, *42*, 3521–3528. [CrossRef]
14. Tollefson, J. IPCC climate report: Earth is warmer than it's been in 125,000 years. *Nature* **2021**, *596*, 171–172. [CrossRef]
15. Taylor, R.G.; Scanlon, B.; Döll, P.; Rodell, M.; Van Beek, R.; Wada, Y.; Longuevergne, L.; Leblanc, M.; Famiglietti, J.S.; Edmunds, M.; et al. Ground water and climate change. *Nat. Clim. Chang.* **2013**, *3*, 322–329. [CrossRef]
16. Ohba, M.; Arai, R.; Sato, T.; Imamura, M.; Toyoda, Y. Projected future changes in water availability and dry spells in Japan: Dynamic and thermodynamic climate impacts. *Weather. Clim. Extremes* **2022**, *38*, 100523. [CrossRef]
17. Fan, Y. Groundwater. How much and how old? *News Views. Nat. Geosci.* **2015**, *9*, 93–94. [CrossRef]
18. Doveri, M.; Menichini, M.; Scozzari, A. Protection of groundwater resources: Worldwide regulations, scientific approaches and case study. In *The Handbook of Environmental Chemistry: Threats to the Quality of Groundwater Resources: Prevention and Control*; Scozzari, A., Dotsika, E., Eds.; Springer-Verlag: Berlin/Heidelberg, Germany, 2016; Volume 40, pp. 13–30.
19. Anderson, M.P.; Woessner, W.W.; Hunt, R.J. *Applied Groundwater Modeling: Simulation of Flow and Advective Transport*; Academic Press: Amsterdam, The Netherlands, 2015; p. 564.
20. Kresic, N.; Mikszewski, A. *Hydrogeological Conceptual Site Model: Data Analysis and Visualization*; CRC Press: Boca Raton, FL, USA, 2013.
21. Reilly, T.E.; Harbaugh, A.W. *Guidelines for Evaluating Ground-Water Flow Models*; U.S. Geological Survey Scientific Investigations Report: Reston, VA, USA, 2004; Volume 5038, p. 30.
22. Grodzka-Łukaszewska, M.; Nawalany, M.; Zijl, W. A Velocity-Oriented Approach for Modflow. *Transp. Porous Media* **2017**, *119*, 373–390. [CrossRef]
23. Ginn, T.R.; Cushman, J.H. Inverse methods for subsurface flow: A critical review of stochastic techniques. *Stoch. Hydrol. Hydraul.* **1990**, *4*, 1–26. [CrossRef]
24. Shapiro, A.M.; Day-Lewis, F.D. Reframing groundwater hydrology as a data-driven science. *Groundwater* **2022**, *60*, 455–456. [CrossRef]
25. Curtis, Z.K.; Li, S.-G.; Liao, H.-S.; Lusch, D. Data-Driven Approach for Analyzing Hydrogeology and Groundwater Quality Across Multiple Scales. *Groundwater* **2017**, *56*, 377–398. [CrossRef] [PubMed]
26. Xu, T.; Valocchi, A.J. Data-driven methods to improve baseflow prediction of a regional groundwater model. *Comput. Geosci.* **2015**, *85*, 124–136. [CrossRef]
27. Bakker, M.; Maas, K.; Schaars, F.; von Asmuth, J.R. Analytic modeling of groundwater dynamics with an approximate impulse response function for areal recharge. *Adv. Water Resour.* **2007**, *30*, 493–504. [CrossRef]
28. Sun, J.; Hu, L.; Li, D.; Sun, K.; Yang, Z. Data-driven models for accurate groundwater level prediction and their practical significance in groundwater management. *J. Hydrol.* **2022**, *608*, 127630. [CrossRef]
29. Gusyev, M.; Haitjema, H.; Carlson, C.; Gonzalez, M. Use of Nested Flow Models and Interpolation Techniques for Science-Based Management of the Sheyenne National Grassland, North Dakota, USA. *Groundwater* **2012**, *51*, 414–420. [CrossRef]

30. Demissie, Y.K.; Valocchi, A.; Minsker, B.; Bailey, B.A. Integrating a calibrated groundwater flow model with error-correcting data-driven models to improve predictions. *J. Hydrol.* **2009**, *364*, 257–271. [CrossRef]
31. Szidarovszky, F.; Coppola, E.A.; Long, J.; Hall, A.D.; Poulton, M.M. A Hybrid Artificial Neural Network-Numerical Model for Ground Water Problems. *Groundwater* **2007**, *45*, 590–600. [CrossRef] [PubMed]
32. Meixner, T.; Manning, A.H.; Stonestrom, D.A.; Allen, D.M.; Ajami, H.; Blasch, K.W.; Brookfield, A.E.; Castro, C.L.; Clark, J.F.; Gochis, D.J.; et al. Implications of projected climate change for groundwater recharge in the western United States. *J. Hydrol.* **2016**, *534*, 124–138. [CrossRef]
33. Menichini, M.; Doveri, M. Modelling tools for quantitative evaluations on the Versilia coastal aquifer system (Tuscany, Italy) in terms of groundwater components and possible effects of climate extreme events. *Acque Sotter. Ital. J. Groundw.* **2020**, *9*, 475. [CrossRef]
34. Hunt, R.J. Ground Water Modeling Applications Using the Analytic Element Method. *Groundwater* **2006**, *44*, 5–15. [CrossRef]
35. Harbaugh, A.W. MODFLOW-2005, the US Geological Survey Modular Ground-Water Model: The Ground-Water Flow Process Reston, VA, USA, 6-A16, 2005: US Department of the Interior, US Geological Survey. Available online: http://pubs.er.usgs.gov/publication/tm6A16 (accessed on 1 May 2022).
36. Rumbaugh, J.O.; Rumbaugh, D.B. *Groundwater Vistas*; Environmental Simulations Inc.: Leesport, PA, USA, 2011; p. 213.
37. Jones, N.L. GMS Reference Manual. In *Aquaveo*; Brigham Young University: Provo, UT, USA, 2014; p. 662.
38. Gallagher, M.; Doherty, J. Parameter estimation and uncertainty analysis for a watershed model. *Environ. Model. Softw.* **2007**, *22*, 1000–1020. [CrossRef]
39. Chen, J.; Brissette, F.P.; Leconte, R. A daily stochastic weather generator for preserving low-frequency of climate variability. *J. Hydrol.* **2010**, *388*, 480–490. [CrossRef]
40. Jones, P.G.; Thornton, P.K.; Heinke, J. Generating Characteristic Daily Weather Data Using Downscaled Climate Model Data from the IPCC's Fourth Assessment; Project Report. 2009, p. 19. Available online: http://dspacetest.cgiar.org/handle/10568/2482 (accessed on 1 May 2022).
41. Menichini, M.; Da Prato, S.; Doveri, M.; Ellero, A.; Lelli, M.; Masetti, G.; Nisi, B.; Raco, B. An integrated methodology to define Protection Zones for groundwaterbased drinking water sources: An example from the Tuscany Region, Italy. *Acque Sotter. Ital. J. Groundw.* **2015**, *4*, 21–27. [CrossRef]
42. Menichini, M.; Doveri, M.; Ellero, A.; Raco, B.; Masetti, G.; Da Prato, S.; Lelli, M.; Nisi, B. Delimitazione delle Zone di Protezione Risorse Idriche destinate al consumo umano. Campo pozzi "Empoli" (FI-ATO2). In *Delimitation of Water Resources Protection Zones for Human Consumption. Empoli" well field (FI-ATO2)*; IGG-CNR Confidential Internal Technical Report n 10988; IGG-CNR: Pisa, Italy, 2013.
43. Da Prato, S.; Doveri, M.; Ellero, A.; Lelli, M.; Masetti, G.; Menichini, M.; Nisi, B.; Raco, B. Integrazioni alla Caratterizzazione geologica, idrogeologica e idrogeochimica dei Corpi Idrici Sotterranei Significativi della Regione Toscana (CISS). 11AR025 Corpo idrico del Valdarno Inferiore e Piana Costiera Pisana-zona Empoli. In *Integrations to the Geological, Hydrogeological and Hydrogeochemical Characterisation of the Significant Underground Water Bodies of the Region of Tuscany (CISS) Valdarno Inferiore and Piana Costiera Pisana Water Body-Empoli Area*; Technical Report IGG n° 10976; IGG-CNR: Pisa, Italy, 2012; p. 18.
44. SIR-Regional Hydrological and Geological Sector. Available online: www.sir.toscana.it (accessed on 1 March 2018).
45. Doveri, M.; Da Prato, S.; Masetti, G.; Menichini, M.; Raco, B.; Vivaldo, G.; Scozzari, A. *Relazione Sulle Attività Svolte Nell'ambito Dell'Accordo di Collaborazione Scientifica AIT-LaMMA-IGG/CNR del 13 March 2017 IGG-CNR Confidential Internal Technical Report n° 12306*; IGG-CNR: Pisa, Italy, 2020; p. 35.
46. SIR-Regional Hydrological and Geological Sector. Available online: www.idropisa.it/consumi_idrici (accessed on 1 March 2018).
47. Cisotto, A.; Rusconi, A.; Baruffi, F. Regional Studies of the North Adriatic Basin Authority on the Aquifers of the Veneto-Friuli Plain. *Mem. Descr. Carta Geol. d'It.* **2007**, *76*, 117–124.
48. Carraro, A.; Fabbri, P.; Giaretta, A.; Peruzzo, L.; Tateo, F.; Tellini, F. Arsenic anomalies in shallow Venetian Plain (Northeast Italy) groundwater. *Environ. Earth Sci.* **2013**, *70*, 3067–3084. [CrossRef]
49. Dal Prà, A.; Veronese, F. Gli acquiferi dell'alta pianura alluvionale del Brenta e i loro rapporti col corso d'acqua. *Atti Istituto Veneto Sc. Lett. Arti.* **1972**, *5*, 189–222.
50. Pilli, A.; Sapigni, M.; Zuppi, G. Karstic and alluvial aquifers: A conceptual model for the plain – Prealps system (northeastern Italy). *J. Hydrol.* **2012**, *464–465*, 94–106. [CrossRef]
51. Sottani, A.; Vielmo, A. Groundwater conservation and monitoring activities in the middle Brenta River plain (Veneto Region, Northern Italy): Preliminary results about aquifer recharge. *Acque Sotter. Ital. J. Groundw.* **2014**, *3*, 3. [CrossRef]
52. Mayer, A.; Sültenfuß, J.; Travi, Y.; Rebeix, R.; Purtschert, R.; Claude, C.; Salle, C.L.G.L.; Miche, H.; Conchetto, E. A multi-tracer study of groundwater origin and transit-time in the aquifers of the Venice region (Italy). *Appl. Geochem.* **2014**, *50*, 177–198. [CrossRef]
53. Bullo, P.; Dal Prà, A. Lo sfruttamento ad uso acquedottistico delle acque sotterranee dell'alta pianura alluvionale veneta. *Geol. Romana* **1994**, *30*, 371–380.
54. ARPAV–Regional Agrncy for Environmental Prevention and Protection of Veneto. Available online: www.arpa,veneto.it (accessed on 1 January 2020).
55. Freund, R.J.; Wilson, W.J. *Regression Analysis: Statistical Modeling of a Response Variable*; Academic Press: New York, NY, USA, 1998.

56. Adnan, N.; Ahmad, M. A comparative study on some method for handling multicollinearity problems. *Matematika* **2006**, *22*, 109–119. [CrossRef]
57. Tracy, N.D.; Young, J.C.; Mason, R.L. Multivariate Control Charts for Individual Observations. *J. Qual. Technol.* **1992**, *24*, 88–95. [CrossRef]
58. Brozzo, G.; Accornero, M.; Marini, L. The alluvial aquifer of the Lower Magra Basin (La Spezia, Italy): Conceptual hydrogeochemical–hydrogeological model, behavior of solutes, and groundwater dynamics. *Carbonates Evaporites* **2011**, *26*, 235–254. [CrossRef]
59. Menichini, M.; Doveri, M.; El Mansoury, B.; El Mezouary, L.; Lelli, M.; Raco, B.; Scozzari, A.; Soldovieri, F. Groundwater vulnerability to climate variability: Modelling experience and field observations in the lower Magra Valley (Liguria, Italy). In *EGU General Assembly Conference Abstracts*; EGU: Vienna, Austria, 2016.
60. El Mezouary, L.; El Mansouri, B.; Kabbaj, S.; Scozzari, A.; Doveri, M.; Menichini, M.; Kili, M. Modélisation numérique de la variation saisonnière de la qualité des eaux souterraines de l'aquifère de Magra, Italie. *Houille Blanche* **2015**, *101*, 25–31. [CrossRef]
61. Mathbout, S.; Lopez-Bustins, J.; Royé, D.; Martin-Vide, J. Mediterranean-Scale Drought: Regional Datasets for Exceptional Meteorological Drought Events during 1975–2019. *Atmosphere* **2021**, *12*, 941. [CrossRef]

 water

Article

Simulation of Heat Flow in a Synthetic Watershed: The Role of the Unsaturated Zone

Eric D. Morway [1,*], Daniel T. Feinstein [2] and Randall J. Hunt [3]

1 U.S. Geological Survey, Nevada Water Science Center, 2730 N. Deer Run Rd. Suite 3, Carson City, NV 89701, USA
2 U.S. Geological Survey, Upper Midwest Water Science Center, Milwaukee Office, 3209 North Maryland Avenue, Milwaukee, WI 53211, USA
3 U.S. Geological Survey, Upper Midwest Water Science Center, 1 Gifford Pinchot Drive, Madison, WI 53726, USA
* Correspondence: emorway@usgs.gov

Abstract: Future climate forecasts suggest atmospheric warming, with expected effects on aquatic systems (e.g., cold-water fisheries). Here we apply a recently published and computationally efficient approach for simulating unsaturated/saturated heat transport with coupled flow (MODFLOW) and transport (MT3D-USGS) models via a synthetic three-dimensional (3D) representation of a temperate watershed. Key aspects needed for realistic representation at the watershed-scale include climate drivers, a layering scheme, consideration of surface-water groundwater interactions, and evaluation of transport parameters influencing heat flux. The unsaturated zone (UZ), which is typically neglected in heat transport simulations, is a primary focus of the analysis. Results from three model versions are compared—one that neglects UZ heat-transport processes and two that simulate heat transport through a (1) moderately-thick UZ and (2) a UZ of approximately double thickness. The watershed heat transport is evaluated in terms of temperature patterns and trends in the UZ, at the water table, below the water table (in the groundwater system), and along a stream network. Major findings are: (1) Climate forcing is the product of infiltration temperatures and infiltration rates; they combine into a single heat inflow forcing function. (2) The UZ acts as a low-pass filter on heat pulses migrating downward, markedly dampening the warming recharge signal. (3) The effect of warming on the watershed is also buffered by the mixing of temperatures at discharge points where shallow and deep flow converge. (4) The lateral extent of the riparian zone, defined as where the water table is near land surface (<1 m), plays an important role in determining the short-term dynamics of the stream baseflow response to heat forcing. Runoff generated from riparian areas is particularly important in periods when rejected infiltration during warm and wet periods generates extra runoff from low-lying areas to surface water.

Keywords: heat transport; watershed modeling; temperature; unsaturated zone

Citation: Morway, E.D.; Feinstein, D.T.; Hunt, R.J. Simulation of Heat Flow in a Synthetic Watershed: The Role of the Unsaturated Zone. *Water* **2022**, *14*, 3883. https://doi.org/10.3390/w14233883

Academic Editor: Cristina Di Salvo

Received: 23 June 2022
Accepted: 17 October 2022
Published: 28 November 2022

1. Introduction

Most future climate forecasts suggest atmospheric warming [1]. As a result, how warming affects aquatic systems is of societal interest. For example, climatic warming is expected to increase the amount of time that humid temperate streams exhibit conditions not suited to cold-water fisheries [2,3]. Such forecasts, however, typically do not fully represent processes that play an important role in how climatic warming is expressed within watersheds. That is, each part of the watershed's subsurface system may alter the extent and timing of how heat is transported within a watershed. From the standpoint of recharge processes, the water table is typically separated from the land surface by an unsaturated zone (UZ) of variable thickness. Because the thickness of the UZ is spatially variable, its combined (or integrated) effect on the amount and timing of recharging water and heat associated with a changing climate signal is highly uncertain. Also, short and long groundwater flow paths commonly discharge to surface-water bodies which host temperature-sensitive plant and animal communities. The combined action of multiple groundwater pathways determines the amount of heat transported through the subsurface and delivered to surface water systems. Thus, accurately forecasting the effects of a warming climate on terminal surface water discharge points and associated ecosystems in a watershed requires a quantitative tool that incorporates both unsaturated and saturated zone processes.

The approaches used here build on two previous publications. Using the Unsaturated-Zone Flow (UZF1) package [4] within MODFLOW-NWT [5], Hunt et al. (2008) [6] demonstrated the importance of the UZ on the distribution, magnitude and timing of recharge at the watershed scale, and showed how each of these are impacted differently by the buffering effects of a variably thick UZ. This study expands on the results presented in Hunt et al. (2008) [6], by considering unsaturated zone heat transport over a range of infiltration signals across a range of water table depths. The second publication, Morway et al. (2022) [7], documents and verifies the mathematical framework for new computationally efficient heat transport capabilities within MT3D-USGS, using a combination of steady and transient flow and transport simulations. However, the examples in Morway et al. (2022) [7] are limited to a one-dimensional profile extending from the top of the UZ to the water table. In this study, we use the revised MT3D-USGS code to simulate heat transport at the watershed scale, following the heat influx from its entry point below the root zone as deep percolation, through the unsaturated and saturated zones, and finally to the terminal surface water discharge points. This work focuses on the groundwater system, however; temperature processes within the surface water discharge points themselves (e.g., heat changes from precipitation and shading, evaporative cooling) are not investigated, to highlight the importance of processes operating in the unsaturated zone.

The watershed used for testing can be considered quasi-hypothetical (see, for example, Anderson and Bowser (1986) [8] for an example in the context of the subsurface propagation of the effects of acid rain). That is, a homogeneous synthetic groundwater model was constructed to allow for control of important system characteristics at a scale typical of a humid temperate watershed (HUC10 size [9]), with realistic land surface and surface water configurations. The imposed transient forcing function, used to account for heat influx under conditions of warming, is also synthetic. In addition, the specified infiltration rates and temperatures vary temporally, to facilitate the exploration of watershed response relations. In effect, our experimental design pairs spatially uniform subsurface properties with temporally variable forcing functions of infiltration rates and temperatures.

Thus, this article has three principal objectives. The first is to extend the method presented in Morway et al. (2022) [7] for simulating unsaturated/saturated heat transport with MODFLOW [5,10] and MT3D-USGS [11] from a one-dimensional column to a three-dimensional surficial aquifer system, with groundwater—surface-water exchange. The second objective assesses the importance of explicitly representing UZ heat transport processes in watershed scale models. Finally, the third objective explores temperature patterns and trends within the UZ, at the water table, within the groundwater (saturated) system, and along the stream network, as the synthetic watershed warms. The warming signal migrating through and being stored in the subsurface is subject to lags (that is, change of phase), to dampening (that is, change of amplitude), and mixing (that is, convergence of flow lines), which jointly giving rise to subsurface thermal buffering. Of particular interest is the effect of the UZ as a low-pass filter, flattening high frequency and high amplitude temperature events before the heat in the percolating water reaches the water table.

2. Methods

The following four sub-sections describe the setup of the MODFLOW and MT3D-USGS models. In order, the sections focus on (1) how the warming climate is represented within the model, (2) a description of the MODFLOW model setup, including various boundary conditions (e.g., streams, lakes, etc.), (3) an in-depth discussion of three different

UZ configurations for elucidating how the role of the UZ as a warming climate may impact groundwater temperatures, and (4) a description of the MT3D-USGS heat transport model. Additional details also may be found in the Supplementary Material Sections S1 and S2.

2.1. Representation of Warming Conditions

The infiltration of "warm" water is an important, if not dominant, pathway by which atmospheric heat enters the subsurface. As a result, changes to watershed heat loading must consider two factors—(1) the amount of infiltration, and (2) the temperature of the infiltration below the bottom of the root zone. Since both are temporally variable, it is the product of these time series that represents a major component of the total heat influx at the top of the UZ. Periods of high infiltration combined with elevated temperatures (e.g., a prolonged warm spring rain or particularly high infiltration event during the hottest part of summer) will significantly increase the influx of heat into the subsurface system. By contrast, if the same elevated atmospheric temperatures that control the temperature of the infiltration are paired with reduced infiltration rates, the influx of heat to the subsurface is reduced.

For the synthetic watershed developed in this study, the top of the UZ corresponds to the bottom of the root zone and extends to the water table. As such, the specified infiltration rates and temperatures correspond to the bottom of the root zone, which is the same as the top of the unsaturated zone. In addition, the specified infiltration rates and temperatures vary on a monthly basis, to capture seasonal cycling. Monthly infiltration rates and associated temperature input into the models include a constant 30-year spin-up period, to ensure a dynamic equilibrium is established prior to the start of a variable 30-year warm-up period.

During the 30-year spin-up period, the monthly infiltration rates specified at the top of the UZ total 8.0 in/year (0.20 m/year) and are held constant in each monthly stress period, at 0.66 in/month (0.02 m/month; Figure 1A). In contrast, the infiltration rate varies monthly during the 30-year warm-up period, commensurate with typical seasonal change, but also includes random noise generated from a uniform distribution. The average annual infiltration rate during the 30-year warm-up period is 8.84 in/year (0.224 m/year) and does not include an underlying trend, although the annual totals do vary. The infiltration rates used during the spin-up and warm-up periods are similar to other modeling efforts investigating climate change impacts in humid temperate watersheds, located in Wisconsin, USA [e.g., Table 3 of Hunt et al. (2016) [3]].

During the spin-up period, the temperatures assigned to the monthly infiltration rates vary, but the same sequence of monthly temperatures are repeated every January in order to generate a constant annual average value (Figure 1B). By contrast, the monthly temperatures assigned to the infiltration rates during the 30-year warm-up period reflect three sources of variability: (1) seasonal oscillations, (2) random noise, and (3) an underlying linear warming trend of 0.0025 °C/month (Figure 1B), which equates to 0.9 °C after 30 years, and is commensurate with predicted warming under a high-emission scenario downscaled for southern Wisconsin for the period 2022–2051 [12].

The use of spin-up and warm-up periods results in relative heat influx values that are in a thermal dynamic equilibrium by the end of the spin-up period, and then transition to an unsteady condition during the warm-up period (Figure 1C). Although the warming trend contained within the infiltration temperature time series of the warm-up period is relatively modest, when combined with the variability of the monthly infiltration rates (absent during the spin-up portion of the synthetic simulations), an appreciable increase in the amount of heat added to the system occurs in some stress periods.

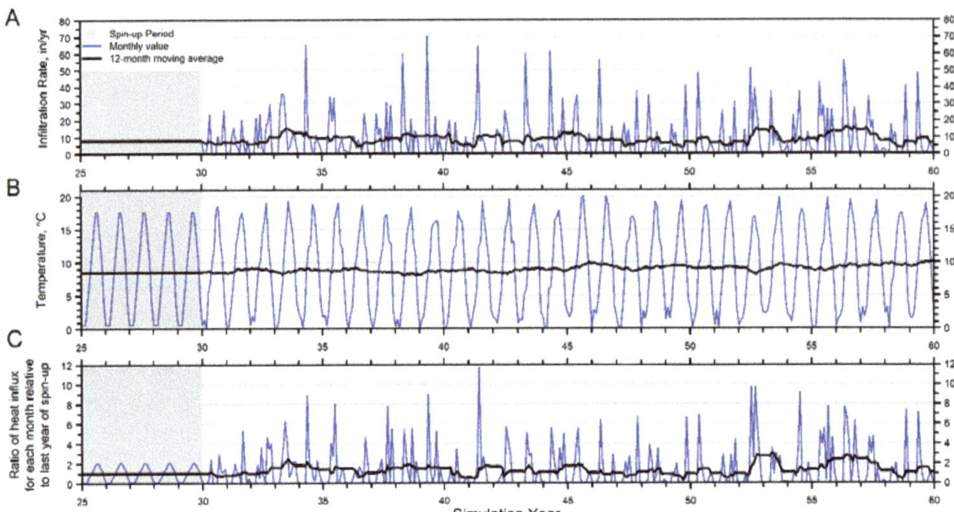

Figure 1. Forcing function applied to a synthetic model for a high emission (RCP8.5) warming scenario, corresponding to representative watershed in southern Wisconsin for warming period 2022–2051 (adapted with permission from Ref. [12]). (**A**) Monthly infiltration rates during the spin-up and warm-up periods. (**B**) Monthly infiltration temperatures during spin-up and warm-up periods. (**C**) A time series of the relative heat influx. Relative heat influx is calculated as the product of the monthly infiltration rates and temperatures for any given month in the warm-up period, divided by the average monthly heat influx during the spin-up period, resulting in a ratio that is referred to as the relative heat influx.

The unsteady heat forcing represented in the model during the warm-up period is calculated as the monthly infiltration rate multiplied by the monthly infiltration temperature. The result, after further multiplying by the heat capacity and density of water, results in units of energy/time. Figure 1C shows the ratio of the heat influx for any month, which is calculated as the heat influx for a given month divided by the average heat influx during the last year of spin-up. Hereafter, this ratio is referred to as the relative heat influx. In Figure 1C, seasonal oscillations around a stationary average are present during the spin-up period. During the warm-up period, the relative heat influx shows a considerably more variable pattern. Monthly episodes of significant forcing, or high relative heat influx, are noted throughout Figure 1C, but especially toward the end of the warming period when high monthly infiltration rates are paired with high monthly temperatures (simulation years 52, 55–56; Figure 1C). Additional discussion of the forcing function, and the ramifications of assumptions used to construct them are given in the Supplementary Material Section S1.

2.2. Model Construction: Groundwater Flow

The quasi-hypothetical watershed-scale model covers an area of about 290 square miles (about 750 square kilometers), corresponding in size to a HUC-10 [9] watershed designation (Figure 2). The domain is conceived as a homogeneous (hydraulic conductivity and specific yield) sandy aquifer, with the water table at variable depth below a spatially varying land surface elevation (resulting in variable UZ thickness from cell to cell). The model grid is 300 rows by 300 columns by 8 layers. Laterally, grid cells are 300 ft (91.4 m) on each side and vary in thickness. Aquifer parameters are uniform throughout the domain (Table 1). Boundary condition parameters, controlling sources and sinks of water, are also spatially homogeneous (Table 1).

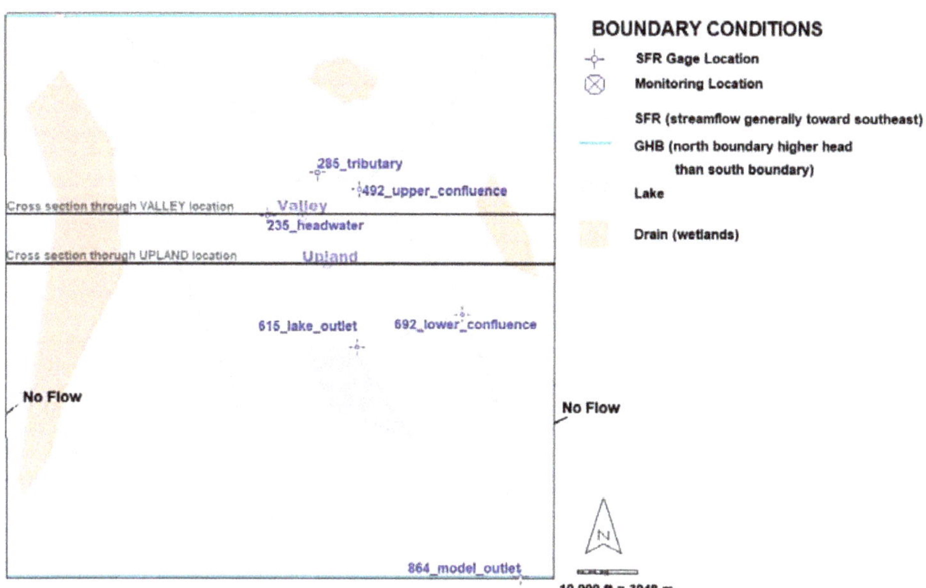

Figure 2. Plan view of synthetic model setup, showing boundary conditions (SFR: streamflow routing package; GHB: general-head boundary package; No Flow: no-flow boundary), locations for monitoring output, and the locations of cross-sections shown in the Supplementary Material (Figures S1–S9). Numbers associated with SFR gage locations correspond to IDs used later in the text.

Table 1. Flow parameter values used in synthetic water model are spatially homogeneous.

MODFLOW-NWT Package	Parameter Name	Value
UPW		
	Horizontal hydraulic conductivity	42.5 ft/day (12.95 m/day)
	Vertical hydraulic conductivity	1 ft/day (0.30 m/day)
	Specific yield	0.26 (unitless)
	Specific storage	1×10^{-5} 1/day
UZF1		
	Vertical hydraulic conductivity	1 ft/day (0.30 m/day)
	Surface infiltration hydraulic conductivity	0.1 ft/day (0.0305 m/day)
	Saturated water content	0.30 (unitless)
	Residual water content	0.04 (unitless)
	Brooks-Corey epsilon	3.87 (unitless)
	Monthly infiltration rate	See Figure 1A
SFR2		
	Channel width	25 ft (7.62 m)
	Channel bed thickness	1 ft (0.30 m)
	Channel bed hydraulic conductivity	20 ft/day (6.10 m/day)
	Channel slope	0.0002 (ft/ft)
	Channel incision (streambed elevation below top of cell)	2.5 ft (0.76 m)

Table 1. *Cont.*

MODFLOW-NWT Package	Parameter Name	Value
DRN		
	Conductance	90,000 ft^2/day (8362 m^2/day)
LAK		
	Lakebed conductance	90,000 ft^2/day (8362 m^2/day)
GHB		
	Conductance	11.37 ft^2/day (1.06 m^2/day)

The General-Head Boundary (GHB) package [10] simulates flow entering the north perimeter boundary and exiting the south perimeter boundary of the model domain. No-flow conditions are imposed along the eastern and western sides of the model at all elevations, and on the bottom of the model. Stream networks are represented by the Streamflow Routing (SFR2) package [13] (Figure 2). Three wetlands are simulated using the Drain (DRN) package [10] and a lake is represented by the Lake (LAK) package [14] (Figure 2). Simulated baseflows within the synthetic model are the sum of (1) direct groundwater discharge to the channel, (2) groundwater discharge to the land surface in riparian areas that subsequently runs off and into the surface-water network, and (3) rejected infiltration resulting from saturation excess in riparian areas that runs off and into the surface-water network. The model is configured so that groundwater–surface-water interaction is one-way as groundwater discharge to the stream; there are no losing reaches. Transient forcing functions and the geometry of surface water sinks, result in appreciable complexity within the flow system, despite the spatially homogeneous parameterization. Additional information on the overarching groundwater model design is provided in Supplementary Material Section S1.

2.3. Representation of Unsaturated Zone Processes

Hunt et al. (2008) [6] demonstrate the importance of including UZ flow processes in steady-state and transient regional-scale groundwater flow models. However, a paucity of data for parameterizing Richards' equation-based approaches, as well as the computational demands of implementing the approach in numerical models, makes alternative approaches such as that of UZF1, highly attractive [15]. At the time of writing, UZF1 is now a common alternative for simulating UZ flow in regional-scale models [2,3,16–22]. Hunt et al. (2008) [6] go on to show how UZ flow and the corresponding changes in UZ storage result in lags between the timing of infiltration at the top of the UZ and recharge to the water table. In addition, because of transient water table elevations, the UZ may at times pinch out as the water table rises, resulting in Dunnian overland flow [23]. This, in turn, reduces net infiltration rates which further alters the individual components of a watershed budget. Thus, approaches that omit UZ processes result in infiltration being transmitted instantaneously to the water table at rates that may not be supported by the system. Such over-simplification can confound the use of head data to estimate recharge and can be problematic where timing of infiltration-related recharge is vital for understanding a process of concern (for example, studies involving solute loading to the water table [24] lagged by the unsaturated zone).

Niswonger et al. (2006) [4] offer a detailed explanation of UZF1. Of particular note is that UZF1 neglects capillary forces, an assumption that provides computational efficiencies for regional-scale models [15]. Moreover, UZF1 is equipped to simulate groundwater discharge to land surface when groundwater heads rise to within a user-specified proximity of the land surface [20]. Additionally, UZF1 simulates rejected infiltration when the specified infiltration rate exceeds the vertical hydraulic conductivity (Hortonian overland flow), or when the UZ becomes saturated (Dunnian overland flow). Options are available for routing both the rejected infiltration and groundwater discharge to land-surface to nearby surface water features, which are often important components of the overall water budget [20]. Although supported by the UZF1 package, evapotranspiration was not simulated from the UZ in the synthetic watershed presented herein; rather, evapotranspiration loss is accounted for in the *net* infiltration rate specified at the top of the UZ. Additional UZF1 information is available in the Supplementary Material Section S1.

Just as with the calculation of infiltration leading to recharge [6], the UZ also acts as a low-pass filter for heat transport, as it migrates downward from near the land surface toward the water table. Equipped with the enhancements described in Morway et al. (2022) [7], MT3D-USGS can simulate heat transport from the top of the UZ, through the UZ and saturated zones, and exchange with surface-water features (represented with different packages). The focus herein is on variably saturated and saturated water temperature, whereas a companion paper focuses on transmission of total energy [25].

Hunt et al. (2008) [6] demonstrated how the thickness of the UZ is a major control on the timing and magnitude of recharge. This investigation goes a step further, and explores how UZ thickness, as well as the parameterization of the UZ, modulate the timing and magnitude of the infiltrating heat signal, prior to becoming recharge. To this end, the synthetic watershed was constructed using three alternative configurations:

- NO_UZ_THK: no UZ processes are simulated by the model. Instead, a monthly infiltration of water and heat is applied directly to the water table, using the Recharge (RCH) [10] and Source/Sink Mixing (SSM) [26] packages, respectively. Note that the NO_UZ_THK and MID_UZ_THK models (explained below) are dimensionally identical (i.e., cell geometries (thicknesses) are the same), meaning that an overlying UZ is present in the NO_UZ_THK, although the grid cells above the water table are inactive. The NO_UZ_THK designation does not imply that the water table is near the land surface throughout the model domain.
- MID_UZ_THK: a model with the same grid cell dimensions as the NO_UZ_THK simulation but simulates UZ processes with the UZF1 and unsaturated zone transport (UZT) [11,27] packages. The UZ average thickness is approximately 11 ft, with a maximum of 62 ft. The land surface slope from surface water features to "upland" locations is low (1.5 ft/300 ft, or 0.005 ft/ft; Figure 3A).
- HI_UZ_THK: the UZ is approximately three times thicker than the MID_UZ_THK setup, averaging approximately 31 ft thick with a maximum thickness of approximately 150 ft. The land surface slope from surface water features is steeper than the MID_UZ_THK model (3.0 ft/300 ft, or 0.01 ft/ft; Figure 3B).

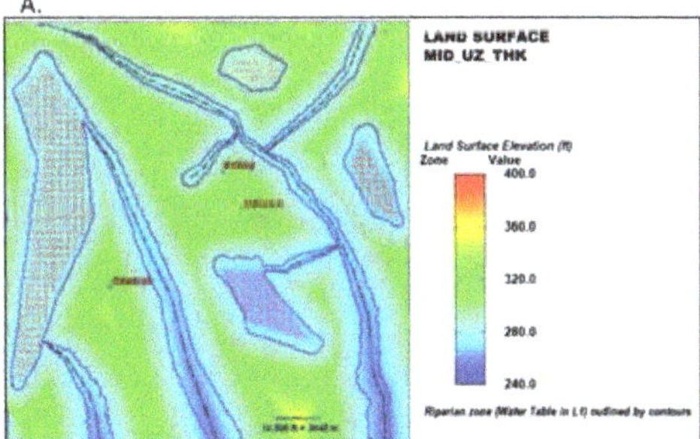

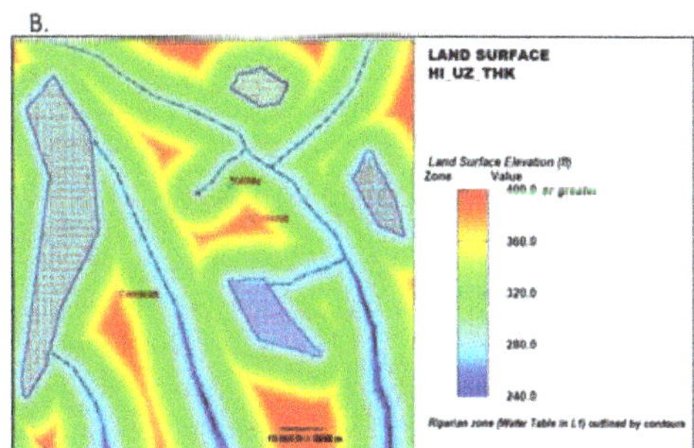

Figure 3. A comparison of the land surface elevations for the (**A**) MID_UZ_THK and (**B**) HI_UZ_THK models. Land surface elevation for each model cell is proportional to its distance from the nearest surface water body. The proportionality constant is 1.5 ft/300 ft (0.005) for the MID_UZ_THK simulation and 3.0 ft/300 ft (0.01) for the HI_UZ_THK simulation. The land surface does not directly enter into the NO_UZ_THK solution, since the UZ is not simulated and therefore no attenuation of the infiltrating signal is realized. The riparian zone contours correspond to conditions at the end of spin-up.

Differences between the MID_UZ_THK and HI_UZ_THK simulations are evidenced by the water table residing in layer 1 across more than 20% of the MID_UZ_THK simulation domain, while residing only in less than 5% of the layer 1 cells in the HI_UZ_THK simulation (Supplementary Material Section S1).

As with any groundwater solute transport simulation, careful consideration of the model layering (i.e., vertical discretization) scheme is an important part of a heat transport simulation. Too many layers can result in overly lengthy model run-times with untenable linker file [28] sizes that do not offer meaningful gains in simulation accuracy. However, too few model layers can misrepresent heat transport processes, most notably conduction. For example, overly thick grid cells cannot accurately account for thermal gradients that drive conduction and therefore the spread of heat.

The simulations in this analysis use a minimum of eight layers, although sensitivity runs with two additional layers dedicated to the UZ were also explored, using both the MID_UZ_THK and HI_UZ_THK configurations (see Supplementary Material). All models employ a 3 ft thick top layer, referred to below as the "receptor" layer, which functionally serves to receive the heat signal in a spatially consistent manner. The alternative of applying a heat signal to layer 1 cells with spatially varying thicknesses would complicate the interpretation of the relative temperatures in layer 1, since the amount of stored heat would also be a function of the thickness of each cell. Furthermore, keeping layer 1 predominately unsaturated (except when adjacent to surface water features), allows a thermal gradient to develop in the upper UZ for simulating conduction between the UZ layers. The logic for deciding the thickness of layers 2 and 3 was determined after a preliminary run of the model. Both layers were assigned a minimum thickness of 6 ft where the water table reached a minimum depth of less than 15 ft sometime during the simulation (the receptor plus the minimum thickness of 6 ft for layers 2 and 3). Where the minimum water table depth was greater than 15 ft, the thickness of layers 2 and 3 was increased, such that the additional UZ thickness was divided equally among them while layer 1 remained a constant 3 ft thick (see for example Figure S3-2). Where the UZ was greater than 15 ft, the water table resided in layer 4. Layers 5 through 8 were fully saturated for the duration of the simulation and were present to enable differing temperature with depth.

2.4. Model Construction—Heat Transport

The mathematical framework and equations for simulating heat transport in the synthetic watershed discussed herein are presented in detail for a one-dimensional system in Morway et al. (2022) [7]. This study adopts a similar approach but applies the methodology at a watershed scale. Table 2 lists the transport and heat flux parameters applied to all three versions of the model. Heat sorption in the matrix is assumed to act instantaneously, portioning the thermal energy between the solid and fluid phases according to a ratio that varies with water content. The partitioning of the total thermal energy between the aqueous and solid phases is commonly referred to as the retardation factor in MT3D-USGS. A retardation factor of 2.0, for example, implies that half of the thermal energy is "sorbed" to the solid phase, and, therefore, the heat moves at half the advective fluid velocity. For unsaturated flow conditions, the retardation factor depends not only on a linear distribution coefficient, but also on the water content. Additional discussion of the treatment of sorption and the selection of parameter values is provided in the Supplementary Material Section S2.

An aspect of heat transport that is often quite different from solute transport is the relative contribution of the mechanical dispersion and molecular diffusion terms that contribute to the overall hydrodynamic dispersion term. In solute transport, mechanical dispersion is generally orders of magnitude greater than molecular diffusion, especially in advection-dominated settings [29]. In heat transport simulations, however, the thermal conduction as represented by the molecular diffusion term can exceed the mechanical dispersion term. In MT3D-USGS, the thermal conduction is a bulk process, representing the movement of heat fronts through both the solid (not explicitly simulated) and fluid phase [7]. In this modeling exercise, a relatively modest longitudinal dispersivity value of 3 ft is assumed. Although advection in the UZ is downward only, as implemented by the kinematic wave approximation within the UZF1 package, conduction and dispersion may still occur in all directions, including upward, in both the saturated and unsaturated zones (see parameters listed for the DSP Package, Table 2).

Table 2. Transport parameter values for synthetic watershed model, spatially homogeneous.

MT3D-USGS Package	Parameter Name	Value
BTN		
	Porosity	0.3 (unitless)
DSP		
	Saturated thermal conductivity	52,669 Joules/(day·ft·°C) [2.0 Joules/(sec·m °C)]
	Residual thermal conductivity	13,167 Joules/(day·ft °C) [0.5 Joules/(sec·m °C)]
	Fluid density	28.3166 kg/ft^3 (1000 kg/m^3)
	Fluid heat capacity	4183 Joules/(kg °C)
	Residual water content	0.04 (unitless)
	Longitudinal dispersivity	3.0 ft (0.91 m)
	Transverse horizontal dispersivity	0.30 ft (0.091 m)
	Transverse vertical dispersivity	0.30 ft (0.091 m)
UZT		
	Monthly infiltration temperature	see Figure 4
RCT		
	Bulk density of solid	51.849 kg/ft^3 (1830 kg/m^3)
	Distribution coefficient	2.68×10^{-3} ft^3/kg (7.59×10^{-5} m^3/kg)
SSM		
	Source temperature	8.55 °C during spin-up (raised 0.03 °C/yr during warm-up)
SFT		
	Initial temperature	8.55 °C
LKT		
	Initial temperature	8.55 °C
Precipitation temperature		see Figure 4 (temperature the same as infiltration, with following exceptions) April: +0.5 °C; May: +1.0 °C; June: +1.5 °C; July: +2.0 °C; August: +1.5 °C; September: +1.0 °C; October: +0.5 °C

Table 3. Additional information pertaining to subplots shown in Figure 4.

Figure 4 Subplot ID	Warming Year	Month	Relative Heat Influx for Current Month	Relative Heat Influx for Proceeding 12 Months	Infiltration Rate (in/mo)	Infiltration Temperature (°C)
A–C	0.00	December	0.07	1.00	0.75	0.02
D–F	2.75	September	3.48	1.30	1.58	12.60
G–I	10.17	February	0.24	1.51	1.13	1.24
J–L	15.17	February	0.00	1.71	0.00	2.62
M–O	24.67	August	0.00	1.20	0.00	17.31
P–R	25.67	August	7.79	2.12	2.25	19.78

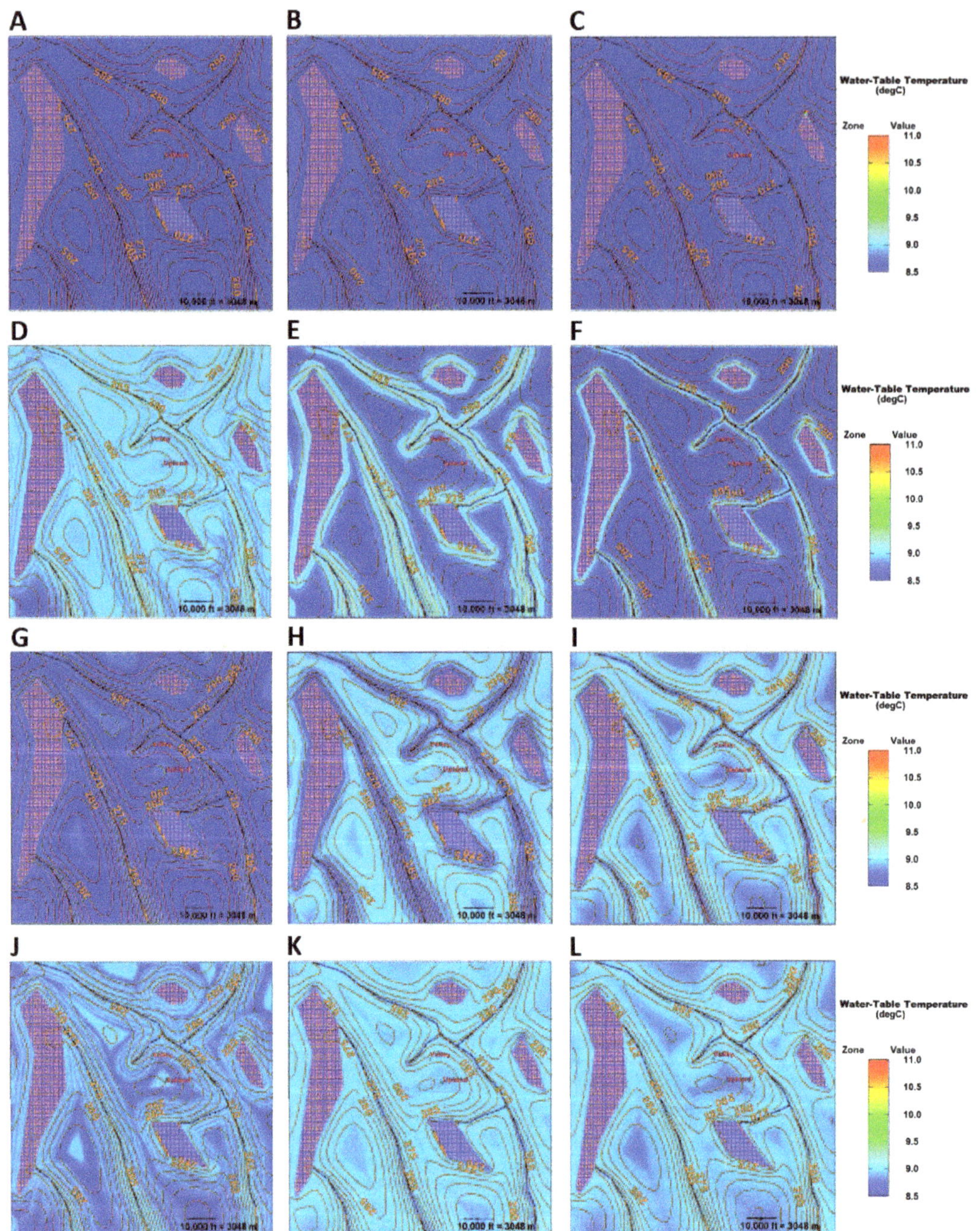

Figure 4. *Cont.*

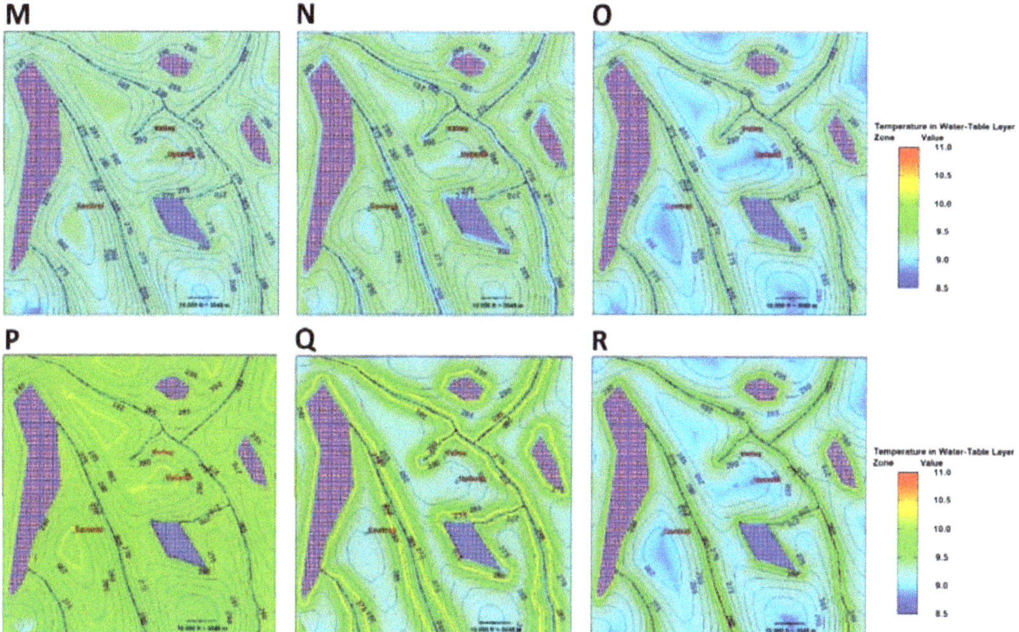

Figure 4. Simulated heads (contour lines) and temperatures (color-filled) for the water table layer through time for the NO_UZ_THK (**A**,**D**,**G**,**J**,**M**,**P**), MID_UZ_THK (**B**,**E**,**H**,**K**,**N**,**Q**) and HI_UZ_THK models (**C**,**F**,**I**,**L**,**O**,**R**). Water table temperatures are shown for (**A–C**) end of spin-up, (**D–F**) after 2.75 years of warm-up, (**G–I**) after 10.17 years of warm-up, (**J–L**) after 15.17 years of warm-up, (**M–O**) after 24.67 years of warm-up, and (**P–R**) after 25.67 years of warm-up. Additional details for each subplot are provided in Table 3.

Within MT3D-USGS, the streamflow transport (SFT) and lake transport (LKT) packages simulate heat transport in the surface water network, including the exchange of heat between surface water and groundwater [11]. SFT solves a 1D advection-dispersion equation for calculating the temperature within each stream reach. In LKT, a single, instantaneously mixed temperature is calculated for each lake interacting with the aquifer. Groundwater discharged directly to surface water features as well as groundwater runoff (i.e., groundwater discharge to land surface adjacent to streams combined with rejected infiltration from the top of UZ, both instantaneously transferred to the nearest stream or lake feature) is routed through the surface-water network in a way that integrates all upstream discharge for any downstream point. As a result, the stream temperature at a particular location may be realistically simulated as higher or lower than the temperature of the ambient groundwater at that same location. The ability to simulate spatially distributed surface water temperatures at specific points within a watershed is increasingly important for resource management.

3. Results

Simulation results are described here and in the Supplementary Material Section S3. Results are grouped under three main subsections that discuss (1) groundwater temperatures near the water table (where recharge occurs) as they relate to the thickness of the UZ, relative heat influx, and time of year, (2) deeper groundwater temperatures, and (3) the cooling influence of groundwater discharge on surface water temperatures after accounting for the temperature changes occurring in the subsurface.

3.1. Water Table Temperatures

A snapshot of simulated water table temperature for the month of December at the end of the 30-year spin-up period shows nearly identical and uniform conditions of 8.55 °C across all three base models (Figure 4A–C). In the 30 years following the spin-up period (i.e., the warm-up period), the effect of the UZ on the relative heat influx results in a complex spatial and temporal temperature response near the top of the saturated groundwater system, as shown by the monthly snapshots of water table temperatures at 2.75 (Figure 4D–F), 10.17 (Figure 4G–I), 15.17 (Figure 4J–L), 24.67 (Figure 4M–O), and 25.67 years (Figure 4P–R). The relative heat influx ratio for each date displayed in Figure 4 are provided in Table 3. A relative heat influx ratio of 1.0 signifies that the heat loading rate for the month or year, depending on which value is considered, is equivalent to the heat loading rate during the last year of spin-up. In Figure 4D–F, for example, the relative heat loading rate for the second year of the 30-year warm-up period (year 2.75) is 1.30. This suggests that 30% more heat flux (infiltration rate multiplied by the temperature of the infiltration) entered the subsurface, relative to the last year of the spin-up period. In general, the monthly heat flux rates vary by an approximate value of 1.0 over the course of a year, while the annual values gradually increase over the 30-year warming period.

The effect of the UZ is demonstrated by comparing the water table temperature maps for the three test models. For the NO_UZ_THK model (recharge is applied directly to groundwater system rather than routed through the UZ) the map for 2.75 years (September) shows mostly homogeneous water-table conditions, averaging a little above 9 °C early in the warm-up period (Figure 4D). In contrast to Figure 4D, the temperatures in the water table layer of the MID_UZ_THK model (Figure 4E) persist at cooler temperatures (~8.55 °C) where the UZ is thick and are prevalent throughout the model domain at the end of the spin-up period (Figure 4B). However, where the UZ is thin and the infiltration quickly converts to recharge (~1 stress period, equivalent to 1 month), simulated warming at the water table is similar in magnitude to the NO_UZ_THK model, for example along the riparian corridor (Figure 4E). The same is true for the HI_UZ_THK model, except that the effect along the riparian corridors is narrower, due to the steeper slope leading away from the streams and water bodies (Figure 4F).

The expression of warming within a particular model layer depends partly on which month is chosen for closer inspection. For example, after 10.17 years of warm-up (relative heat influx of 1.51), which corresponds to February (relative heat influx of 0.24), the water table temperatures in the NO_UZ_THK model (Figure 4G) remain mostly homogeneous, although the water table temperature has cooled, relative to the temperatures at 2.75 years (Figure 4D). The overall cooling between 2.75 and 10.17 years reflects the direct input of colder water to the water table rather than mixing with warmer water through a thicker UZ, as simulated by the UZT package. For the MID_UZ_THK model (Figure 4H) at 10.17 years, the water table temperatures generally increased from year 2.75. However, the groundwater temperatures are inverted compared to what they were at 2.75 years—the riparian corridor is now cooler than the non-riparian corridor areas (i.e., compare Figure 4E to Figure 4H). Similar temperatures are exhibited in the HI_UZ_THK model (i.e., compare Figure 4F to Figure 4I). In another February snapshot from five years later (15.17 years; Figure 4J–L), all three models show similar patterns, as seen at 10.17 years of generally warmer water table temperatures, because the warming trend applied to the temperature of the infiltration begins to affect the overall ambient temperature of the water table.

At 24.67 years into warming, corresponding to August, the NO_UZ_THK and MID_UZ_THK results (Figure 4M,N) show a similar and fairly uniform water table temperature across the model domain. For the same simulated period, the water table temperatures in the HI_UZ_THK model (Figure 4O) appreciably depart from the NO_UZ_THK and MID_UZ_THK model results. That is, the water table temperature is less uniform—areas below a thicker UZ are cooler, for example, at the UPLAND location. The final set of water table temperature maps (Figure 4P–R), also corresponding to August, show the groundwater temperature at 25.67 years. This period is characterized by the highest monthly relative heat influx during the entire warming period (7.79; Figure 1C) as well as a very high average annual relative heat influx (2.12). The NO_UZ_THK model shows warmer temperatures over most of the domain, at nearly 10 °C (Figure 4P). In contrast, the MID_UZ_THK model shows warmer riparian corridors and cooler temperatures under the uplands (Figure 4Q). The HI_UZ_THK model for year 25.67 (Figure 4R) is cooler under the uplands and along the river corridors than either of the other two models. A comparison of the water table temperatures after 24.67 and 25.67 years of warming shows that appreciable warming occurred during the elapsed year in the NO_UZ_THK and MID_UZ_THK models (i.e., comparing Figure 4M to Figure 4P, and Figure 4N to Figure 4Q, respectively) and only a small amount of warming in the HI_UZ_THK model (i.e., comparing Figure 4O to Figure 4R). Given that the only difference between these models is the thickness of the UZ, the cooler temperatures in the HI_UZ_THK model suggest meaningful thermal buffering in the UZ before infiltrating heat reaches the water table.

3.2. Groundwater System Temperatures

Time series plots of the simulated temperature at the UPLAND location show markedly different behavior for each layer of the three base models (Figure 5). For example, the temperature response in the water table layer (layer 4) of the NO_UZ_THK model exhibits higher frequencies and amplitudes compared with the MID_ and HI_UZ_THK models. This result is expected, since the UZ is not simulated and therefore unable to buffer the infiltrating heat signal. Thus, the layer 4 response at the UPLAND location is notably flashier in the NO_UZ_THK model (Figure 5A). The temperature response in layer 4 for the other two models with thicker unsaturated zones is much smoother, and the total warm-up in layer 4 of the MID_UZ_THK (Figure 5B) model is approximately 0.3 °C less by the end of the simulation, compared with the HI_UZ_THK model (Figure 5C). The simulated temperatures also trend upward at the VALLEY location in all layers Figure 5D–F), albeit with different behaviors. For example, the amplitude of the temperature swings in layer 1 at the UPLAND location is greater than at the VALLEY location; however, larger amplitudes are seen at the VALLEY location in layers 2 and 4, compared with the UPLAND location. Additional cross-sectional results are provided in the Supplementary Material Section S3.

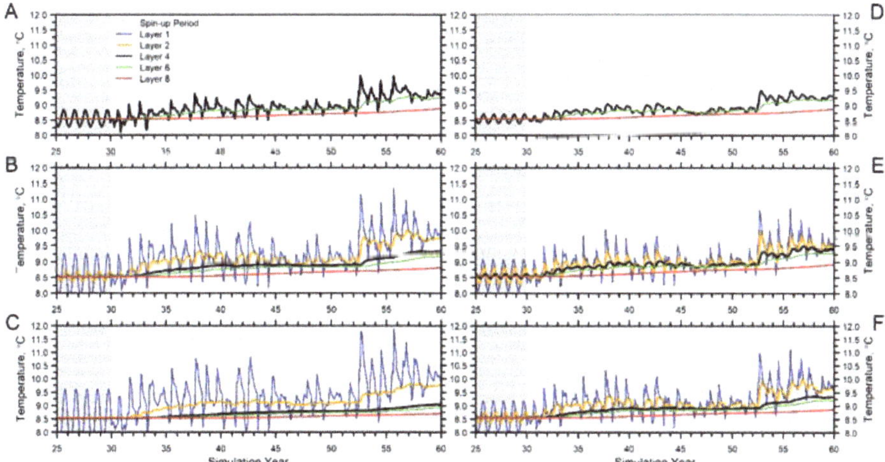

Figure 5. Simulated temperature hydrographs by model layer at the (**A–C**) UPLAND well and (**D–F**) VALLEY well locations for the spin-up and warm-up periods. Groundwater temperature hydrographs are further organized as follows: (**A,D**) NO_UZ_THK, (**B,E**) MID_UZ_THK, and (**C,F**) HI_UZ_THK models. Depths of the various layers for both locations are shown in the Supplementary Material Section, Figures S1–S9.

3.3. Stream Baseflow Temperatures

Groundwater discharge plays an import role in cooling streamflow temperatures, particularly during late summer, when its proportional contribution to streamflow is greatest (i.e., baseflow conditions). Because groundwater discharge is comprised of an ensemble of subsurface flow paths, streamflow temperature during baseflow largely reflects the (1) thermal buffering that occurred within the UZ prior to the infiltrating water becoming recharge, (2) buffering within the saturated zone as groundwater flow paths of different temperature converge, (3) thermal buffering that occurred within the UZ prior to it becoming recharge, and (4) other processes not specifically addressed in this paper, including direct atmospheric effects on surface water. Heat in storm runoff from the land surface, not addressed here but incorporated in a companion paper by Feinstein et al. (2022) [25], does not affect baseflow temperatures, but rather acts on total streamflow temperatures. The key point is that simulated baseflow temperatures for a given location within a stream network represent a composite of heat accumulated from upstream in the watershed. In particular, the baseflow temperature responds to dampened heat flows through the UZ and the saturated system, along with the undampened heat contribution from groundwater runoff (by way of rapid transfers from groundwater discharge to the land surface plus rejected infiltration from the top of the unsaturated zone).

The hypothetical stream gage locations used in this investigation to describe baseflow conditions are divided into two groups of three (Figure 6): the first group consists of upgradient gages (Figure 2) corresponding to a headwater location (site 235), a tributary outlet (site 285), and an upper confluence location (site 492), while the second group of downgradient gages consists of a lake outlet (site 615), a lower confluence (site 692), and the model outlet (site 864). Temperatures in the lake outlet gage reflect a single temperature computed for a well-mixed lake through time. Temperatures at the model outlet gage reflect the integrated response of an entire upgradient surface-water network over time.

For the upgradient gages, results of the NO_UZ_THK model (Figure 6B) show a seasonal temperature frequency and a rising magnitude trend, but with little temperature separation at the three gage locations (Figure 6A). The MID_UZ_THK model (Figure 6C) shows seasonality and rising trends in stream temperatures similar to the NO_UZ_THK model (Figure 6D), but with increased separation among the thermal hydrographs. For example, the 95th percentile temperature increase is greatest at the tributary location and lowest at the upstream headwater location. In contrast to the other simulations, stream temperatures generated by the HI_UZ_THK model are virtually identical across the three upgradient locations, and the upward trend is notably dampened, compared with the NO_UZ_THK and MID_UZ_THK models.

For the three downgradient gage locations (Figure 7A), the streamflow temperature response is different from that of the upgradient locations. For example, in all three test models, the lake outlet gage (Figure 7A) shows a pronounced yearly oscillation superimposed on the rising trend. Although the codes used in this work do not simulate all components of the lake temperature budget, the lake outlet results have heuristic value. Annual temperature swings of roughly 0.3 °C are simulated at the lake outlet, where the lake acts as a well-mixed reservoir, integrating discharge from its groundwater contributing area, which largely consists of areas with little UZ thickness. That is, the simulated streamflow temperature at the lake outlet reflects the solitary temperature simulated for the entire lake. The lower confluence location (ID 692) is somewhat dampened for all three base model versions. However, the results at the model outlet gage are less flashy in the NO_UZ_THK model (Figure 7B) compared to the flashier temperatures in the MID_UZ_THK model (Figure 7C), with the 95th percentile stream baseflow temperature increase for the moderately thick UZ model exceeding 0.6 °C, and an excursion (maximum minus minimum) exceeding 1.0 °C. Results of the HI_UZ_THK model (Figure 7D), by contrast, show a dampened response in the streams for both the lower confluence and model outlet locations.

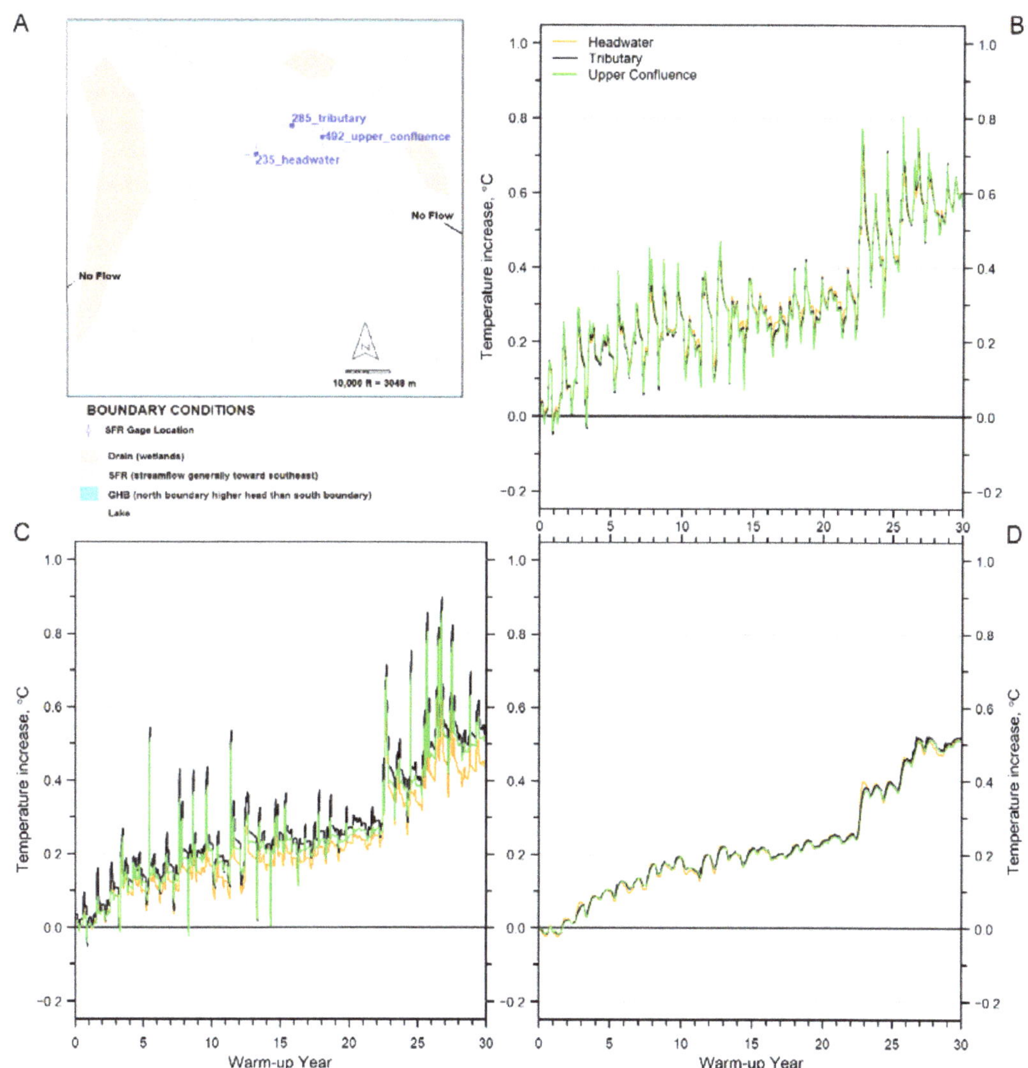

Figure 6. Simulated stream baseflow temperature response to warming (**A**) at the upgradient gage locations for the (**B**) NO_UZ_THK, (**C**) MID_UZ_THK, and (**D**) HI_UZ_THK model simulations.

For the NO_UZ_THK and MID_UZ_THK simulations, the flashiest stream temperature response is at the model outlet gage (Figure 7B,C respectively)—an initially counterintuitive result, considering that this location integrates contributions from the largest portion of the watershed. However, because the model outlet is flanked by riparian areas (the water table resides in the top 3-foot-thick layer), there is minimal UZ buffering for the MID_UZ_THK model, and no buffering for the NO_UZ_THK model; therefore, direct runoff contributed by precipitation will immediately (that is, within the same model monthly time step) influence the stream temperature. The extent of the riparian area varies within the contributing area of each gage, where a rough trend of increasing riparian area at more downstream gage locations is observed (Table 4). For the NO_UZ_THK model, the specified recharge volumes (as opposed to simulating infiltration with UZF) result in elevated water levels near the streams and (unrealistic) high groundwater gradients, which, in turn, facilitate rapid lateral groundwater flow to nearby streams that discharge appreciable amounts of flow and heat in a short period of time. In the MID_UZ_THK model, an extensive riparian area exists within the contributing area, that is, between the lower confluence (gage 692) and model outlet (gage 864). In this circumstance, a thin UZ associated with a water table near the land surface facilitates rejected infiltration; that is, the ability of the groundwater system to accept infiltration is significantly reduced and it is therefore shunted as runoff to the nearby stream. Conversely, the reduced riparian area in the HI_UZ_THK simulation, along with more dampening of a thick UZ, results in a smoother stream temperature response.

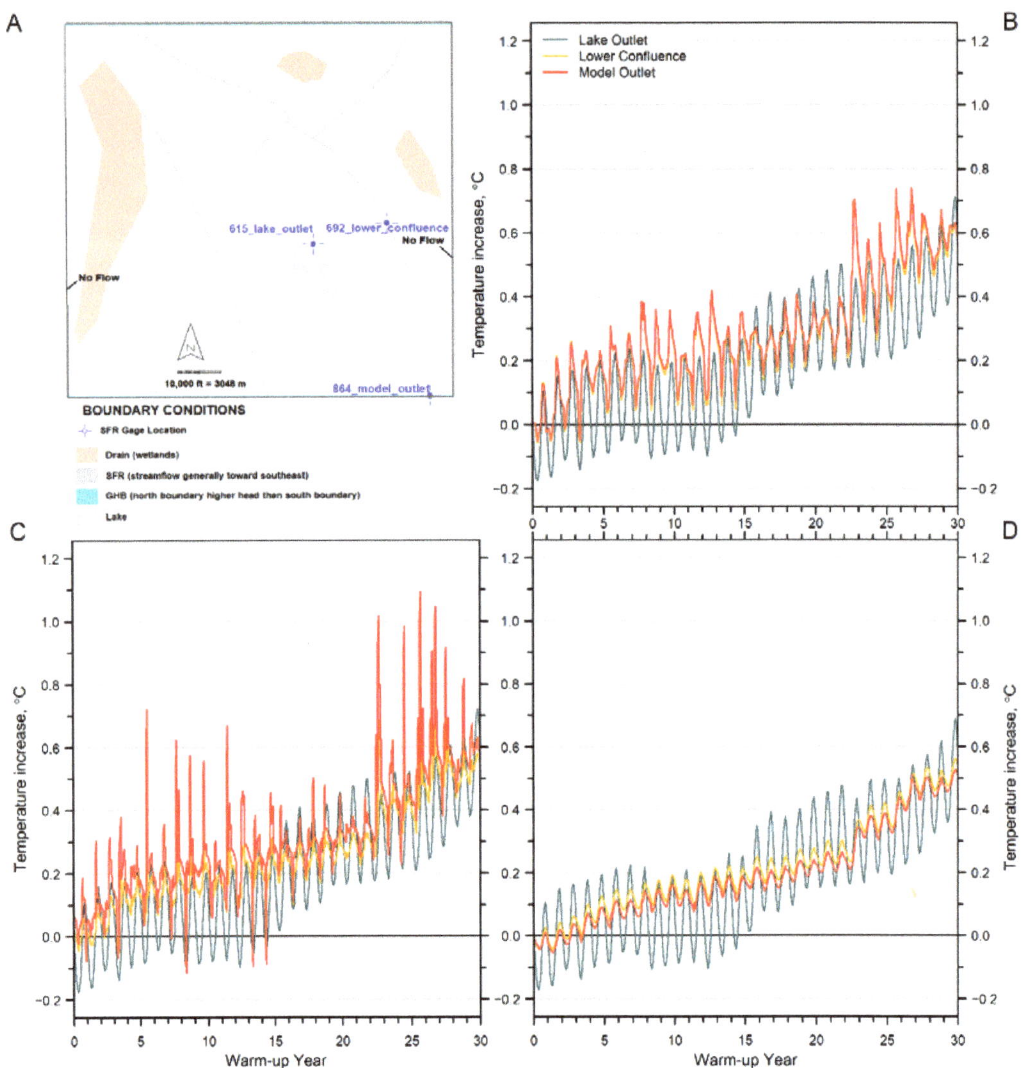

Figure 7. Simulated stream baseflow temperature response to warming (**A**) at the downgradient gage locations for the (**B**) NO_UZ_THK, (**C**) MID_UZ_THK, and (**D**) HI_UZ_THK models.

Table 4. Percent of the contributing area above each stream gage (Figure 2) that is classified as riparian zone.

Base Model Version	Stream Gage	Contributing Area (Fraction of Model Domain)	Riparian Area (Percent of Contributing Area)
NO_UZ_THK	285-tributary	0.042	20.6%
	492-upper confluence	0.202	25.4%
	692-lower confluence	0.353	27.6%
	864-model outlet	0.447	31.8%
MID_UZ_THK	285-tributary	0.042	16.9%
	492-upper confluence	0.202	22.1%
	692-lower confluence	0.353	22.6%
	864-model outlet	0.447	24.9%
HI_UZ_THK	285-tributary	0.042	0.8%
	492-upper confluence	0.202	1.6%
	692-lower confluence	0.353	3.2%
	864-model outlet	0.447	4.4%

By the end of the 30-year warm-up period (i.e., the end of the simulations), an overall increase in the stream baseflow temperature of approximately 0.5 °C is simulated at all three of the downgradient gage locations (Figure 7). This increase is roughly half of the 1.0 °C rise in the simulated water table temperature at the VALLEY well location (the shallowest layer for each UZ model; Figure 5D–F), an area with a similarly shallow water table. Thus, there is a thermally dampened response in the stream temperatures relative to the groundwater system, which is suggestive of groundwater mixing—the upwelling of cooler groundwater from deeper groundwater flow paths combining with shallower and warmer flow paths—before discharging into the stream. It is important to emphasize that at the end of the 30-year warm-up period, simulated temperatures throughout the system have not reached a new dynamic equilibrium. In other words, the UZ continues to buffer the underlying warming signal applied to the infiltration during the last 30 years of the simulation. Additionally, cooler groundwater from deeper parts of the aquifer mixes with the warmer groundwater near the water table to further dampen the effect of the warming signal on the stream temperatures. Finally, longer flow paths, unaffected by 30 years of warming, may begin to show signs of more significant warming, given enough time. For example, the simulated groundwater temperatures in layer 8 do show signs of warming by the end of the simulations (Figure 5), although it is the most muted response across all layers. Therefore, the overall watershed residence time, and the distribution of residence times within a watershed, influence the thermal resiliency of a watershed subject to warming.

Additional UZ layers for further resolving UZ flow and transport had little effect on the final temperatures, indicating that the kinematic wave approximation within the UZF1 package provides sufficient information to capture lags in the infiltrating heat flux. However, including at least one completely unsaturated layer above the water table enables MT3D-USGS to simulate lags in heat reaching the water table, since MT3D-USGS instantaneously mixes the unsaturated and saturated temperatures (i.e., "concentrations") in cells containing the water table [11]. A parameter sensitivity analysis (Supplementary Material Section S3) showed that the simulated water table temperatures responded more strongly to perturbations than the stream temperatures. Table S3-1 lists the parameters that were adjusted. Parameters related to flow of water (UZ vertical hydraulic conductivity and saturated water content) had modest sensitivity, while heat transport-related parameters (i.e., distribution coefficient, thermal conductivity) were more sensitive. However, a highly reduced

UZ vertical hydraulic conductivity did appreciably reduce the amount of groundwater recharge, which was balanced by an increase in rejected infiltration, leading to an increase in the amount of overland flow to surface water, which in turn affected the heat balance of the system.

4. Discussion and Implications for Watershed Heat Transport Modeling

The simulated temperatures throughout the watershed may be evaluated in terms of how the infiltrating heat signal's amplitude, frequency, and phase are modified first by the UZ, and secondly by the saturated zone. For example, seasonal swings in the average simulated temperature of layer 1 can be as high as 2.5 °C (Figure 5B,C,E,F), although they are frequently less than that. When the heat signal reaches the bottom of the UZ (represented by layers 1–3), the amplitudes of the seasonal swings in temperatures have almost entirely disappeared, although small seasonal swings in the groundwater temperature are still evident at the VALLEY location in layer 4 of the MID_UZ_THK simulation (Figure 5E). The existence of some seasonality in temperature for layer 4 in the MID_UZ_THK model (Figure 5E) compared with the HI_UZ_THK model (Figure 5F) further demonstrates the dampening effect of the UZ. Thus, the temperature swings assigned to the infiltration at the top of the UZ (Figure 1) are largely smoothed by the unsaturated and saturated zones.

By the end of the warm-up period, the simulated average temperature increase in layer 4—representative of the shallow part of the groundwater system—is approximately 0.75 °C at the UPLAND location in the MID_UZ_THK model (Figure 5B). At the VALLEY location, the average temperature of layer 4 increased by nearly 1.0 °C (Figure 5E). The average temperature increase in the deeper aquifer, represented by layer 8, was only approximately 0.25 °C and 0.40 °C at the UPLAND (Figure 5B) and VALLEY (Figure 5E) locations, respectively, in the MID_UZ_THK model. In general, layer 8 is representative of groundwater temperatures roughly 100 ft (30 m) below the water table.

The behavior of stream baseflow temperatures during warming is shown for downstream gages in Figure 7. At the end of the 30-year warm-up period, the stream temperatures rose between 0.5–0.6 °C in the three models, compared with the end of the spin-up period. Thus, the model, as expected, simulates less overall warm-up in the stream temperatures compared with the amount of warm-up applied to the infiltrating water (2 °C, Figure 1B). The dampened stream temperature response is sustained by the discharge of colder groundwater from deeper in the aquifer mixing with the groundwater discharge. Moreover, the effect of UZ thickness on stream temperatures also is likely evident in the results; for the upgradient locations, the HI_UZ_THK stream temperatures (Figure 6D) are much smoother and considerably more muted, compared with the MID_UZ_THK stream temperatures (Figure 6C).

A final consideration in evaluating infiltrating heat in a watershed is the phase, or lag time, between the forcing boundary condition (i.e., the temperature of the infiltration) and the downgradient response. The effects of lag time are most clearly seen during periods of high heat inflow, where the response is felt relatively quickly in the MID_UZ_THK simulation (i.e., warmer temperatures below the UPLAND area in Figure 4Q), whereas cooler temperatures persist for the same location in the HI_UZ_THK simulation (Figure 4R).

These findings have implications for watershed heat transport simulations in humid temperate climates. They are:

1. A potential effect of warming climate on groundwater temperatures in a watershed depends on the relative heat flux—the product of infiltration rate and associated temperature—that determines the amount of heat entering the subsurface. For example, if the temperature of infiltrating water increases during a warming climate, the net change in groundwater temperature may be lessened if drought conditions cause the rate of infiltration to be reduced;
2. The UZ acts as a low-pass filter. Both the magnitude and timing of water and heat pulses entering the subsurface and migrating downward to the water table, are attenuated by the UZ. Neglecting the UZ from a model simulation (as in the NO_UZ_THK

version of the synthetic model) effectively "short circuits" the dampening and lag time influences of the UZ;

3. The effect of a warming climate is buffered in a watershed by the total thickness of the groundwater system. A relatively thick groundwater system gives rise to mixed water temperatures at natural discharge points where shallow and deep flow lines converge. The convergence of flow lines dampens the heat signal carried by recharge that eventually discharges as baseflow to surface water;
4. The spatial extent of riparian zones plays an important role in determining the flashiness of a stream's response to heat forcing. That is, the riparian zone sheds (or shunts) precipitation to the surface water network, without the low-pass filtering of the UZ;
5. Additional vertical discretization to more accurately simulate the movement of wetting and heat fronts did not change simulation results. However, omission of the UZ and its effects on heat transport in a watershed-scale model produces erroneous results;
6. A sensitivity analysis of the flow and heat transport parameters showed an appreciable influence on simulated temperatures in both the saturated and unsaturated components of the subsurface, as well as on simulated stream temperatures (Supplementary Material Section S3).

5. Limitations of the Methodology

A discussion of the limitations and assumptions used in this work are provided in Supplementary Material Section S3, with a brief summary here.

- Root zone processes (i.e., evapotranspiration) are neglected; therefore, the infiltration rate is equated with the water that drains out of the root zone and enters the top of the UZ.
- As noted above, the UZF1 package in MODFLOW implements simplifying assumptions that neglect capillary forces. As a result, UZF1 simulates downward-only gravitational flow. This simplification is generally considered acceptable at a watershed scale [30].
- With the UZF1 package active in MODFLOW, one of three potential states is simulated for any active cell. They are either (1) unsaturated (i.e., partially-saturated over the entire thickness of the active grid cell), (2) a mix of unsaturated and saturated conditions (i.e., the water table is present within the cell), or (3) fully saturated. For water table cells, a single water content value is calculated that is equivalent to a volume average of both the unsaturated and saturated portions of the cell. Ambiguity arising from this mixed condition appears to have minimal effect on the heat flux solution, insofar as refined layer discretization hardly changes model results (Supplementary Material Section S3).
- Although conduction occurs through the matrix material of an aquifer and may transport heat more rapidly than in the fluid phase in low convection environments, MT3D-USGS simulates a single "bulk" diffusion term that approximates heat transport through both phases. In other words, the conductive propagation of heat through the solid and fluid phases is represented as a conjoined movement that is slower than thermal diffusion through a pure solid but faster than thermal diffusion through a pure fluid. In a predominantly horizontal flow-field, the upward or downward thermal diffusion is generally secondary, compared with the convection in humid temperate climates [25]. It is conceivable that the conductive flux through matrix material is dominant when the temperature gradient is unusually strong.
- The methods applied in this study were designed for temperate climate regions. It neglects processes such as mountain-front recharge in settings with deep water tables (>30 m), long flow paths (>2–3 km), and long UZ residence times, which are characteristic of arid and semi-arid regions.

- The effects of changes in viscosity owing to temperature changes are not considered in this study. However, variations in viscosity over the relatively small temperature changes simulated in the model are expected to be small.

A second group of limitations that are not related to the methodology chosen, but instead arise from the way the synthetic watershed model was constructed, include:

- Temporal smoothing of system dynamics via the use of a monthly climate forcing.
- Simplification of the thermal influence of storm runoff to streams was ignored/not simulated: that is, the restriction of the simulation to monthly average baseflow conditions.
- Inadequate representation of lake energy budget considerations as important for lake temperature. For example, neglecting the formation of ice during the winter months, energy changes related to evaporation, and lake thermal stratification.

Finally, it is important to note that the heat forcing function used to represent watershed warming in this study was designed to illustrate the components of watershed heat transport rather than represent an expected future condition. A companion paper, Feinstein et al. (2022) [25], incorporates a heat forcing function derived from predictions of climate trends.

6. Conclusions

This study developed a methodology for simulating watershed scale heat transport in a humid temperate climate. Beyond the use of the modified MT3D-USGS code described in Morway et al. (2022) [7], the applied methodology relies on two aspects of the model design:

- Whereas specification of infiltration is critical for representative groundwater flow models, specification of the heat forcing function, represented by the product of the infiltration rate added to the top of UZ multiplied by the infiltration temperature (the relative heat influx) is critical for developing a representative heat transport model.
- Heat transport in watershed models stand to benefit from a discretization scheme with at least one unsaturated layer. This approach enables the simulation to store, dampen, and/or lag the heat pulse before it is mixed with an underlying water table cell.

By the end of each simulation, the increase in the stream baseflow temperature (approximately 0.5 °C) is approximately half of the temperature increase at the water table (approximately 1.0 °C). Even with simulating monthly average conditions, the spatial extent of the riparian zone (water table < 1 m deep) plays an important role in determining the temperature 'flashiness' of the stream response to heat forcing. Thin UZs in riparian areas are more likely to generate rejected infiltration (runoff), which effectively short-circuits the dampening effects of a thicker UZ.

The methods applied in this study of a synthetic watershed highlight the importance of including the UZ in heat transport models. The UZ acts a low-pass filter that dampens the simulated effect of an infiltrating heat signal over time. That is, the thickness of the UZ can modify the amplitude, frequency, and phase change of the infiltrating heat signal as it migrates down to the water table. Moreover, because the thickness of the UZ varies across the active model domain, explicit representation of the UZ within a watershed model better captures the spatially-varying effect of the UZ on heat fluxes delivered to the water table. Equipped with a spatially and temporally refined recharging heat flux simulated by the model, the subsequent heat-buffering effects of the groundwater (saturated) system on a migrating heat signal are better accounted for. For example, as the shallow and deep flow paths converge near discharge points, the respective temperatures associated with each flow path mix. In this way, the cumulative and combined effects of the unsaturated and saturated zones on the temperature of the discharge to surface water features is more accurately simulated. Thus, heat transport models that consider the unsaturated and saturated zones are better equipped to evaluate the impacts of a changing climate on ecologically sensitive endpoints such as stream habitats.

Supplementary Materials: The following supporting information can be downloaded at https://www.mdpi.com/article/10.3390/w14233883/s1. Figure S1-1: Monthly and average yearly infiltration rates over warming period; Figure S1-2: Average yearly infiltration rate over warming period; Figure S1-3: Monthly temperature signal for spin-up; Figure S1-4: Forcing function component: Monthly temperature; Figure S1-5: Forcing function component: Relative monthly heat influx; Figure S1-6: Relative heat influx: 12-month averages during warming; Figure S1-7: Average relative heat influx by month for 30-year spin-up and 30-year warming; Figure S1-8: Synthetic model; Figure S1-9: Model layering and water-table elevation for UPLAND and VALLEY locations at end of spin-up period; Figure S1-10: Depth to water table at end of spin-up; Figure S2-1: Linear relations between thermal conductivity and volumetric moisture content; Figure S3-1: Head hydrographs for base model layers 1-8 at UPLAND and VALLEY locations; Figure S3-2: Temperature and water-table elevation in cross section through UPLAND location for three base models, layers 1-8, at selected times during warming period; Figure S3-3: Stream baseflow temperature for three base models in response to warming at selected gages; Figure S3-4: Cross sections showing layering for 8-layer base model version and 10-layer revised base model version of MID-UZ-THK; Figure S3-5: Cross sections showing layering for 8-layer base model version and 10-layer revised base model version of HI-UZ-THK; Figure S3-6: Temperature hydrographs comparing results for 8-layer and 10-layer model versions for the UPLAND location; Figure S3-7: Temperature hydrographs comparing results for 8-layer and 10-layer model versions for VALLEY location; Figure S3-9: Sensitivity results for MID_UZ_THK model as a percentage of model domain for August, by year, with water-table temperature at or above 9.5°C; Figure S3-10: Sensitivity results for MID_UZ_THK model for stream temperatures at the model outlet gage during warming period; Table S1-1: Infiltration rates over warming period in inches/year; Table S1-2: Vertical distribution of water-table elevations and corresponding temperatures at selected times during warming. Units for average temperature are degrees Celsius; Table S2-1: Spatially homogeneous transport and heat flux parameters; Table S3-1: Summary of changes to parameter values for sensitivity simulations using the MID_UZ_THK base model. References [31–34] at the end of the reference list are cited in the Supplementary Materials.

Author Contributions: E.D.M., D.T.F. and R.J.H. shared equally in the conceptualization of the material.; E.D.M. and D.T.F. split much of the writing, with help from R.J.H.; E.D.M. and D.T.F. collaborated on the figures; E.D.M. took the lead on the response to reviewers, with contributions from coauthors. All authors have read and agreed to the published version of the manuscript.

Funding: The authors would like to acknowledge support from the U.S. Geological Survey Land Change Science/Climate Research and Development for this work.

Data Availability Statement: The model executables, input and output files associated with the example models described in this manuscript are available in an online model archive located at: https://doi.org/10.5066/P99NUKIX (Morway et al., 2022b) [35].

Acknowledgments: Any use of trade, firm, or product names is for descriptive purposes only and does not imply endorsement by the U.S. Government. The authors wish to thank Dan Bright and Ramon Naranjo, both from the U.S. Geological Survey's Nevada Water Science Center, as well as anonymous reviewers selected by the journal, for their helpful reviews.

Conflicts of Interest: The authors declare no conflict of interest.

References

1. IPCC. *Climate Change 2022: Impacts, Adaptation, and Vulnerability, Technical Summary*; Cambridge University Press: Cambridge, UK, 2022.
2. Hunt, R.J.; Walker, J.F.; Selbig, W.R.; Westenbroek, S.M.; Regan, R.S. *Simulation of Climatechange Effects on Streamflow, Lake Water Budgets, and Stream Temperature Using GSFLOW and SNTEMP, Trout Lake Watershed, Wisconsin*; U.S. Geological Survey Scientific Investigations Report 2013-5159; U.S. Geological Survey Scientific: Reston, VA, USA, 2013; p. 118. Available online: http://pubs.usgs.gov/sir/2013/5159/ (accessed on 15 January 2022).
3. Hunt, R.J.; Westenbroek, S.M.; Walker, J.F.; Selbig, W.R.; Regan, R.S.; Leaf, A.T.; Saad, D.A. *Simulation of Climate Change Effects on Streamflow, Groundwater, and Stream Temperature Using GSFLOW and SNTEMP in the Black Earth Creek Watershed, Wisconsin*; 2328-0328; U.S. Geological Survey, Scientific Investigations Report 2016-5091; U.S. Geological Survey Scientific: Reston, VA, USA, 2016; p. 117. [CrossRef]

4. Niswonger, R.G.; Prudic, D.E.; Regan, R.S. *Documentation of the Unsaturated-Zone Flow (UZF1) Package for Modeling Unsaturated Flow between the Land Surface and the Water Table with MODFLOW-2005*; 2328-7055; U.S. Geological Survey Scientific: Reston, VA, USA, 2006; p. 62. [CrossRef]
5. Niswonger, R.G.; Panday, S.; Ibaraki, M. *MODFLOW-NWT, a Newton Formulation for MODFLOW-2005*; U.S. Geological Survey Scientific: Reston, VA, USA, 2011; p. 44. [CrossRef]
6. Hunt, R.J.; Prudic, D.E.; Walker, J.F.; Anderson, M.P. Importance of unsaturated zone flow for simulating recharge in a humid climate. *Groundwater* **2008**, *46*, 551–560. [CrossRef]
7. Morway, E.D.; Feinstein, D.T.; Hunt, R.J.; Healy, R.W. New capabilities in MT3D-USGS for Simulating Unsaturated-Zone Heat Transport. *Groundwater* **2022**. [CrossRef] [PubMed]
8. Anderson, M.P.; Bowser, C. The role of groundwater in delaying lake acidification. *Water Resour. Res.* **1986**, *22*, 1101–1108. [CrossRef]
9. USDA-NRCS. HUC10: USGS Watershed Boundary Dataset of Watersheds. Available online: https://datagateway.nrcs.usda.gov (accessed on 25 May 2022).
10. Harbaugh, A.W. *MODFLOW-2005, the U.S. Geological Survey Modular Ground-Water Model: The Ground-Water Flow Process*; U.S. Geological Survey Scientific: Reston, VA, USA, 2005. [CrossRef]
11. Bedekar, V.; Morway, E.D.; Langevin, C.D.; Tonkin, M.J. *MT3D-USGS Version 1: A U.S. Geological Survey Release of MT3DMS Updated with New and Expanded Transport Capabilities for Use with MODFLOW*; 2328-7055; U.S. Geological Survey Scientific: Reston, VA, USA, 2016. [CrossRef]
12. USGS. National Climate Change Viewer. Available online: https://www2.usgs.gov/landresources/lcs/nccv/maca2/maca2_watersheds.html (accessed on 31 January 2022).
13. Niswonger, R.G.; Prudic, D.E. *Documentation of the Streamflow-Routing (SFR2) Package to Include Unsaturated Flow Beneath Streams—A Modification to SFR1*; 2328-7055; U.S. Geological Survey, Techniques and Methods 6-A13; U.S. Geological Survey Scientific: Reston, VA, USA, 2005. [CrossRef]
14. Merritt, M.L.; Konikow, L.F. *Documentation of a Computer Program to Simulate Lake-Aquifer Interaction Using the MODFLOW Ground-Water Flow Model and the MOC3D Solute-Transport Model*; U.S. Geological Survey Water-Resources Investigations Report 00–4167; U.S. Geological Survey Scientific: Reston, VA, USA, 2000; p. 146. Available online: https://pubs.er.usgs.gov/publication/wri004167 (accessed on 15 May 2020).
15. Niswonger, R.; Prudic, D.E. Comment on "evaluating interactions between groundwater and vadose zone using the HYDRUS-based flow package for MODFLOW" by Navin Kumar, C. Twarakavi, Jirka Simunek, and Sophia Seo. *Vadose Zone J.* **2009**, *8*, 818–819. [CrossRef]
16. Markstrom, S.L.; Niswonger, R.G.; Regan, R.S.; Prudic, D.E.; Barlow, P.M. *GSFLOW-Coupled Ground-Water and Surface-Water FLOW Model Based on the Integration of the Precipitation-Runoff Modeling System (PRMS) and the Modular Ground-Water Flow Model (MODFLOW-2005)*; U.S. Geological Survey Techniques and Methods 6-D1; U.S. Geological Survey Scientific: Reston, VA, USA, 2008; p. 240. Available online: https://pubs.usgs.gov/tm/tm6d1/pdf/tm6d1.pdf (accessed on 15 May 2020).
17. Morway, E.D.; Gates, T.K.; Niswonger, R.G. Appraising options to reduce shallow groundwater tables and enhance flow conditions over regional scales in an irrigated alluvial aquifer system. *J. Hydrol.* **2013**, *495*, 216–237. [CrossRef]
18. Woolfenden, L.R.; Nishikawa, T. *Simulation of Groundwater and Surface-Water Resources of the Santa Rosa Plain Watershed, Sonoma County, California*; 2328-0328; U.S. Geological Survey, Scientific Investigation Report 2014-5052; U.S. Geological Survey Scientific: Reston, VA, USA, 2014. [CrossRef]
19. Haserodt, M.J.; Hunt, R.J.; Fienen, M.N.; Feinstein, D.T. *Groundwater/Surface-Water Interactions in the Partridge River Basin and Evaluation of Hypothetical Future Mine Pits, Minnesota*; 2328-0328; U.S. Geological Survey, Scientific Investigations Report 2021-5038; U.S. Geological Survey Scientific: Reston, VA, USA, 2021. [CrossRef]
20. Feinstein, D.T.; Hart, D.J.; Gatzke, S.; Hunt, R.J.; Niswonger, R.G.; Fienen, M.N.J.G. A simple method for simulating groundwater interactions with fens to forecast development effects. *Groundwater* **2020**, *58*, 524–534. [CrossRef] [PubMed]
21. Khadim, F.K.; Dokou, Z.; Bagtzoglou, A.C.; Yang, M.; Lijalem, G.A.; Anagnostou, E. A numerical framework to advance agricultural water management under hydrological stress conditions in a data scarce environment. *Agric. Water Manag.* **2021**, *254*, 106947. [CrossRef]
22. Kitlasten, W.; Morway, E.D.; Niswonger, R.G.; Gardner, M.; White, J.T.; Triana, E.; Selkowitz, D. Integrated Hydrology and Operations Modeling to Evaluate Climate Change Impacts in an Agricultural Valley Irrigated with Snowmelt Runoff. *Water Resour. Res.* **2021**, *57*, e2020WR027924. [CrossRef]
23. Dunne, T.; Black, R.D. An experimental investigation of runoff production in permeable soils. *Water Resour. Res.* **1970**, *6*, 478–490. [CrossRef]
24. Muldoon, M.A.; Madison, F.M.; Lowery, B. Variability of Nitrate Loading and Determination of Monitoring Frequency for a Shallow Sandy Aquifer, Arena, Wisconsin. 1998. Available online: https://wgnhs.wisc.edu/catalog/publication/000810/resource/wofr199809 (accessed on 25 January 2022).
25. Feinstein, D.T.; Hunt, R.J.; Morway, E.D. Simulation of heat flow in a synthetic watershed: Lags and damping across multiple pathways under a climate-forcing scenario. *Water* **2022**, *14*, 2810. [CrossRef]

26. Zheng, C.; Wang, P.P. *MT3DMS: A Modular Three-Dimensional Multispecies Transport Model for Simulation of Advection, Dispersion, and Chemical Reactions of Contaminants in Groundwater Systems*; Documentation and User's Guide; Contract Report SERDP-99-1; U.S. Army Engineer Research and Development Center: Vicksburg, MS, USA, 1999; Available online: http://hydro.geo.ua.edu/mt3d (accessed on 22 June 2022).
27. Morway, E.D.; Niswonger, R.G.; Langevin, C.D.; Bailey, R.T.; Healy, R.W. Modeling variably saturated subsurface solute transport with MODFLOW-UZF and MT3DMS. *Groundwater* **2013**, *51*, 237–251. [CrossRef] [PubMed]
28. Zheng, C.; Hill, M.C.; Hsieh, P.A. *MODFLOW-2000, the U.S. Geological Survey Modular Ground-Water Model: User Guide to the LMT6 Package, the Linkage with MT3DMS for Multi-Species Mass Transport Modeling*; 2331-1258; U.S. Geological Survey Scientific: Reston, VA, USA, 2001. [CrossRef]
29. Zheng, C.; Bennett, G.D. *Applied Contaminant Transport Modeling*, 2nd ed.; John Wiley & Sons, Inc.: New York, NY, USA, 2002.
30. Harter, T.; Hopmans, J.W. (Eds.) *Role of Vadose-Zone Flow Processes in Regional-Scale Hydrology: Review, Opportunities and Challenges*; Kluwer Academic Publishers: Wageningen, The Netherlands, 2004; pp. 179–208.
31. Healy, R.W.; Ronan, A.D. *Documentation of Computer Program VS2DH for Simulation of Energy Transport in Variably Saturated Porous Media—Modification of the U.S. Geological Survey's Computer Program VS2DT*; U.S. Geological Survey Water-Resources Investigations Report 96-4230 36; United States Environmental Protection Agency: Washington, DC, USA, 1996. [CrossRef]
32. Van Duin, R.H.A. The Influence of Soil Management on the Temperature Wave Near the Soil Surface. Wageningen University & Research, Technical Bulletin no. 29, Institute for Land and Water Management Research. 1963. Available online: https://edepot.wur.nl/417072 (accessed on 22 June 2022).
33. Westenbroek, S.M.; Engott, J.A.; Kelson, V.A.; Hunt, R.J. *SWB Version 2.0—A Soil-Water-Balance Code for Estimating Net Infiltration and Other Water-Budget Components*; Book 6; U.S. Geological Survey Techniques and Methods: Lakewood, CO, USA, 2018; Chapter A59; p. 118. [CrossRef]
34. Zheng, C. MT3DMS V5.3: A Modular Three-Dimensional Multispecies Transport Model For Simulation Of Advection, Dispersion, And Chemical Reactions Of Contaminants In Groundwater Systems: Supplemental User's Guide. 2010. Available online: https://hydro.geo.ua.edu/mt3d/index.htm (accessed on 25 August 2021).
35. Morway, E.D.; Feinstein, D.T.; Hunt, R.J. MODFLOW-NWT and MT3D-USGS Models for Appraising Parameter Sensitivity and Other Controlling Factors in a Synthetic Watershed Accounting for Variably-Saturated Flow Processes. Available online: https://doi.org/10.5066/P99NUKIX (accessed on 25 January 2022).

Article

Evaluation of Fresh Groundwater Lens Volume and Its Possible Use in Nauru Island

Luca Alberti , Matteo Antelmi *, Gabriele Oberto, Ivana La Licata and Pietro Mazzon

Civil and Environmental Engineering Department, Politecnico di Milano, Piazza L. da Vinci 32, Milano 20133, Italy
* Correspondence: matteo.antelmi@polimi.it; Tel.: +39-02-2399-6668

Abstract: A proper management of fresh groundwater lenses in small islands is required in order to avoid or at least limit uncontrolled saltwater intrusion and guarantee the availability of the resource even during drought occurrences. An accurate estimation of the freshwater volume stored in the subsoil is a key step in the water management decision process. This study focused on understanding the hydrogeological system behaviour and on assessing the sustainable use of the groundwater resource in Nauru Atoll Island (Pacific Ocean). A first phase, concerning the hydrogeological characterization of the island, highlighted the occurrence of few drought-resilient freshwater lenses along the seashore. The second part of the study focused on the characterization of a freshwater lens found in the northern coastal area and identified such area as the most suitable for the development of groundwater infrastructures for water withdrawal. The characterization activities allowed quantifying the freshwater lens thickness and volume in order to assess the capability to satisfy the population water demand. A geo-electrical tomography survey was carried out, and a 3D density-dependent numerical model was implemented in SEAWAT. The model results demonstrated that in small islands freshwater can unexpectedly accumulate underground right along the seashore and not in the centre of the island as is commonly believed. Furthermore, the model can constitute a useful tool to manage the groundwater resources and would allow the design of sustainable groundwater exploitation systems, avoiding saltwater intrusion worsening.

Keywords: small island; groundwater storage; groundwater management; geo-electrical survey; density-dependent model; SEAWAT; water security

Citation: Alberti, L.; Antelmi, M.; Oberto, G.; La Licata, I.; Mazzon, P. Evaluation of Fresh Groundwater Lens Volume and Its Possible Use in Nauru Island. *Water* **2022**, *14*, 3201. https://doi.org/10.3390/w14203201

Academic Editors: Cristina Di Salvo and Thomas M. Missimer

Received: 6 September 2022
Accepted: 6 October 2022
Published: 11 October 2022

1. Introduction

The management of freshwater reserves is nowadays becoming increasingly important. Freshwater stored in coastal aquifers is particularly vulnerable to degradation because of its proximity to seawater. Coastal aquifers often must face environmental problems related to seawater intrusion as the result of indiscriminate and unplanned groundwater exploitation for fulfilling the freshwater need of the growing global population [1–5].

Small islands, consisting of extremely small surface areas [6] and maximum elevations approaching only a few metres [7], have a particular physical structures and unique hydrological systems. This category includes the small coral islands of the Caribbean Sea and the coral atolls of the Pacific and Indian Oceans, where the surface water does not exist in an exploitable form, and fresh groundwater resources are limited. On these islands, conventional options for freshwater supplies are limited to groundwater development and rainwater harvesting [8]. The groundwater in small islands is located into the subsoil in the form of lenses, relatively thin layers of freshwater floating above the denser seawater. The mixing zone separating freshwater and seawater influences the depth at which freshwater is available [9]. Usually, in small islands' subsoil, the mixing zone contains a gradual transition in salinity from freshwater to seawater as well [6,10,11], and its thickness is related to the tidal fluctuations [12,13]. Buddemeier and Oberdorfer [14] suggested that the mixing zone, defined by salinities between 2.5% and 95% of seawater concentration, can occupy a significant volume of atoll island aquifers compared to the freshwater component. Water infiltrates into the ground and becomes recharged to groundwater that moves from the inner parts of the island towards the coast and discharges along the coastline, where the outflow is usually not regular but varies according to changes in permeability and aquifer thickness [8,15,16]. Fresh groundwater lenses' salinization due to seawater intrusion is a main issue and has a great impact on groundwater quality and can even prevent the utilization of the underground resource [6–8,17]. Werner et al. [18] found that the combined effects of complex geology, tides, episodic ocean events, strong climatic variability, and human pressures strongly influence the extension of fresh groundwater lenses. Thus, these underground water resources are particularly vulnerable to saline contamination but, on the other hand, are the unique water supply sources for many island populations. Furthermore, the freshwater supply security is threatened by the sea level rise due to the effects of climate change [19,20]. A correct knowledge of the hydrogeological system and the fresh groundwater lens availability is fundamental for the water resources management of many small islands [2,21,22]. Numerical modelling is a very useful tool and able to solve many hydrogeological issues: from the interpretation of the contamination sources to their path in the aquifer [23] to the study of the best remediations strategies [24,25]. In this case, numerical modelling is a tool used to quantify freshwater availability range, starting from simple empirical relationships to three-dimensional density-dependent models [18,26]. Using a variable-density numerical model, Underwood et al. [27] discussed the effect of various hydrogeological parameters, already identified by Falkland [8], on the size of the fresh groundwater lenses on small islands. Bailey et al. [28] also started from Falkland's study [8] and carried out a sensitivity analysis of atoll fresh groundwater lens thickness, providing an insight into some of the fresh groundwater lens-controlling factors. Alberti et al. [15] implemented a 2D numerical density-dependent model to understand the phenomenon of freshwater accumulation in the sandy part of Nauru coastal belt. Similarly, Babu et al. [29] presented a quasi-3D sharp interface finite element model to investigate the saltwater intrusion dynamics in Tongatapu Island, Kingdom of Tonga, Polynesia. As an example of groundwater resource management in island aquifers using numerical models, Coulon et al. presented first a parameter estimation framework [30] and then a numerical model to optimize the pumping rates while avoiding saltwater intrusion in the Magdalen Islands, Quebec, Canada [31].

Vulnerability studies focused on many small islands in the Pacific: Houghton et al. [32] showed that the costs of general infrastructure for water supply (e.g., reverse osmosis (RO) desalination plant) and protection are often well beyond the financial possibilities of most small island states. The need to implement measures for reducing vulnerability and increasing resilience of systems to climate variability in small islands was highlighted by Nurse and Moore [33]. The objective for small islands is then to look at alternative sources of water in the long run [34–36].

Nauru is an isolated raised coral-limestone island located 41 km south of the Equator in the central Pacific Ocean, standing 4300 m above the ocean floor. The National Sustainable Development Strategy (NSDS) for Nauru was prepared in November 2005. In the document the government identified sustainable supply of power and water as fundamental for supporting the island economic growth. In the past, Nauru imported water from the neighbouring islands (Marshall Islands and Solomon Islands), but this is very expensive and is no longer an option. At present, Nauru's main sources of water are rainfall and desalination (RO). These water supplies are enough in normal rainfall year but become insufficient in case of prolonged drought or when the desalination plants are not in operation. On the other hand, the RO is expensive in terms of energy consumption and needs expertise for its maintenance. Thus, the best option is to look for alternative water resources such as groundwater, which could turn out to be a potential source of water for Nauru and that has not yet been exploited accordingly to its possibilities. Several studies were focused on the groundwater potential for Nauru, such as Falkland's study [37] that reports the activities needed to inspect this water resource in the island and make it available for the people.

In this context, Politecnico di Milano was in charge of a project concerning the assessment of the practicability of infrastructural actions for sustainable usage of groundwater resources in the island [38]. The project consisted of three steps: (1) development of the conceptual site model (CSM) through the geological and hydrogeological characterization of the island; (2) implementation of 2D and 3D density-dependent flow and transport groundwater numerical models focused on the island's part more suitable for the groundwater development; and (3) study and preliminary design of infrastructural actions for groundwater sustainable exploitation. Alberti et al. [15] presented the first step of the project, describing both the island geological structure, its climate, the activities carried out to identify the hydrogeological setting, the seawater intrusion occurrence in Nauru aquifer, and, at last, implemented the 2D numerical model. The present paper mainly focused on the second project step concerning the 3D modelling and the survey carried out. The main challenges concerned: (a) better understanding the behaviour of the hydrogeological system and why groundwater accumulates along the coast rather than in the centre of the island and (b) quantifying the thickness and volume of the freshwater lens in the northern sector of the island to evaluate the possibility, aiming to satisfy, at least in part, the water demand of the population. The study shows, as a proper combination of hydrogeological investigations and density-dependent groundwater modelling, all the necessary steps to correctly manage groundwater resources on small islands prone to saltwater intrusion and long drought periods.

2. Materials and Methods

The Republic of Nauru (Figure 1) is an isolated, uplifted limestone island located 41 km south of the equator (0°32′ S, 166°56′ E). This island represents a raised atoll standing 4300 m above the ocean floor, with a maximum land-surface altitude of 70 m above the sea level (cross section in Figure S1 in Supplementary Materials). The total land area of Nauru is only 22 km^2, surrounded by a fringing coral reef between 150 and 250 m wide. The land area consists of a narrow coastal plain (Bottomside) 100 to 300 m wide and with an elevation ranging from 0 to 10 m a.s.l., which encircles a limestone escarpment rising some 30 m to a central plateau (named Topside). The Bottomside consists of a sandy or rocky beach on the seaward edge and a beach ridge or foredune, behind which are either relatively flat ground or, in some places, low-lying small lagoons filled by brackish water. The high plateau consists of a matrix of coral-limestone pinnacles and limestone outcrops, between which lie extensive deposits of soil and high-grade tricalcic phosphate rock, which was extensively mined in the past century. On the southwest-central part of the island, there is a wide and fertile depression (about 120,000 m^2), where a brackish water lake, known as Buada Lagoon, is located (Figure 1). This is the only surface water body (38,000 m^2) existing on the entire island. Jacobson and Hill [39], in 1987, led the first hydrogeological survey in Nauru, basing all their measurements on a reference point called the reduced level (RL). Groundwater level measurements of all the following hydrogeological studies on the island [6,15] were referred to as RL [40].

Most of the wells and monitoring wells in Nauru are located along the coastline [15], where most people live, and the groundwater is mainly exploited. Methods used for water extraction are pumping or bailing. In detail, along the coastline, there are about 350 domestic wells, 126 of which are used as monitoring wells [15,41]; referring to the measurements carried out by Nauru government, most of these wells present electrical conductivity (EC) values definitely higher than 2200 μS/cm, which is the drinkable limit (1.5 g/L of TDS) defined by Nauru government [42]. Only in the northern part of the island, in Anetan and Ewa districts, the EC values in multipipe monitoring well S1 (5 pipes) and S18 (4 pipes) are lower than the drinkable limit in the shallow part (Figure 2a,b). The surveys show that in this zone, freshwater depth ranges between 6 m (June 2008) and 8 m

(April 2010) below the ground level. In other areas, the common EC values are always higher, as shown in Figure 2c,d for monitoring wells S21 and S23.

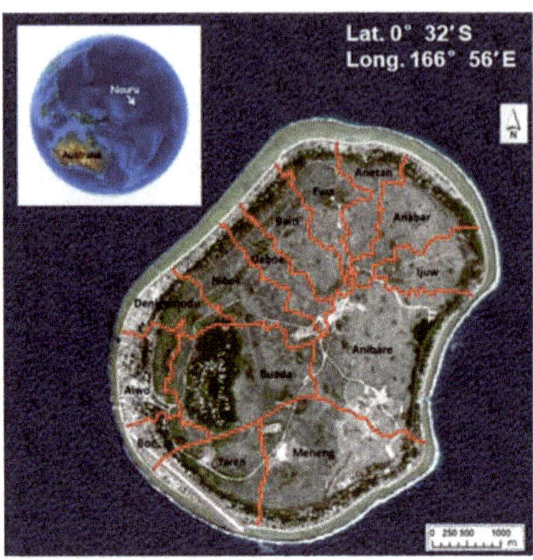

Figure 1. Nauru position at 0°32′ S and 166°56′ E and aerial photo of the island in 2009 with its districts.

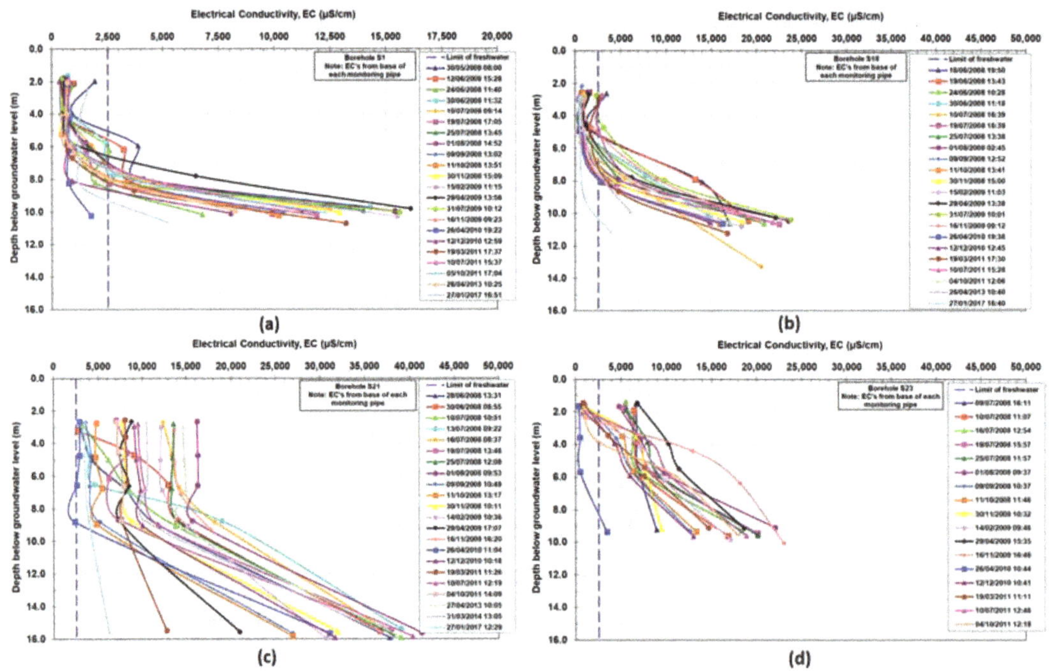

Figure 2. EC values over the depth of the following monitoring wells: (**a**) well S1, (**b**) well S18, (**c**) well S21, and (**d**) well S23.

For a better understanding of the geological and hydrogeological setting of the island, Alberti et al. [15] discussed in detail its hydrogeological structure and the activities carried out to identify the related properties and the seawater intrusion occurrence in Nauru aquifer.

2.1. Hydraulic Tests

During a first characterization phase, Politecnico's researchers tried to perform some pumping tests, but the small size of the pumps available in Nauru did not allow the adequate stressing of the aquifer. Consequently, none of these provided satisfactory results, and during the last visit to Nauru, 12 pneumatic slug tests were performed focusing on the northern sector of the island. The tests were carried out in situ, pushing the Geoprobe systems ®rods in the aquifer to the desired depth and pumping air into the rods to lower the water level. Once reached, the system equilibrium air was released, and the sensor registered the groundwater recovery inside the rods. The duration of each test was about 20 min. The slug tests allowed obtaining hydraulic conductivity values in the proximity of the piezometers S1 and S18. For the slug tests, the method used for the interpretation was derived from Bouwer and Rice [43], whereas for the pumping test, the interpretative approach was derived from the adaptation for unconfined aquifer of Neumann [44]. In detail, the test carried out 10 m north of the piezometer S1 provided a hydraulic conductivity value of 11.98 m/d at 5 m depth, while the slug tests carried out next to the piezometers S1 and S18 showed the common feature of having a greater permeability with increasing depth. Close to S1, the hydraulic conductivity increases from 3.46 m/d to a depth of 7 m from the ground level and up to 29 m/d at a depth of 15 m from the ground level; next to S18 instead, the measured hydraulic conductivity is 0.38 m/d at 14 m depth and increases up to 7.08 m/d at a depth of 17 m. The tests allowed to achieve the hydraulic conductivity values for the Bottomside for the Anetan district. All the tests performed are summarized in Table 1.

Table 1. Hydraulic conductivity value measured through slug tests.

Slug Test On	Depth (m)	Hydraulic Conductivity (m/d)
S18	10.2–13	0.4
S18	13.1–15	7.1
Near to S18	4.8–5.1	12.0
S1	6.7–7.7	3.5
S1	7.9–9.7	32.0
S1	9.8–11.2	12.6
S1	11.3–13.6	12.3
S1	13.8–15	30.0
S21	33.08–35.11	26.0
S21	35.11–36.98	42.5
S21	36.98–39.17	37.7
S21	39.17–46.08	14.0
Average value		19.2
Median value		13.3
Std. deviation		13.8

2.2. Geo-Electrical Investigation

For the Nauru project [38], the researchers of Politecnico di Milano, in partnership with the National Research Council (Consiglio Nazionale delle Ricerche—CNR), carried out a geoelectrical survey in order to define in detail the thickness and quantity of the freshwater lens present in the northern area of Nauru, in the Capelle zone belonging to Anetan and Ewa districts (Figure 2). The results of the Nauru project's first phase [15] show that this area presents a freshwater lens that maintains significant thickness over the years and has been resilient even in drought periods. The 2D numerical model [15] allowed to interpret the cause of the freshwater presence in the proximity of the seashore.

The geoelectrical prospections are part of the geophysical survey methods and, through an indirect way, allow detecting and characterizing the shape, dimensions, and physical properties of the underground structures. The electrical resistivity tomography (ERT), through vertical electrical sounding (VES), is the most used technique. A georesistivimeter "PASI E3 Digit" (produced by Pasi s.r.l.) was used for the geo-electrical investigation, while the software "Earthimager 2D" (Advanced Geosciensces Inc.) was used for the tomographic inversion. It is based on the observation of the potential difference created by the introduction of electrical current into the ground and on the evaluation of the electrical resistivity. As this parameter is very sensitive to the presence of water in pores and to salt concentration in water, its spatial variations give a hint of the freshwater thickness in the subsoil. However, the data interpretation is difficult when the study is carried out in thin aquifers affected by saltwater intrusion. Indeed, assuming these conditions, a sharp surface separating the freshwater unit from the salty one does not exist, and a transition zone is present, characterized by variable thickness and salt concentration (related also to tidal fluctuations [12]). In the Nauru case, a clear evaluation of the freshwater thickness of the transition zone and of the saltwater occurrence was carried out by means of the integration of geoelectrical data with EC (electrical conductivity) and head measurements in wells close to geo-electrical investigations. Tidal fluctuations turned out to affect the tomographic data because their acquisition occurred in a time frame comparable with the one of sea level variations. Thus, the collected hydrogeological data were used to link, at known depth, the EC values to the geophysical resistivity data.

On April 2013, during the geophysical survey, 20 VES and 8 ERT were carried out in the northern part of the island (Figure 3). The survey was mainly focused on the Bottomside, where 16 VES were carried out at 8–10 m depth from the ground level, and 4 on the Topside at a depth ranging from 17 to 27 m from the ground level. All the VES were performed by means of the Schlumberger quadrupole using an AB/2 spacing of about 50 m. Differently, all the ERT were carried out on the Bottomside along 4 cross-sections having a length ranging from 100 to 150 m.

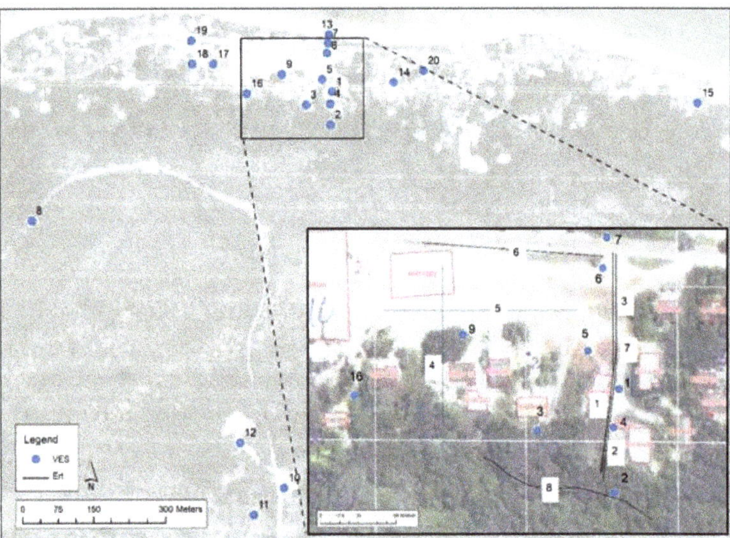

Figure 3. VES (blue points in map) and ERT (black lines) carried out in the Capelle area located in the northern part of Nauru island at the border between Anetan and Ewa districts.

2.3. 3D Numerical Model Implementation

The evaluation of the available freshwater volume in the Nauru aquifer is the starting point for the sustainable use of the groundwater resource. The next step of the study is the implementation of the 3D numerical model. This is the tool to evaluate the effect on the characterized fresh groundwater lens related to the infrastructural actions for groundwater sustainable exploitation that is planned at point 3 of the Nauru project. Infrastructures for groundwater supply indeed have to be designed with great consideration of the system's impact on the natural groundwater resource. Pumping water infrastructure often involves large-scale ground water withdrawals, which is not feasible in small islands such as Nauru, where the saltwater intrusion phenomenon could be worsened by groundwater extraction for the well-known up-coning effect. Infiltration galleries, horizontal wells, or skimming wells could be appropriate methods of groundwater abstraction from small coral islands aquifers: they avoid the problem of saline intrusion by spreading the impact of pumping (head loss) over a larger area of the freshwater lens than wells normally do [45,46]. Nauru, as many small islands, is vulnerable to water shortage, and the use of horizontal drains is thought to be a suitable tool for a sustainable groundwater management.

The 2D model discussed in detail by Alberti et al. [15] was used as the basis for the 3D numerical model. A great number of layers was needed to simulate the saltwater intrusion phenomenon. Since the freshwater occurs mostly in the northern zone, only the northern half of the island was modelled (Figure 4). The model was created using the MODFLOW-2000 and SEAWAT-2000 finite difference codes [47,48], by means of the graphical user interface Groundwater Vistas. SEAWAT-2000 is a previous release of the SEAWAT computer program for simulation of three-dimensional, variable-density, transient ground-water flow in porous media. SEAWAT-2000 was designed by combining a modified version of MODFLOW-2000 and MT3DMS into a single computer program. SEAWAT-2000 contains all the processes distributed with MODFLOW-2000 and also includes the variable-density flow process and the integrated MT3DMS transport process.

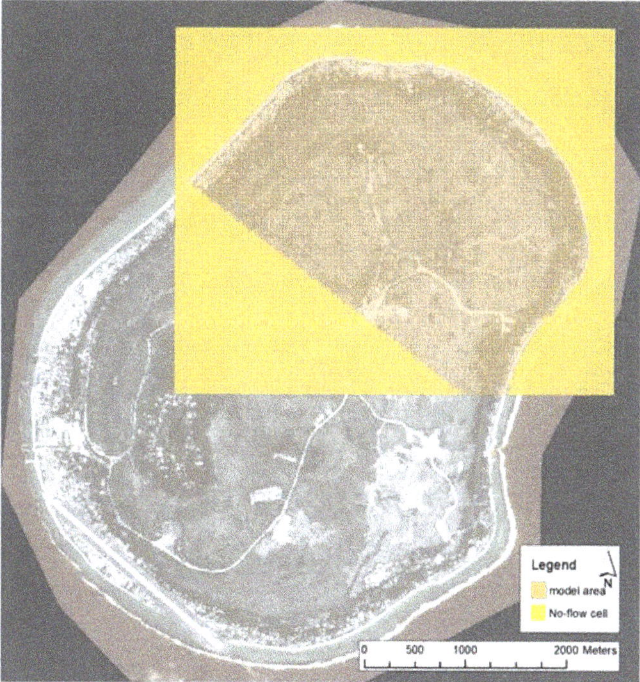

Figure 4. Modelling domain of the density-dependent 3D numerical model.

The model consisting of 117 rows and 144 columns has a uniform spacing of cells (30×30 m) that cover a model domain of about 15 km^2 (Figure 4). No-flow boundary conditions were applied to the cells outside the area of interest (yellow cells in Figure 4), reducing to 9 km^2 the model domain covered by active cells. Over the vertical direction, the domain was subdivided by 22 layers (north-south section in Figure 5a), for which thickness increases from the surface to the bottom, ranging from 0.5 m to 5 m. The choice to reproduce the shallow layers through a strong refinement is justified since the concentration data in monitoring wells showed greater changes in the upper part of the aquifer [15]. The bottom of the model is set at the −70 m depth related to the reference level (RL) in order to have a stable configuration of the saltwater wedge.

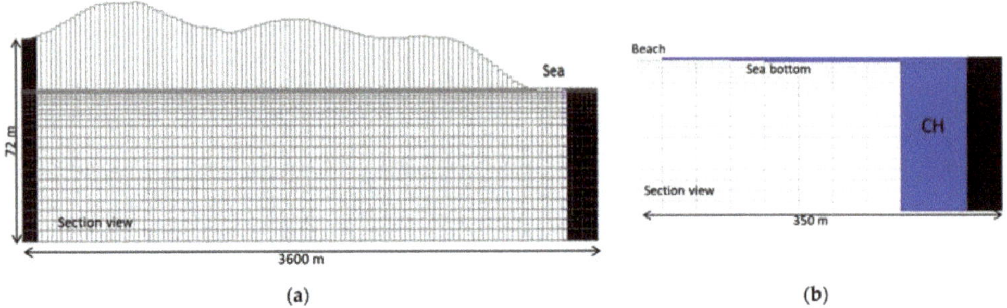

Figure 5. (**a**) Vertical discretization of the model domain (black cells are no-flow boundary conditions) and (**b**) zoom on the Bottomside area, where blue cells are Dirichlet boundary conditions having the same head and concentration values; the first seven thin layers were used to represent the sea bottom deepening and the shallow lagoon.

Similar to the 2D numerical model [15], constant head and concentration (Dirichlet) boundary conditions representing the sea were assigned in the cells surrounding the island considering the sea bottom slope from the coastal line to the reef. In order to reproduce the sea bottom slope, the first 7 layers were used assigning them the Dirichlet condition at an increasing depth (Figure 5b). Then, at the border of the model domain from the first layer to the deepest one, the same boundary conditions was assigned using the following values (Figure S2 in Supplementary Materials): a constant head of 1.59 m above RL (corresponding to average sea level [4]) and a concentration of 35.7 kg/m^3 of total dissolved solids (TDS) [49].

The hydrogeological and physical parameters values (i.e., hydraulic conductivity, porosity, dispersivity, etc.) initially implemented in the 3D numerical model were the same values implemented in the 2D numerical model [15] and reported here (Table 2).

Table 2. Hydrogeological parameters initially assigned to the 2D models [15].

Hydrogeological Parameter	Value	
Hydraulic conductivity (m/d)—horizontal and vertical	800 for limestone (zone 1) 40 for sand (zone 1)	80 for limestone (zone 1) 4 for sand (zone 1)
Porosity	0.3	
Specific storage (1/m)	0.0003	
Specific yield	0.3	
Longitudinal dispersivity (m)	50 for limestone 2 for sand	
Transverse dispersivity (m)	5 for limestone 0.2 for sand	
Vertical dispersivity (m)	0.2 for limestone 0.008 for sand	
Recharge (mm/y)	540	
Molecular diffusion (m^2/d)	8.64×10^{-6}	
Sea water TDS concentration	35.7 kg/m^3	

Before starting the calibration process, some of these values were changed considering the hydrogeological investigations result and the specific hydrological data for each year considered in the 3D simulation.

Concerning the hydraulic conductivity, the hydrogeological characterization and the 2D numerical model [15] highlighted (in many areas of the Bottomside) the presence of sand sediments covering the karstified limestone for a depth ranging from 1 to 15 m below the ground level depending on the direction of the oceanic currents. These sediments were probably originated because of the erosion operated by the atmospheric agents and the waves during the last phase of the island formation, when the sea level decreased, and the cliff was created. Based on 67 available stratigraphic logs, the subsoil setting was implemented into the numerical model through two different zones of hydraulic conductivity and dispersivity. To the first zone, representing the sandy sediment of the Bottomside (layers 1–12), were assigned low values of dispersivity and hydraulic conductivity (Figure S2) as in the 2D simulation [15] but decreasing in the latter (from 40 to 20 m/d) based on hydraulic tests average results (Table 1); to the second zone, representing the Topside and the deep limestone of the Bottomside, higher values of the two parameters (800 m/d and 50 m) were assigned, maintaining the same values used in the 2D simulation.

The parameters calibration of the density-dependent flow model was carried out through the simulation of 3 different transient conditions (Figure 6).

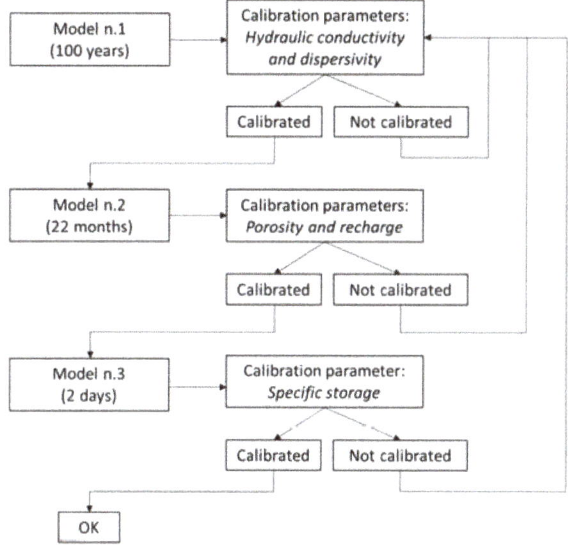

Figure 6. Diagram scheme of the numerical modelling calibration phases.

Clearly, for each simulation, the recharge value was updated considering the specific period of simulation. The calibration process followed a trial-and-error approach using 29 concentration and heads data collected at different depths in 12 monitoring wells shown in Figure 7. Among these monitoring wells, 5 are multipipe (S1, S2, S18, S21, and S23) and allow to detect the concentration at different aquifer depths.

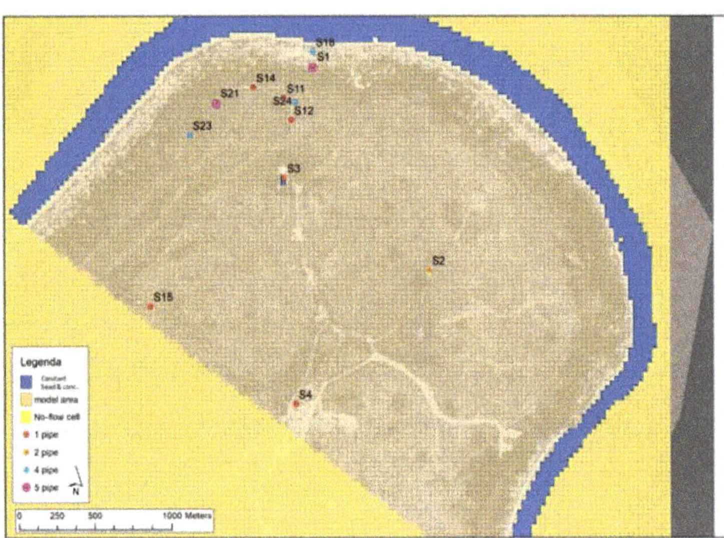

Figure 7. Available target points location for the model calibration.

The first phase (model n.1) provided a calibration of the hydraulic conductivity and dispersivity parameters for a quasi-steady state simulation: a long transient simulation (100 years) with the aim to recreate the heads and initial total dissolved solids (TDS) concentration distribution observed in the model area during the November 2009 survey (provided by CNR).

For model n.1, a recharge value slightly different from Ghassemi et al. [49] was used: 590 mm/year instead of 540 mm/year was calculated for the Topside (Thornthwaite method) considering the total precipitation (2000 mm) and temperatures measured by Nauru Government from December 2008 to November 2009. In the coastal zone, where the large part of the population lives, the value was increased by 11% (reaching 657 mm/year). This because Nauru's villages are not equipped with sewer systems, and the habitants dispose the used freshwater into sinkholes, which directly leak in groundwater. This additional term of recharge is constituted both by the desalted water and the rainwater stored in the harvesting systems the houses are usually equipped with. The study carried out by Bouchet and Sinclair [41] highlighted that the population with access to a pumping well used averagely about 114 l/d/pc of water (Table 3). However, in Anetan and Ewa districts (the ones included in the model domain), only the 36% of the population has the access to a well. For the remaining habitants, an average water consumption of about 88 l/d/pc was estimated, supplied by domestic rainwater harvesting and the desalting plant.

Table 3. Water usage estimates 2010 for people in Nauru.

Parameter	Minimum (l/d/pc)	Maximum (l/d/pc)	Average (l/d/pc)
Groundwater	68	121	94
Drinkable water (desalination or rainwater)	20	20	20
Total water needs	88	141	114

Part of this water, disposed in leaking sinkholes, represents an additional recharge for the coastal aquifer system. From the data reported by SOPAC [41], 20 l/d/pc of water was considered for food purposes. Consequently, the remaining and 68 l/d/pc was evaluated to potentially infiltrate in the subsoil, respectively, for people with access to the well and without access. The additional recharge rate (91 m^3/d) was assessed considering the number of inhabitants (i.e., 1180) of Ewa and Anetan and the percentage of people with/without well access. With the urbanised area in the model domain comprising about 300,000 m^2, a maximum additional recharge of about 111 mm/year was assessed. During the calibration process, this value was considered as the upper limit because few houses discharge waste waters directly into the sea, and the rainwater collected through the harvesting systems is partially subtracted to the natural meteoric recharge. Finally, the calibrated value was 67 mm/year and was applied to the entire urbanized part of the Bottomside.

Once the numerical model n.1 satisfactorily reproduced the measured hydraulic heads and concentration distributions, a 22-month unsteady-state simulation (model n.2) was run to reproduce the TDS concentration variation and compare results with the TDS distribution measured through 6 characterization surveys carried out from December 2009 to September 2011 (Figure S3). The model was implemented using 22 stress periods with 10 time steps each and a multiplier of 1.2. The sea boundary condition was changed, assigning to each SP the head value corresponding the average sea level in each specific month, while the TDS concentration remained the same. The recharge was updated using rainfall and temperature daily collected by the Nauru Rehabilitation Company. Along the Bottomside, the same average additional recharge was maintained to consider house water infiltration. In this second phase, the calibration process mainly focused on storage parameters (Sy and Ss) to which the saline intrusion was shown to be more sensitive. Further, an update of the previously determined hydraulic conductivity was needed, and because of this, the transient model n.1 was run again.

Finally, the calibrated model resulting from the second step of the calibration (Figure S4) process was applied for a 2-day unsteady-state simulation (model n.3) with the aim to prove the model capacity in reproducing the tidal signal recorded during the 4–5 October 2011 survey [15]. In this case, there were no rainy events during groundwater and sea level collection; consequently, the adopted recharge was null except for the Bottomside, where only the additional recharge was maintained. Differently, as the sea level fluctuated during the simulation, the head of the boundary condition representing the sea was changed using a time variant constant head (Figure S5). Hence, hourly stress periods were adopted (total 48 SP) with 5 time steps each and a multiplier of 1.2. The boundary concentration was maintained 35.7 kg/m^3. In this modelling phase, the calibration process mainly focused on specific yield values; nevertheless, some small changes of the previously determined parameters were needed, and this meant to modify transient model n.2 to match again the salinity distribution in the aquifer.

At the end of the calibration process achieved through the 3 models, the hydrogeological parameters adopted are reported in Table 4.

Table 4. Hydrogeological parameters after the calibration processes.

Hydrogeological Parameter	Topside	Bottomside
Hydraulic conductivity (m/d)	800	10–15
Effective porosity (-)	0.02	0.15
Specific yield (-)	0.02	0.15
Specific storage (1/m)	1×10^{-5}	3×10^{-4}
Longitudinal dispersivity (m)	80	2
Transverse dispersivity (m)	5	0.2
Vertical dispersivity (m)	0.2	0.008
Recharge (mm/year)	590	657
Molecular diffusion (m^2/d)	8.64×10^{-6}	8.64×10^{-6}

3. Results

The assessment of the freshwater thickness was completed through two steps. Initially, thanks the characterization activities (hydrogeological and geophysical surveys) focused on the northern side of Nauru, the freshwater volume present under the Capelle area was estimated.

Then, by means of the calibrated model, the freshwater thickness in a more extended area surrounding the Capelle area (i.e., Ewa and Anetan districts) and in general in the northern side of the island was evaluated.

3.1. Freshwater Thikness in Capelle Area

The analysis of the collected data (VES, ERT, and hydrogeological data) allowed to evaluate the thickness and extension of the freshwater lens present in the Capelle area. Ten VESs and eight ERTs (Figure 8) were performed in this zone and compared with the collected hydrogeological data (EC and wells stratigraphic logs). The result of this analysis showed a constant freshwater thickness from west to east, whereas it lessened northward because of the presence of the saltwater edge. For instance, ERT 7 (Figure 9) indicates fresh-brackish transition) showed the thickness decrease of freshwater moving from the cliff toward the sea: from south to north, the thickness reduced from an average value of 6 m to 0 m. The west-east ERT (N° 5, 6, and 8) showed a more constant thickness of freshwater (Figures S6 and S7), for which the average values range from 1–2 m in the western part to 5 m on the eastern side close to the cliff, where it seems to deepen. The maximum thickness (12 m) was detected near well S1 (ERT N° 1, 2, 7 and VES 3, 4).

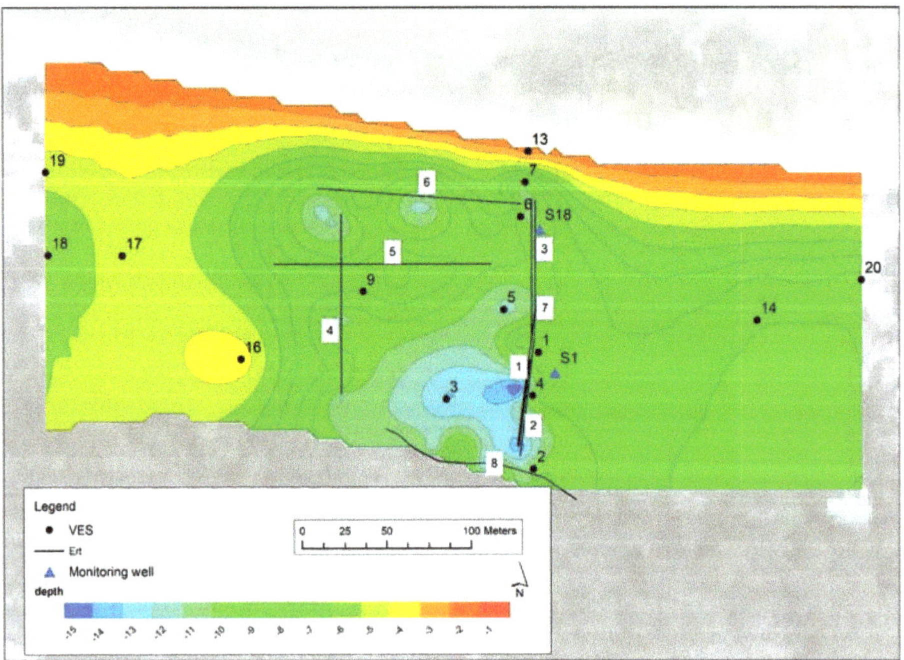

Figure 8. Depth distribution of the freshwater lens in Capelle area based on the geophysical survey results.

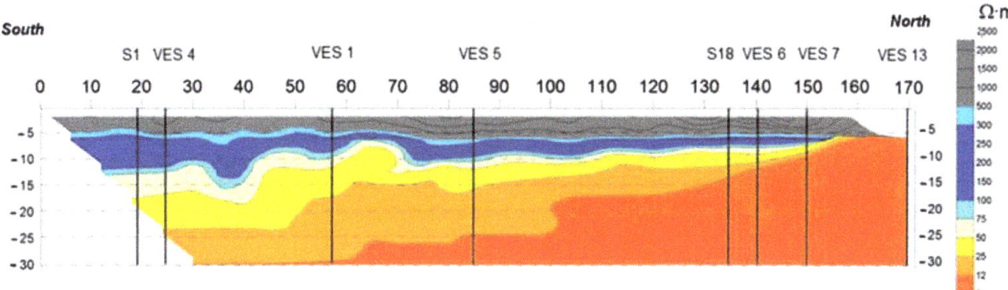

Figure 9. ERT n.7 section interpretation with VES monitoring wells positions; the 75 Ω·m value (light blue) indicates fresh-brackish transition.

From the ERT, it also resulted that the wells used by the population in this area do not substantially affect the freshwater distribution in the subsoil: the electrical data do not show up-coning phenomena related to wells activity. Unfortunately, the four VES performed on the Topside did not result as useful for this analysis because the thickness of the unsaturated aquifer (about 30 m) and the limestone resistivity value (30,000 Ω·m) were too high for the instruments used for the survey.

Finally, the combined analysis of VES, ERT, and hydrogeological data collected in monitoring wells allowed to determine the volume of freshwater in subsoil. The scarp was chosen as southern limit (given that the VESs on the Topside are not usable), and the coastline was defined as the northern limit. Based on the EC values detected in the monitoring wells, for the geoelectric investigations, a resistivity value of 75 Ω·m was considered as the fresh-brackish water transition.

Figure 9 shows the freshwater bottom in the Capelle area. The depth where this surface lays is significantly reduced moving westward (VES 17-18-19), while it reaches the maximum depth close to VES 3 and 4. Moving eastwards, the freshwater thickness increases (Figure S7), but the available data did not allow to identify if and where the freshwater lens pinchouts.

For the evaluation of freshwater volume in the area, the hydraulic head measured in multipipe monitoring well S1 and S18 (the only one where the tube elevation was known) was used. Through the difference between the piezometric surface and the bottom of freshwater, a lens volume equal to 300,000 m^3 was evaluated: it led to a value of 45,000 m^3 of freshwater considering an effective porosity of 15%.

3.2. Numerical Model Results

Figures 10 and 11 and Table 5 report concentration results for model n.1 (Figure S8): the simulated concentration values are similar to the measured ones during the survey of November 2009, with an absolute concentration mean of residuals equal to 1.87 kg/m^3, corresponding to a 5% error (i.e., standardized RMSE).

Table 5. Model n.1: statistical parameters between observed and simulated concentration data.

Statistic	Value
Residual mean (kg/m^3)	1.34
Absolute residual mean (kg/m^3)	1.87
Residual standard deviation (kg/m^3)	1.94
RMSE	1.15

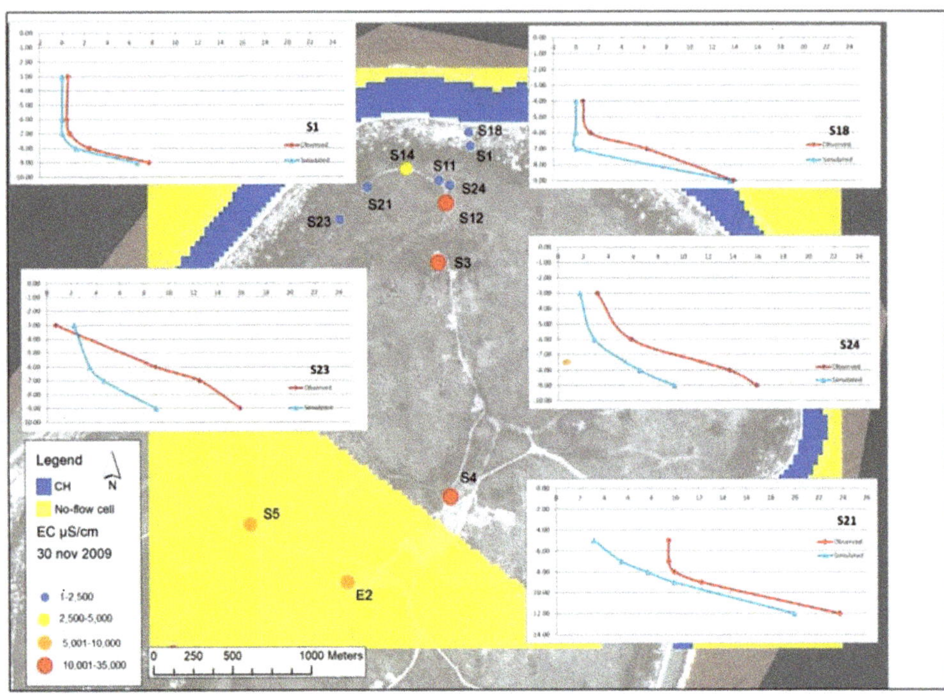

Figure 10. Comparison between observed (red profiles) and simulated (blue profiles) concentration in the targets of the model n.1.

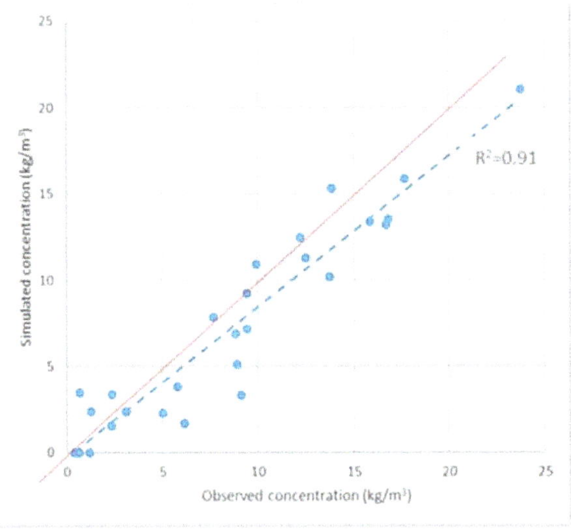

Figure 11. Observed vs. simulated concentrations for model n.1.

The observed vs. simulated graph (Figure 11) shows a good alignment of the points on the 45° line with a slight tendency to underestimate the concentrations and a R^2 coefficient of 0.90 for the trend line (Figure 11). The 3D saltwater distribution simulated by the calibrated model is shown in Figure S9, where along the coast, a freshwater lens (deep blue colour) is almost present everywhere. This lens in the Capelle area (Figure S9b) reaches an average thickness of 6 m, confirming the results of EC and geophysical investigations. Importing the model results in the Leapfrog software, a fresh groundwater volume stored in November 2009 under the Capelle area was approximately evaluated equal to 120,500 m^3.

While the model n.1 results represent an instantaneous picture (November 2009) of the saltwater concentrations, the model n.2 allowed to understand how the concentrations change over the time under different recharge patters. The results confirmed the hints presented in the previous work [15]: they demonstrated that the freshwater can also be stored on Topside during wet periods, but it suddenly disappears (Figure S10). This is because it mixes with deeper salt waters and at the same time rapidly flows towards the ocean by means of the high hydraulic conductivity of the limestone. Consequently, a few months after the rainy period, the groundwater on the Topside becomes salty with concentrations greater than 6 g/L from a depth greater than 2 m from the water table (Figure S11 in Supplementary Materials, S23pipe2). Differently, the freshwater remains stored in a strip located almost all along the entire coastline (Figure S11, S1pipe3). Along the coast of the Anibare district (eastern coast), the lens is narrower and sometime almost disappears because of the reduced width and thickness of the sand deposits (Figure S10). Only between the Anetan and Anabar districts is the lens completely absent due to the presence of a limestone outcrop that did not allow the deposition of sand.

The ability of the model in simulating the Nauru's groundwater system and the saltwater intrusion behaviour is also demonstrated by the results of model n.3. In Figure S12, at the Topside and Bottomside monitoring wells, the comparison between measured and simulated groundwater levels during a two-day transient simulation is shown. The tide influence is evident, with a faster and higher groundwater heads variation on Bottomside and a slower and more contained one on Topside. The model, on both sides, is able to correctly simulate the groundwater levels variation (i.e., efficiency of the tide) and the tidal lag [15].

4. Discussion

In Nauru, there are three main water supply sources: (1) groundwater, which supplies the not-potable water needs for the population having access to a well (about the 36% of the total population); (2) rainwater harvesting in tanks placed on house roofs, which furnish the population with water used for potable and not-potable purposes; and (3) desalination, which provides high-quality drinking water. Past experiences in Nauru demonstrated that the current supply of drinking water is not able to meet the population needs when extended droughts periods overlap with RO plant failures or power shortage. Due to the island's economic fragility, economic crises are frequent, and during these events, the government cannot adequately maintain the RO plant or buy the necessary fuel for its operation. In the past, such issues caused frequent water scarcity (e.g., in the period 2000–2010), considerably increasing the reliance on groundwater for the water supply. From this comes the need to properly assess the fresh groundwater resource on the island in order to increase Nauru's resilience to climate change effects.

The results of previous investigation [15] highlighted in the Nauru aquifer the existence of freshwater lenses stored close to the coastline that can be exploited sustainably. The present study demonstrates that it is possible to quantify the freshwater stored underground by properly combining hydrogeological investigations and density-dependent numerical modelling. A general hydrogeological characterization of the island [15] and a more detailed investigation in a smaller area, by means of geophysical and EC survey, allowed to properly calibrate the numerical model in quasi-steady-state conditions (model n.1) and in unsteady-state conditions (model n.2 and n.3). This led to obtain a general evaluation of groundwater resources for the northern part of Nauru and a good assessment of the freshwater stored in the Capelle area.

The fresh groundwater present in the subsoil of Capelle was assessed by means of the model n.2 to range between 190,000 and 88,300 m^3, respectively, in April 2010 (SP5) and April 2011 (SP17), corresponding to the end of a rainy period and of a dry period (Figure S13). At the end of the simulation (SP22), after few rainy months (1086 mm in 6 months), the volume was the same (83,400 m^3). The assessment is in compliance with the geophysical investigations results that provided 45,000 m^3 in April 2013 after a relative dry period (885 mm in the previous 6 months). The difference is linked to three aspects: (a) before April 2013, the rain was 20% less; (b) the model is unable to simulate the reduced thickness west of Capelle probably due to some pumping well where the pumping rate is unknown; and (c) as shown above, the model is inclined to underestimate the salt concentrations.

Nevertheless, these results demonstrate that the model is an appropriate tool to obtain a good first assessment of the fresh groundwater resource in Nauru, but once the suitable areas are individuated, a detailed hydrogeological investigation is always necessary to improve the model simulation capacity and to correctly design the water abstraction system.

In addition to the estimation achieved for Capelle, an assessment for the remaining of the Bottomside and for the Topside represented in the model can be evaluated. The total fresh groundwater stored along the Bottomside from April 2010 to September 2011 changed between 1,208,100 and 510,390 m^3, with a decrease of about 58% during a drought period of 1.5 years. This result is similar to the Capelle area, where the decrease was 56%. Differently, in the same period, on the Topside, the decrease was about 99%, with a variation from 5,040,500 to 570 m^3, confirming the complete disappearance of freshwater in 18 months.

The amount of fresh groundwater present under the Capelle area could be sustainably exploited to redistribute the water resource to satisfy the population demand in the surrounding Anetan and Ewa districts and/or reduce the RO plants operation. From the SOPAC report [41], the water usage estimated in Nauru for the population having access to a private well (about the 36% of the total population) ranges from 88 l/d/pc to 141 l/d/pc (Table 3). Actually, groundwater supplies not-potable water needs, whereas the RO plant and domestic rainwater harvesting provide the needed amount of drinking water (estimated volume equal to 20 l/d/pc in Table 3). The inhabitants of the Ewa and Anetan districts are approximately 1180 (i.e., 12% of the population of the entire island). Keeping a conservative point of view, it is possible to assume an amount of 45,000 m^3 for the fresh groundwater lens present in Capelle area and to hypothesize it to entirely satisfy the Ewa and Anetan water demand. Assuming a groundwater consumption pro-capita equal to 68 or 121 l/d, the calculations estimated, respectively, the depletion time of the groundwater resource in approximately 1.5 or 1 years in case of complete absence of rainfall for that period. Differently, considering a low-recharge period and a volume of 83,000 m^3, the maximum depletion time would, respectively, increase up to 1.6 and 2.8 years.

The reported example is a simple calculation with the aim to show that even if a strong groundwater abstraction is unsustainable in a long-term perspective, the groundwater lenses stored along the Bottomside can play an important role in the case of emergency. However, in greater generality, groundwater should be considered among the available water resources in Nauru in order to implement a new water management system able to increase the island's resilience to climate change effects.

However, the water derived from the freshwater lenses has lower quality than the desalted water or the rainwater because of the leakage from domestic sewage pits into the subsoil and of the salt concentration increase, therefore potentially leading to the worsening of health and environmental issues. For this reason, alternative strategies for a possible sustainable use of fresh groundwater must be accompanied by sewage infrastructure design, monitoring activities [50], and wellhead protection area definition.

Furthermore, fresh groundwater usage in a coastal area is a difficult issue, and an accurate infrastructure design is needed to avoid the worsening of groundwater due to the saltwater intrusion phenomenon. Because of their small extension and low elevation above the sea level, the oceanic atolls are particularly vulnerable to saltwater intrusion, which can be worsened in the future by the expected sea level rise due to the climate change. Groundwater exploitation in small limestone atoll islands occurs generally through hand-dug wells, which are about 1 m deeper than the groundwater level. This kind of wells is used just for domestic purposes and usually has very modest pumping rates [6]. On the other side, the groundwater extraction using vertical pumping wells in small islands could lead to saltwater up-coning, causing the deterioration of groundwater quality. Therefore, alternative infrastructures such as scavenger wells or infiltration galleries need to be developed in such fragile environments, and density-dependent modelling represents a useful tool for their design.

5. Conclusions and Future Perspectives

In small islands, groundwater can represent an important resource of freshwater. However, fresh groundwater should not be exploited without a proper management, especially in those systems where aquifer vulnerability is high, such as small islands' aquifers, consisting of shallow and thin groundwater lenses. In these contexts, uncontrolled saltwater intrusion phenomena and the depletion of the groundwater lenses due to a potential overexploitation of the resource necessitate a proper water management supported by monitoring activities, a detailed study of the aquifer system behaviour, and an accurate estimation of groundwater availability. In the present study, these objectives were fulfilled through hydrogeological/geophysical investigations and numerical modelling.

Nauru is a very small island located in the Pacific Ocean, and the single use of groundwater as freshwater supply is not adequate to satisfy the population demand. The Nauru Water Plan suggested an integration of groundwater with other water sources: currently, rainwater and desalted water in Nauru are used to satisfy the drinkable water demand, whereas the groundwater is mainly used for non-potable purposes because it is often contaminated. Improving the use of this last resource is essential for Nauru island to guarantee future water security even during drought occurrences or in case of desalination plant failure.

Previous studies investigated the groundwater system in the island, also considering possible future issues such as climatic variations, potential sea level risings, and the increase of extreme climate conditions. Nevertheless, a thorough understanding of the hydrogeological system behaviour and the definition of groundwater availability in Nauru was still required. This study confirms the conceptual model proposed by Alberti L. et al. [15]: fresh groundwater lenses are hosted into the sandy sediments of the coastal zone close to the seashore, which is a result in contrast with the previous studies that had identified freshwater lenses in the limestone, forming the internal part of the island. This conceptual site model has been verified through three unsteady-state, 3D, density-dependent numerical models, which allowed to understand the system behaviour: low hydraulic conductivity of sand makes the flow of groundwater slow down toward the sea, thus allowing freshwater storage where salt water is expected to penetrate more easily. Furthermore, the collected data and model results confirmed the resilience of those lenses in drought conditions.

In particular this study pointed out the presence, in the Bottomside of the northern part of the island, of a fresh groundwater lens that could potentially represent a sustainable solution in meeting the Nauru population's water demand. The hydrogeological investigations allowed the assessment of the freshwater volume stored in this part of the island, and this led to suggest some possible use of this resource. The unsteady-state, 3D, density-dependent model represents a useful tool available to assess the possible sustainable fresh groundwater exploitation to prevent saltwater up-coning occurrences. The implemented model has the capability to provide key information regarding the design and optimization of new suitable groundwater abstraction systems, also assessing their impacts on fresh groundwater availability and quality under different climate conditions. This ability allows to perform an optimal design of the infrastructures by considering several parameters (i.e., number, position, depth, and pumping rates) and supporting public decision makers to set up actions and plans to fulfil the goal of a sustainable groundwater management.

Future research developments should mainly concern two aspects: (a) the extension of the survey and modelling activities to the southern part of Nauru to reach a full assessment of freshwater resources in the island and (b) the extension of a similar survey approach to other smaller islands to find confirmation that even in different hydrogeological conditions, freshwater can accumulate not only in the centre of the islands but in areas where the hydrogeological properties promote a slow groundwater flow and a reduced saltwater mixing process.

Supplementary Materials: The following are available online at https://www.mdpi.com/article/10.3390/w14203201/s1, Figure S1: Hydrogeological section profile of Nauru island, Figure S2: Plan and 3D representation of the hydraulic conductivity distribution in the numerical model, Figure S3: Recharge values (m/d) adopted in the model n.2 for each monthly Stress Period and the 6 monitoring campaigns (red lines), Figure S4: Measured hydraulic heads compared with simulated hydraulic heads after calibration process, Figure S5: Sea levels (RL) attributed at the boundary condition for the model n.3 in order to reproduce the sea level variations, Figure S6: 3D representation of resistivity values detected trough geoelectrical investigations at Capelle area, Figure S7: Freshwater thickness distribution in Capelle area considering 75 m as separation limit between fresh and brackish water, Figure S8: Observed concentration values compared with simulated ones, Figure S9: Model n.1, 3D representation of salt concentrations in groundwater in November 2009 (a) in the model domain and (b) along a cross section passing through Capelle area, Figure S10: Model n.2, 3D concentration representation (g/l for unsteady-state simulation run from December 2009 (SP1) to September 2011 (SP22), Figure S11: Model n.2, simulated versus observed concentrations from December 2009 to September 2011 in (a) S1 and (b) S23 monitoring wells/respectively Bottomside and Topside, Figure S12: Observed and simulated groundwater levels (RL) resulting from the 2 days simulations in monitoring wells S1 (a) and S3 (b), Figure S13: Representation through Leapfrog of the model n.2 unsteady state results, in blue the fresh groundwater body along the Bottomside and in green under the Topside; the view is from the bottom of the model and 1.5 Kg/m^3 is the represented concentration surface that shows the shrinking of the fresh groundwater in 1.5 years.

Author Contributions: Conceptualization, L.A.; methodology, G.O.; software M.A. and G.O.; validation, I.L.L. and G.O.; investigation, L.A. and G.O.; data curation, M.A. and I.L; writing—original draft preparation, I.L.L., P.M. and G.O.; writing—review and editing, L.A. and M.A.; visualization, M.A. and P.M.; supervision, M.A.; project administration, L.A.; funding acquisition, L.A. All authors have read and agreed to the published version of the manuscript.

Funding: This study was supported by the Nauru Project (2010–2015), which was funded by Milano Municipality and related to EXPO 2015. Authors did not receive funds for covering the costs to publish in open access.

Institutional Review Board Statement: Not applicable.

Informed Consent Statement: Not applicable.

Acknowledgments: The authors thank the Nauru Rehabilitation Company (NRC) and the technicians of Ministry of Commerce, Industry, and Environment (CIE), for carrying out the field activities in Nauru Island in collaboration with the researchers of Politecnico di Milano. The authors would like to thank Daniel Feinstein (U.S. Geological Survey—Wisconsin) for his help and his assistance in the model implementation and run. Finally, the authors would like to thank Alfredo Lozej (CNR) for his help during geophysical surveys and Louis Bouchet for being the connection between the researchers of Politecnico di Milano and Nauru Island.

Conflicts of Interest: The authors declare no conflict of interest.

Nomenclature

CSM	Conceptual site model
CNR	National research council
EC	Electrical conductivity
ERT	Electrical resistivity tomography
RL	Reduced level
RO	Reverse osmosis
SP	Stress periods
TDS	Total dissolved solids
VES	Vertical electrical sounding

References

1. Sharan, A.; Lal, A.; Datta, B. A review of groundwater sustainability crisis in the Pacific Island countries: Challenges and solutions. *J. Hydrol.* **2021**, *603*, 127165. [CrossRef]
2. Leoni, B.; Zanotti, C.; Nava, V.; Rotiroti, M.; Stefania, G.A.; Fallati, L.; Soler, V.; Fumagalli, L.; Savini, A.; Galli, P.; et al. Freshwater system of coral inhabited island: Availability and vulnerability (Magoodhoo Island of Faafu Atoll—Maldives). *Sci. Total Environ.* **2021**, *785*, 147313. [CrossRef]
3. Eissa, M.A.; Thomas, J.M.; Pohll, G.; Shouakar-Stash, O.; Hershey, R.L.; Dawoud, M. Groundwater recharge and salinization in the arid coastal plain aquifer of the Wadi Watir delta, Sinai, Egypt. *Appl. Geochem.* **2016**, *71*, 48–62. [CrossRef]
4. Yaqoob, A.A.; Parveen, T.; Umar, K.; Ibrahim, M.N.M. Role of nanomaterials in the treatment of wastewater: A review. *Water* **2020**, *12*, 495. [CrossRef]
5. Balzan, M.V.; Potschin-Young, M.; Haines-Young, R. Island ecosystem services: Insights from a literature review on case-study island ecosystem services and future prospects. *Int. J. Biodivers. Sci. Ecosyst. Serv. Manag.* **2018**, *14*, 71–90. [CrossRef]
6. White, I.; Falkland, T. Management of freshwater lenses on small Pacific islands. *Hydrogeol. J.* **2010**, *18*, 227–246. [CrossRef]
7. Wheatcraft, S.W.; Buddemeier, R.W. Atoll Island Hydrology. *Groundwater* **1981**, *19*, 311–320. [CrossRef]
8. Falkland, A.C.; Custodio, E.; Diaz Arenas, A.; Simler, L. Hydrology and Water Resources of Small Islands: A Practical Guide. In *Studies and Reports in Hydrology*; Unesco: Paris, France, 1991; Volume 49, ISBN 9231027530.
9. Vacher, H.L.; Quinn, T.M. *Geology and Hydrogeology of Carbonate Islands*; Elsevier: Amsterdam, The Netherlands, 2004; Volume 54, ISBN 0-444-51644-1.
10. Falkland, A.C. Hydrology and Water Management on Small Tropical Islands. 1993. Available online: https://iahs.info/uploads/dms/iahs_216_0263.pdf (accessed on 5 September 2022).
11. Mastrocicco, M. Studies on water resources salinization along the Italian coast: 30 years of work. *Acque Sotter.—Ital. J. Groundw.* **2021**, *10*, 7–13. [CrossRef]
12. Licata, I.L.; Langevin, C.D.; Dausman, A.M.; Alberti, L. Effect of tidal fluctuations on transient dispersion of simulated contaminant concentrations in coastal aquifers. *Hydrogeol. J.* **2011**, *19*, 1313–1322. [CrossRef]
13. Mulligan, A.E.; Langevin, C.; Post, V.E. Tidal Boundary Conditions in SEAWAT. *Ground Water* **2011**, *49*, 866–879. [CrossRef]
14. Buddemeier, R.W.; Oberdorfer, J.A. Internal Hydrology and Geochemistry of Coral Reefs and Atoll Islands: Key to Diagenetic Variations. In *Reef Diagenesis*; Schroeder, J.H., Purser, B.H., Eds.; Springer: Berlin/Heidelberg, Germany, 1986; pp. 91–111.
15. Alberti, L.; Licata, I.L.; Cantone, M. Saltwater Intrusion and Freshwater Storage in Sand Sediments along the Coastline: Hydrogeological Investigations and Groundwater Modeling of Nauru Island. *Water* **2017**, *9*, 788. [CrossRef]
16. Nakada, S.; Umezawa, Y.; Taniguchi, M.; Yamano, H. Groundwater Dynamics of Fongafale Islet, Funafuti Atoll, Tuvalu. *Ground Water* **2012**, *50*, 639–644. [CrossRef]
17. Gingerich, S.B.; Voss, C.I.; Johnson, A.G. Seawater-flooding events and impact on freshwater lenses of low-lying islands: Controlling factors, basic management and mitigation. *J. Hydrol.* **2017**, *551*, 676–688. [CrossRef]
18. Werner, A.D.; Sharp, H.K.; Galvis, S.C.; Post, V.E.A.; Sinclair, P. Hydrogeology and management of freshwater lenses on atoll islands: Review of current knowledge and research needs. *J. Hydrol.* **2017**, *551*, 819–844. [CrossRef]
19. Oberle, F.K.J.; Swarzenski, P.W.; Storlazzi, C.D. Atoll groundwater movement and its response to climatic and sea-level fluctuations. *Water* **2017**, *9*, 650. [CrossRef]
20. Lentini, A.; De Caterini, G.; Cima, E.; Manni, R.; Ventura, G. Della Resilience toclimate change: Adaptation strategies for the water supply system of Formia and Gaeta (Province of Latina, Central Italy). *Acque Sotter.—Ital. J. Groundw.* **2021**, *10*, 35–46. [CrossRef]
21. Karatzas, G.P.; Dokou, Z. Optimal management of saltwater intrusion in the coastal aquifer of Malia, Crete (Greece), using particle swarm optimization. *Hydrogeol. J.* **2015**, *23*, 1181–1194. [CrossRef]
22. Thiéry, D. Saltwater Intrusion Modelling with an Efficient Multiphase Approach: Theory and Several Field Applications. In *18th Salt Water Intrusion Meeting: 18 SWIM*; Instituto Geologico y Minero de España: Madrid, Spain, 2004.
23. Colombo, L.; Alberti, L.; Mazzon, P.; Antelmi, M. Null-Space Monte Carlo Particle Backtracking to Identify Groundwater Tetrachloroethylene Sources. *Front. Environ. Sci.* **2020**, *8*, 142. [CrossRef]
24. Antelmi, M.; Renoldi, F.; Alberti, L. Analytical and numerical methods for a preliminary assessment of the remediation time of pump and treat systems. *Water* **2020**, *12*, 2850. [CrossRef]

25. Antelmi, M.; Mazzon, P.; Höhener, P.; Marchesi, M.; Alberti, L. Evaluation of mna in a chlorinated solvents-contaminated aquifer using reactive transport modeling coupled with isotopic fractionation analysis. *Water* **2021**, *13*, 2945. [CrossRef]
26. Alberti, L.; Francani, V.; La Licata, I. Characterization of salt-water intrusion in the lower Esino Valley, Italy using a three-dimensional numerical model. *Hydrogeol. J.* **2009**, *17*, 1791–1804. [CrossRef]
27. Underwood, M.R.; Peterson, F.L.; Clifford, I.V. Groundwater Lens Dynamics of Atoll Islands. *Water Resour. Res.* **1992**, *28*, 2889–2902. [CrossRef]
28. Bailey, R.T.; Jenson, J.W.; Olsen, A.E. Numerical Modeling of Atoll Island Hydrogeology. *Groundwater* **2009**, *47*, 184–196. [CrossRef]
29. Babu, R.; Park, N.; Yoon, S.; Kula, T. Sharp interface approach for regional andwell scale modeling of small island freshwater lens: Tongatapu island. *Water* **2018**, *10*, 1636. [CrossRef]
30. Coulon, C.; Pryet, A.; Lemieux, J.M.; Yrro, B.J.F.; Bouchedda, A.; Gloaguen, E.; Comte, J.C.; Dupuis, J.C.; Banton, O. A framework for parameter estimation using sharp-interface seawater intrusion models. *J. Hydrol.* **2021**, *600*, 126509. [CrossRef]
31. Coulon, C.; Lemieux, J.; Pryet, A.; Bayer, P.; Young, N.L.; Molson, J. Pumping Optimization under Uncertainty in an Island Freshwater Lens Using a Sharp-Interface Seawater Intrusion Model Water Resources Research. *Water Resour. Res.* **2022**, *58*, 1–18. [CrossRef]
32. Houghton, J.T.; Ding, Y.; Griggs, D.J.; Noguer, M.; van der Linden, P.J.; Dai, X.; Maskell, K.; Johnson, C.A. *IPCC Climate Change 2001: The Scientific Basis*; The Press Syndacate of the University of Cambridge: Cambridge, UK, 2001.
33. Nurse, L.; Moore, R. Adaptation to Global Climate Change: An Urgent Requirement for Small Island Developing States. *Rev. Eur. Community Int. Environ. Law* **2005**, *14*, 100–107. [CrossRef]
34. Birawida, A.B.; Ibrahim, E.; Mallongi, A.; Rasyidi, A.A.A.; Thamrin, Y.; Gunawan, N.A. Clean water supply vulnerability model for improving the quality of public health (environmental health perspective): A case in Spermonde islands, Makassar Indonesia. *Gac. Sanit.* **2021**, *35*, S601–S603. [CrossRef]
35. Doorga, J.R.S. Climate change and the fate of small islands: The case of Mauritius. *Environ. Sci. Policy* **2022**, *136*, 282–290. [CrossRef]
36. Boojhawon, A.; Surroop, D. Impact of climate change on vulnerability of freshwater resources: A case study of Mauritius. *Environ. Dev. Sustain.* **2021**, *23*, 195–223. [CrossRef]
37. Falkland, A. *Nauru Water Management Visit Report 28th Oct–1st Nov 2002*; Ecowise Environmental; Australian Agency for International Development: Canberra, Australia, 2002; 48p.
38. Nauru Project. Available online: http://nauru.como.polimi.it (accessed on 19 March 2022).
39. Jacobson, G.; Hill, P.J. Hydogeology and groundwater resources of Nauru Island, Central Pacific Ocean. *Groundwater* **1988**, *12*, 85.
40. Alberti, L.; Cantone, M.; Oberto, G.; Sampietro, D. GNSS Static Suvey Report. Published by Politecnico di Milano (DIIAR Department) for Nauru Project. Available online: http://nauru.como.polimi.it/activities-report-november-2010/gnss-nauru-survey-report-oct-2011 (accessed on 19 March 2022).
41. Bouchet, L.; Sinclair, P. *Assessing the Vulnerability of Shallow Domestic Wells in Nauru*; SOPAC Technical Report 435; SOPAC Secretariat: South Orange, NJ, USA, 2010.
42. Jacobson, G.; Hill, P.J.; Ghassemi, F. Geology and Hydrogeology of Nauru Island. In *Geology and Hydrogeology of Carbonate Islands*; Vacher, H.L., Quinn, T.M., Eds.; Elsevier: Amsterdam, The Netherlands, 1997; pp. 707–742. ISBN 0444815201.
43. Bouwer, H.; Rice, R.C. A slug test method for determining hydraulic conductivity of unconfined aquifers with completely or partially penetrating wells. *Water Resour. Res.* **1976**, *12*, 423–428. [CrossRef]
44. Neuman, S.P. Effect of partial penetration on flow in unconfined aquifers considering delayed gravity response. *Water Resour. Res.* **1974**, *10*, 303–312. [CrossRef]
45. Falkland, T. Water resources issues of small island developing states. *Nat. Resour. Forum* **1999**, *23*, 245–260. [CrossRef]
46. Falkland, A.C. *Vaipeka Water Gallery Extension Project, Aitutaki, Cook Islands*; Australian Agency for International Development: Canberra, Australia, 1995.
47. Harbaugh, A.W. *MODFLOW-2005, The U.S. Geological Survey Modular Ground-Water Model—The Ground-Water Flow Process*; US Department of the Interior, US Geological Survey: Reston, VA, USA, 2005; p. 253. [CrossRef]
48. Langevin, C.D.; Guo, W. MODFLOW/MT3DMS-based simulation of variable-density ground water flow and transport. *Ground Water* **2006**, *44*, 339–351. [CrossRef] [PubMed]
49. Ghassemi, F.; Jakeman, A.J.; Jacobson, G.; Howard, K.W.F. Simulation of seawater intrusion with 2D and 3D models: Nauru Island case study. *Hydrogeol. J.* **1996**, *4*, 4–22. [CrossRef]
50. Kim, Y.; Yoon, H.; Lee, S.H. Freshwater-salt water interface dynamics during pumping tests. *Acque Sotter.—Ital. J. Groundw.* **2019**, *8*, 35–39. [CrossRef]

 MDPI

Article

ORGANICS: A QGIS Plugin for Simulating One-Dimensional Transport of Dissolved Substances in Surface Water

Rudy Rossetto [1,*], Alberto Cisotto [2], Nico Dalla Libera [2], Andrea Braidot [2], Luca Sebastiani [1], Laura Ercoli [1] and Iacopo Borsi [3]

1 Crop Science Research Center, Scuola Superiore Sant'Anna, 56122 Pisa, Italy
2 Distretto Idrografico Alpi Orientali, 30121 Venezia, Italy
3 TEA SISTEMI S.p.A., 56122 Pisa, Italy
* Correspondence: rudy.rossetto@santannapisa.it

Citation: Rossetto, R.; Cisotto, A.; Dalla Libera, N.; Braidot, A.; Sebastiani, L.; Ercoli, L.; Borsi, I. ORGANICS: A QGIS Plugin for Simulating One-Dimensional Transport of Dissolved Substances in Surface Water. *Water* **2022**, *14*, 2850. https://doi.org/10.3390/w14182850

Academic Editor: Cristina Di Salvo

Received: 5 August 2022
Accepted: 8 September 2022
Published: 13 September 2022

Abstract: Surface water in streams and rivers is a valuable resource and pollution events, if not tackled in time, may have dramatic impacts on aquatic ecosystems. As such, in order to prepare pollution prevention plans and measures or to set-up timely remedial options, especially in the early stages of pollution incidents, simulation tools are of great help for authorities, with specific reference to environmental protection agencies and river basin authorities. In this paper, we present the development and testing of the ORGANICS plugin embedded in QGIS. The plugin is a first attempt to embed surface water solute transport modelling into GIS for the simulation of the concentration of a dissolved substance (for example an organic compound) in surface water bodies including advection dispersion and degradation. This tool is based on the analytical solution of the popular advection/dispersion equation describing the transport of contaminants in surface water. By providing as input data the concentration measured at the entry point of a watercourse (inlet boundary condition) and the average speed of the surface water, the model simulates the concentration of a substance at a certain distance from the entry point, along the profile of the watercourse. The tool is first tested on a synthetic case. Then data on the concentration of the pharmaceutical carbamazepine monitored at the inlet and outlet of a vegetated channel, in a single day, are used to validate the tool in a real environment. The ORGANICS plugin aims at popularizing the use of simple modelling tools within a GIS framework, and it provides GIS experts with the ability to perform approximate, but fast, simulations of the evolution of pollutants concentration in surface water bodies.

Keywords: water pollution; solute transport modelling; Geographic Information System (GIS); pollution prevention plans; pharmaceuticals; carbamazepine; longitudinal dispersion coefficient; decay rate coefficient

1. Introduction

Surface water in streams and rivers is a valuable resource, and pollution events, if not tackled in time, may have dramatic impacts on aquatic ecosystems [1–4]. As such, in order to prepare pollution prevention plans or to set up timely remedial options, especially in the early moment of pollution incidents, simulation tools are of great help for authorities, with specific reference to environmental protection agencies and river basin authorities. Modelling of solute transport in surface water is then a valuable and common option [5–9], especially by means of analytical solutions simplifying system description and reducing complexity [10–12], providing fast, even approximate, answers, particularly in cases of insufficient data availability for the implementation of more complex numerical models.

In order to advance the use of modelling tools and to support the digitalization of the technical sector these tools must be user-friendly, and built around free and open-source codes. An open code may guarantee the reproducibility and the reliability of the analyses performed [13,14] and their early deployment and impact [15,16].

Geographic Information Systems (GISs) are worldwide mainstreaming tools and methodologies for storing, managing/analyzing and visualizing large temporal, spatial and non-spatial datasets (for both geometric and alphanumerical data). Hence, they are widely applied to support environmental modelling [17,18]. The possibility of including modelling tools in Geographic Information Systems (i.e., via plugins), and the existing community of users and developers in constant growth, is potentially the most relevant strength of such tools [18–20]. For thirty years GIS and water modelling codes/applications have been successfully integrated. Vieux [21] presented an application of the GIS, ARC/INFO and the finite element solution to the kinematic wave equations to process the spatially variable terrain in a small watershed using a Triangular Irregular Network for the solution of overland flow. Oliveira et al. [22] presented ArcGIS-SWAT, a geodata model and GIS interface for the Soil and Water Assessment Tool (SWAT; [23]). Becker and Jiang [24] developed a computationally efficient method for predicting contaminant mass flux to a specified boundary, carrying out the method in a GIS and taking full advantage of widely available digital hydrologic data. Akbar et al. [25] showed a GIS-based modelling system called ArcPRZM-3 for the spatial modelling of pesticide leaching potential from soil towards groundwater. Rossetto et al. [26] integrated a suite of groundwater modelling tools in the gvSIG GIS application [27]. Oliveira and Martins [28] developed an application for the preliminary characterisation of the river boundary condition for a MODFLOW [29] finite difference groundwater flow numerical model. Bittner et al. [30] developed the LuKARS GIS-based model for simulating the hydrological effects of land use changes on karst systems.

Among GIS desktop applications, QGIS [31] is probably the most popular free geospatial software. Rosas-Chavoya et al. [20] conducted a bibliometric analysis on the acceptance of this application on documents published in Scopus from 2005 to 2020, considering 931 manuscripts. They observed a favorable trend in the acceptance of QGIS across the world and the development of large collaborative networks.

Several plugins have been developed for QGIS within the hydrological or aquatic domain. In particular, Nielsen et al. [32] developed the Water Ecosystems Tool, a workflow implemented (as a plugin) in QGIS, for the application and evaluation of aquatic ecosystem models. Ellsäßer et al. [33] developed the QWaterModel as an easy-to-use tool to make evapotranspiration predictions available to broader audiences. The QWaterModel is a QGIS plugin compatible with all versions of QGIS3. Dile et al. [34] developed an open-source user interface for the SWAT [23], QSWAT, using various functionalities of the QGIS application. Rossetto et al. [18] presented the FREEWAT plugin for managing the groundwater resource, including tools for the management of hydrochemical data [35] and nitrate leaching assessment [36].

In this paper, we present the development and testing of the ORGANICS plugin as a first attempt to embed surface water solute transport modelling into GIS. This tool allows users to simulate the concentration of a dissolved substance (for example an organic compound) in surface water bodies by applying an analytical solution of the advection dispersion equation, which includes also a first-order degradation term. By providing as input the stepwise time-variant concentration measured at the entry point of a watercourse, along with the related average water velocity, the concentration of a substance at a certain distance from the entry point along the profile of the watercourse, is simulated. A sketch of the presented problem is shown in Figure 1.

After presenting the theoretical and modelling approach, we show an example application of the plugin (to be used as a tutorial), and then a real case study application to simulate carbamazepine concentration in a vegetated channel collecting poorly treated wastewater.

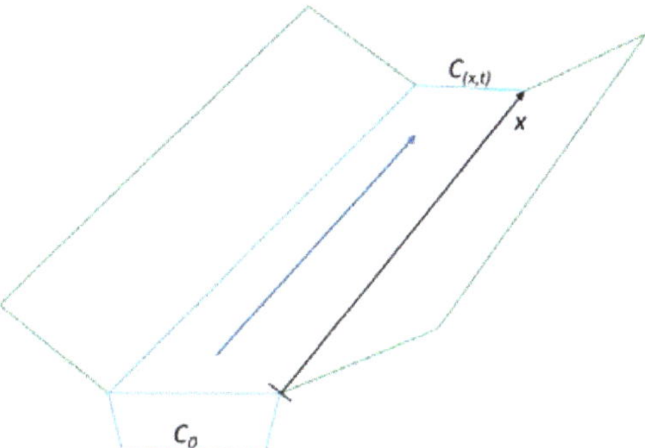

Figure 1. Schematic draw of contaminant movement along a surface water reach.

2. Materials and Methods

The model used in the ORGANICS plugin is based on the analytical solution of the popular advection/diffusion/decay equation, in one-dimensional form, taken from [10,37]: where:

$$\frac{\partial C}{\partial t} + \frac{\partial v_x C}{\partial x} = E\frac{\partial^2 C}{\partial x^2} - kC \quad (1)$$

C: is the solute concentration expressed as mass per unit volume of water [M/L^3],

v_x: longitudinal fluid flow velocity is the input velocity [L/T].

x: is the longitudinal coordinate [L],

t: is time [T],

E: is the longitudinal dispersion coefficient accounting for the combined effects of ionic or molecular diffusion and hydrodynamic dispersion [L^2/T].

k: is a first-order decay rate [T^{-1}].

Using a constant concentration boundary condition at $x_0 = 0$ [L] (the inlet point of a surface water body reach) at the initial time ($t_0 = 0$) [T], for each $x > 0$ and $t > 0$ the Equation (1) may be reduced to the following analytical solution (2) [10,37].

$$C(x,t) = \frac{C_0}{2}\left\{\exp\left[\frac{Ux}{2E}(1-\Gamma)\right]\mathrm{erfc}\left(\frac{x-Ut\Gamma}{2\sqrt{Et}}\right) + \exp\left[\frac{Ux}{2E}(1+\Gamma)\right]\mathrm{erfc}\left(\frac{x+Ut\Gamma}{2\sqrt{Et}}\right)\right\} \quad (2)$$

$$\Gamma = \sqrt{1+4\frac{kE}{U^2}} \quad (3)$$

and U is the velocity module [L/T].

This approach entails the following assumptions:

(a) the flow is one-dimensional and oriented according to the main direction of the flow in the surface water body;

(b) the concentration at the inlet remains constant for the specified simulated time interval (first kind boundary condition);

(c) chemical interactions between different dissolved substances are not considered, nor are reactive geochemical processes simulated with other components (i.e., riverbed matrix);

(d) the morphology of the bed of the surface water body does not affect the solution;

(e) no sorption processes or production terms are considered;

(f) at $t_0 = 0$ [T] the initial condition is $C_{(x,0)} = 0$ [M/L^3] along all the simulated domains.

The developed code uses Equation (2) to calculate the concentration value along the line input provided by the user to represent the selected surface water body. This input can be given as a linear vector layer (for instance the common ESRI. shp file). The code calculates the solution (concentration value) at nodes at homogeneous lengths from the inlet point as specified by the user, and at selected homogeneous time-steps also defined by the user.

2.1. *Software Development*

The plugin was developed in Python3 [38] language, using the Qt5 [39] graphics libraries and the QGIS [40] Application Program Interface. The solute transport code is embedded in the QGIS desktop application. This means a full integration at programming language level, with models using GIS data format included as full component of the host GIS application [41,42].

The plugin can be used as an add-on to the QGIS desktop application, version 3.4 or higher, once it has been installed on a PC. During the development phase, some specific Python libraries were used, but each of them is already included in the official QGIS desktop distribution. This choice allows the user to use the plugin without requiring further software updates.

2.2. *Data Needs*

In order to use the plugin, the user has to prepare a set of input files. These are:

- a *.csv file specifying the water average longitudinal velocity (U) and concentration values at the inlet of the watercourse (constant concentration boundary condition), and the time these data refer to. The file must comply with the template format defined for the plugin. In particular, the file must contain data relating to (at least) one dataset, specifying:
    - starting date and time, in YYYY-MM-DD HH: MM: SS format;
    - average flow velocity in the surface water body, in $\text{m} \cdot \text{s}^{-1}$. This value will be used at all the node of the surface water body;
    - the concentration of the source at the inlet point.

  When considering time-varying boundary condition (that is the concentration input changing with time), the user must specify for each different time all the above information on consecutive lines of the. csv file.
- an ESRI linear Shapefile representing the surface water body. The file may consist of one or more segments. The line must be digitized towards the flow direction. When more segments are used, the topology must be respected (all the lines must be connected).

An example of the required files to run a first test are provided in the *template_files* folder of the plugin itself as Supplementary material.

2.3. *Model Implementation and Run*

Once data are prepared in the form of the required files, the first operation consists in loading the Shapefile geographic layer of the line into the QGIS view.

By clicking on *ORGANICS* in the Plugin menu, the main window opens. This is divided in four sections: (1) *Run*; (2) *Plot Results*; (3) *Help*; (4) *About*. By entering the *Run* section, the user input the following data (Figure 2):

(a) the *.csv file. Upon the file selection, the drop-down menus will automatically update. Through these menus the user must select the name of the fields in the *.csv file corresponding to the required information;

(b) the linear shapefile representing the watercourse. At this step the user must specify the length of the homogeneous reaches at whose ends the concentration values will be calculated;

(c) the value of the first-order decay rate coefficient (s^{-1});

(d) the value of the coefficient of longitudinal dispersion (m^2/s);

(e) the length of the timestep (in minutes) at which the solution will be calculated over a time interval (in minutes from the start of the simulation) at each point of the reach;

(f) the name of the output file (*.shapefile) and the directory where the file will be saved. Should this field be left blank, the output will be saved as a temporary layer (memory layer) in QGIS.

Although time data are input both in seconds and in minutes, all the calculations are internally run in seconds, while results are provided in minutes or within hours.

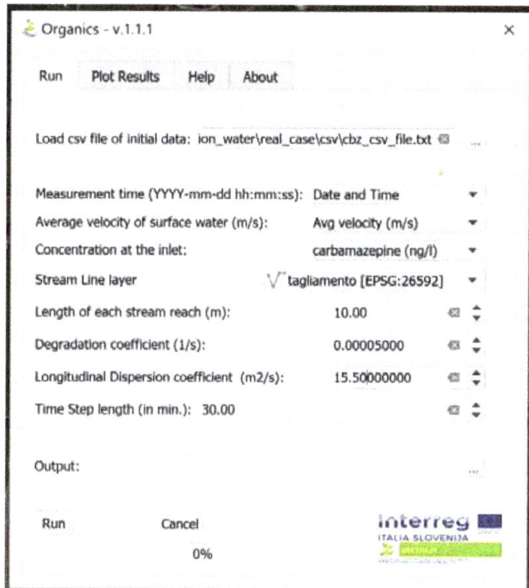

Figure 2. The ORGANICS plugin main window.

Once the simulation is run, the code will: (i) divide the line in several nodes, according to the length specified by the user, and (ii) calculate the solution at each node at the end of a reach, at each time step as specified by the user. The result consists in a *.shapefile point layer in which the simulated concentration values at different times are saved at each point. This layer can be used to visualize the results using the tools in QGIS. For example, it can be themed with color scales depending on the level of the selected solute concentration. Animations to visualize the evolution of the concentration in the various points may be produced by applying the *TimeManager* plugin (plugin, downloadable from the QGIS *PluginManager*).

A *.csv file of the output will also be saved in the previously defined destination folder. This file can be used by the user to conduct further analyses externally to the plugin and/or from QGIS (i.e., using spreadsheet).

Graphs of the solutions may be drawn opening the *Plot Results* section, where a number of options for producing solution plots or further customizing the draw and to save it in image format (i.e., as *.png file) are provided (Figure 3).

The user must choose the output layer to process. Graph drawing can be performed at any time, even after the execution of the model, by selecting the output file from the drop-down menu. However, in order for the desired layer to appear in the ORGANICS menu, the layer must be loaded in the QGIS layer panel. Once the layer is selected, the drop-down menus below will automatically update.

The following options for creating the graphs are available:

- *Select a position* (distance in m from the entry point): this option will create a graph displaying the concentration trend in a point defined by the user at a certain distance from the starting point, as a function of time (Figure 4);
- *Use the selected position on the layer*: this option allows to view the same result as above, but in this case the position is provided by selecting, using the classic selection tools on the map, one or more points of the output layer (Figure 5);
- *Select a time*: this graph will display the concentration values, at a given simulated time, as a function of the distance from the inlet (Figure 6). The times available for selection correspond to the discretization obtained with the time step chosen in input by the user.

Figure 3. The ORGANICS plugin *Plot results* section.

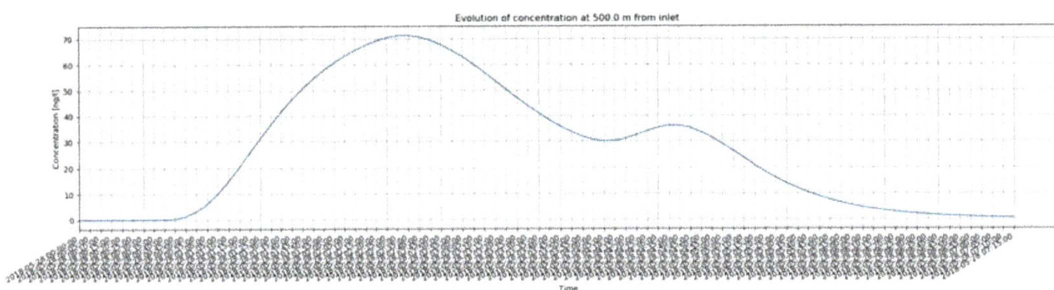

Figure 4. Example graph: solution at a selected point as function of time.

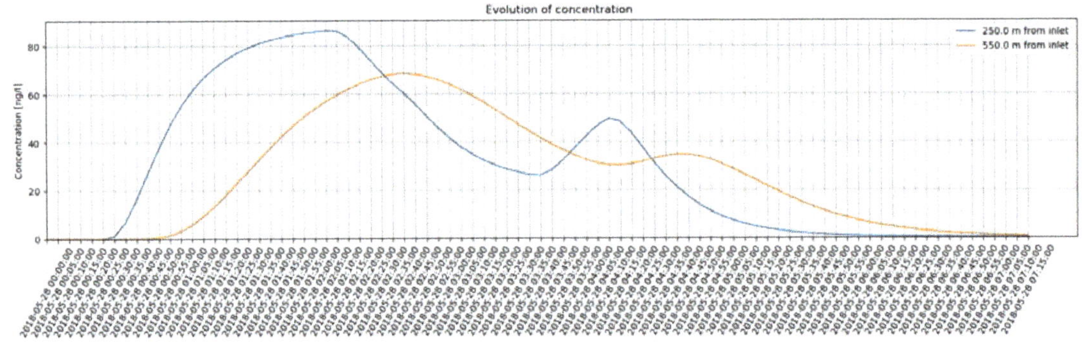

Figure 5. Example graph: solution at a selected points as function of time. Two points selected by using the GIS selection tools.

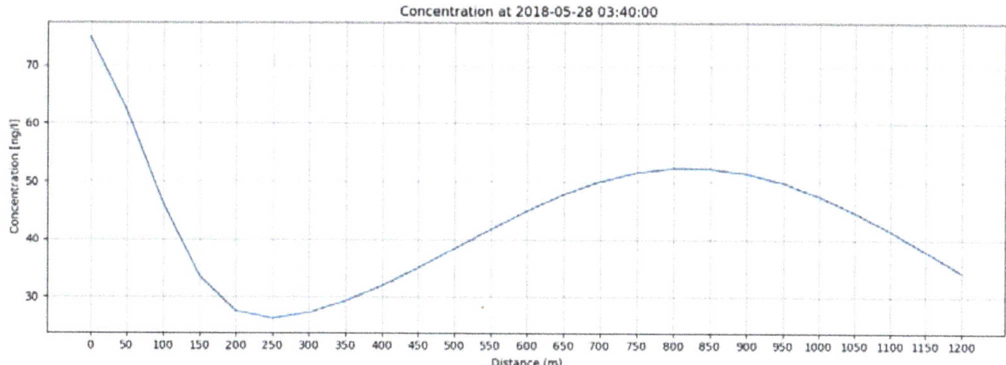

Figure 6. Example graph: solution at a selected time as function of distance from the inlet point.

3. Results and Discussion

3.1. Model Validation

We successfully validated the model implemented within the tool by simulating the same case described by the analytical solution shown in [10]. The simulation results (Figure 7) are obtained using the following parameters:

$U = 1.0$ m/s

$k = 0$ s^{-1} (no decay is simulated)

$E = 5.0$ m^2/s

where:

$C_{(0,0)} = C_0$, and

$C_{(x,0)} = 0$ at $x > 0$

The solution (concentration value) is provided at 100, and 1000 m from the source.

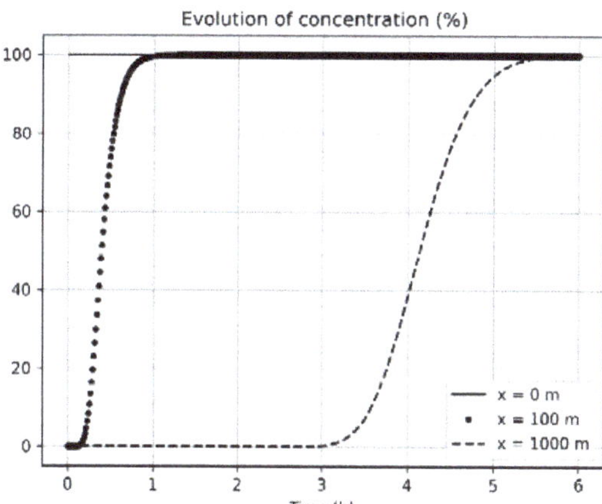

Figure 7. Computed simplified analytical solution implemented in Organics in time at $x = 100$ m and $x = 1000$ m.

3.2. Example Problem

In this section, three tests are presented in order to show the behavior of the calculated solution in different scenarios. The cases tested are:

(a) C_0 mass injection, constant over time;

(b) time-limited pulse C_0 mass injection;

(c) C_0 mass input, variable over time (multi-pulse input condition).

Concerning the cases with time-limited or time-varying input (cases b and c, respectively), we implemented a solution using the principle of superimposition of the individual analytical solutions at each stress period, as reported in [10,43,44]. The parameter values used in the example problem are presented in Table 1. The .csv file and the .shp file are provided in the Supplementary Material.

Table 1. Parameters values used in the example problem.

Parameter	Value	Units
Total length of the reach	1200	m
Simulation length	50	m
Time step	10	min
Velocity (*U*)	0.1	m/s
First order decay rate (*k*)	0.00005	s^{-1}
Longitudinal dispersion (*E*)	5.0	m^2/s
Initial time	28 May 2018 00:00	dd/mm/yyyy hh:mm

3.2.1. C_0 Mass Injection, Constant over Time

In this test, we simulated the release of a mass with concentration C_0 = 100 ng/L constant over time. The solution is presented at the end of the simulation time, which is the time needed for the mass to reach the outlet.

Figure 8 displays the solution at the beginning of the reach (x = 0 m), at x = 500 m, and approximately at the end of the reach (x = 1100 m). In the middle of the reach (x = 500 m) the solution tends to an asymptotic value, which is less than 100 ng/L due to the effect of the simulated decay process.

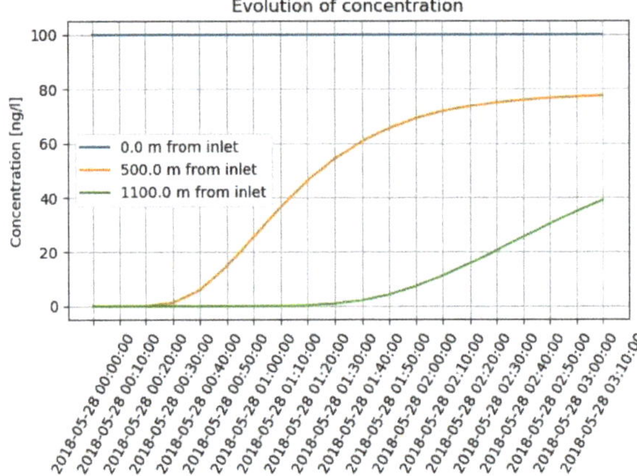

Figure 8. C_0 mass injection, constant over time: concentration values at selected points of the reach.

3.2.2. Time-Limited Pulse C_0 Mass Injection

In this test we simulated a time-limited pulse C_0 = 100 ng/L mass injection for a 2 h duration. This time-limited pulse case, at constant concentration, can be implemented by defining in the *csv file an initial period (of known duration, with concentration C_0; first line in the *csv file) followed by a second period with zero concentration (second line in the *csv file). In this test, the second input is then two hours long with concentration set at C_{2h} = 0 ng/L. The solution is then displayed in Figure 9 at an infinite time (that is, the time needed for the dissolved substance to reach the outlet of the water course considered). The solution is presented at x = 500 m (Figure 9). At this distance, mass arrival is recorded after 30 min from the beginning of the simulation.

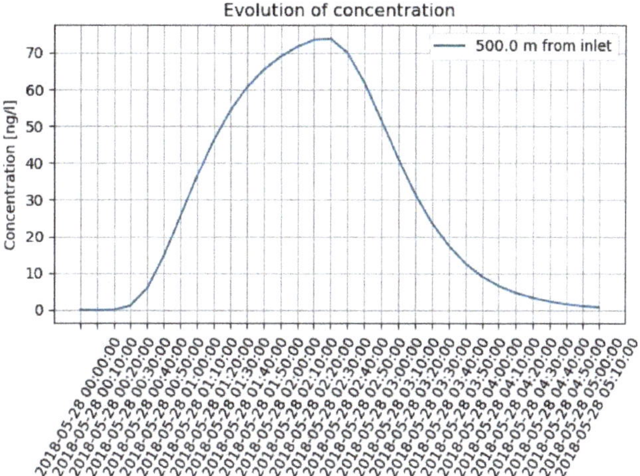

Figure 9. Time-limited pulse C_0 mass injection: concentration values at 500 m from the inlet.

3.2.3. C_0 Mass Input, Variable over Time (Multi-Pulse Boundary Condition)

In this test, we simulated C_0 mass input, variable over time (multi-pulse input condition) according to data presented in Table 2. The global solution works as the superposition of the several pulses, each one having a constant condition for a specified time interval. Superposing each "pulse-solution" makes the model able to consider time-dependent boundary conditions. Results are shown at $x = 400$ m, $x = 800$ m, and $x = 1200$ m from the inlet point for time step length of:

- 20 min (Figure 10);
- 10 min (Figure 11);
- 5 min (Figure 12);

from the beginning of the simulation, in order to present the impact of the different time discretization on the solution (Figures 10–12).

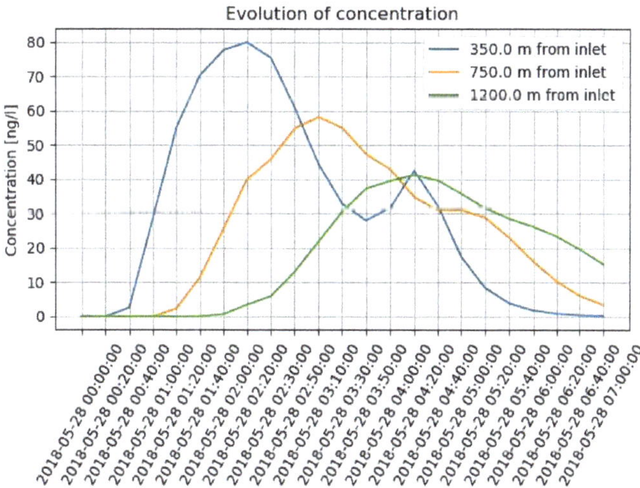

Figure 10. Simulated concentration at $x = 400$ m, x = 800 m, and $x = 1200$ m from the inlet point with 20 min time step length.

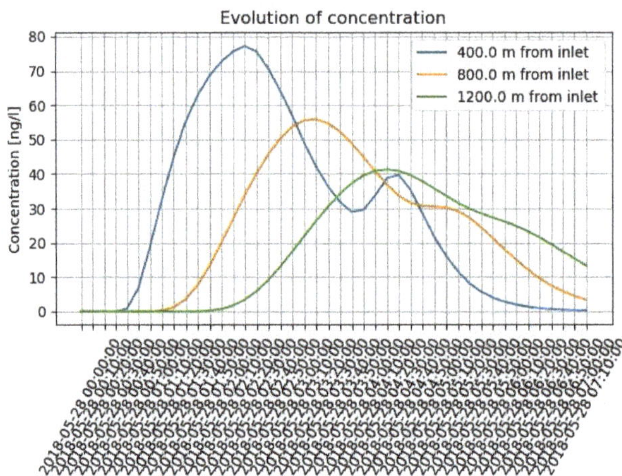

Figure 11. Simulated concentration at $x = 400$ m, $x = 800$ m, and $x = 1200$ m from the inlet point with 10 min time step length.

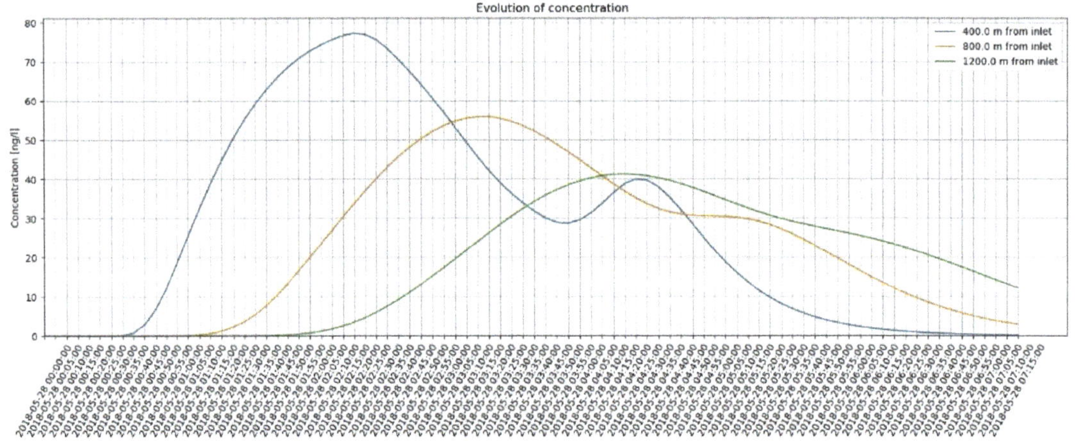

Figure 12. Simulated concentration at $x = 400$ m, $x = 800$ m, and $x = 1200$ m from the inlet point with 5 min time step length.

Table 2. Data used in the test presented in Section 3.2.3.

Date and Time	C_0 (ng/L)
28 May 2018 0:00	0
28 May 2018 0:20	100
28 May 2018 2:00	50
28 May 2018 2:30	25
28 May 2018 3:30	75
28 May 2018 4:00	0

3.3. Case Study Application

The ORGANICS plugin was then applied to compute the concentration of the pharmaceutical compound carbamazepine at a reach of a vegetated channel receiving poorly treated wastewater, in low-flow conditions, in the Pisa municipality (Tuscany, Italy, Figure 13). Carbamazepine (CBZ) is an anticonvulsant or anti-epileptic drug commonly found in poorly treated wastewater and consequently in surface- and ground-water [45,46]. Removal efficiency in secondary wastewater treatment is typically less than 10% for CBZ [47]. Composite samples (2 volumes of 0.5 L every 30 min representative for one hour) were collected approximatively every two hours during an experiment run on 28 May 2018 at the inlet (point PSM_w = 0 m) and the outlet (point PSM_z = 420 m) of the channel reach (Figure 13) in low-flow conditions. Analytical determinations were performed following the method described in [48]. Mean longitudinal flow velocities were measured by means of an acoustic digital current meter (OTT Messtechnik GmbH, Kempten; Germany). Data for CBZ and mean longitudinal flow velocities are presented in Table 3.

Figure 13. Case study location, investigated channel reach, and monitored points.

Table 3. Carbamazepine concentrations at the inlet (PSMw) and outlet (PSMz) points of the vegetated channel on 28 May 2018, and simulated results.

Time	Inlet (PSM_w) (ng/L)	Outlet (PSM_z)(ng/L)	Flow Velocity (m/s)	Simulated Value ($PSM_{z\ sim}$) (ng/L)
07:20	123	-	0.025	-
09:50	181	-	0.026	-
11:00	-	105	-	112
12:20	162	-	0.026	-
13:10	-	112	-	116
14:20	162	-	0.021	-
15:40	-	115	-	119
16:50	150	-	0.029	-
18:00	-	125	-	122

A multi-pulse boundary condition was prepared by exploiting data from the five monitoring points. Figure 14 shows the simulated carbamazepine concentration at the outlet point, PSM_z, 420 m from the inlet point. The *Simulated value* (PSM_{z_sim}) column in Table 3 presents the simulated value against the measured ones (*Outlet* (PSM_z) *column*). In Figure 13 CBZ simulated concentrations are themed with color scales depending on the concentration value.

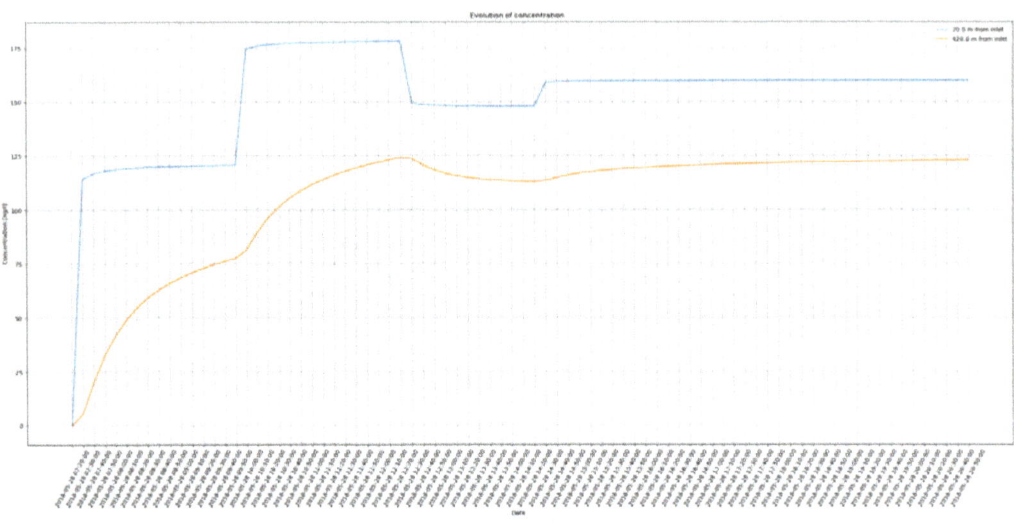

Figure 14. Simulated concentration at x = 20 m, and x = 420 m from the inlet point with 10 min time step length.

The first-order decay rate and the longitudinal dispersion coefficients are relevant parameters in our analyses, and, more in general, in surface water quality modelling [49,50]. No experimental data were available for these parameters. As such, the model was calibrated to get the best fit (R^2 = 0.95) between simulated and measured concentrations with values of the longitudinal dispersion coefficient of 35 m^2/s and decay rate equal to $3 \cdot 10^{-5}$ s^{-1}. Good fit (R^2 > 0,9) was also obtained varying these two parameters within the range of 30 and 35 m^2/s for longitudinal dispersivity and between $2.5 \cdot 10^{-5}$ and $3 \cdot 10^{-5}$ s^{-1} for the decay rate coefficient. The values of the longitudinal dispersion coefficient are coherent with values found in [49,50] for similar open channels. The calibrated values of the decay rate are slightly higher than calculated values from half-life time data for CBZ reported in [51].

4. Conclusions

The developed open-source and free plugin allows simulating transport of dissolved substances in water courses following advection/dispersion, and degradation processes. The present formulation combining a simple analytical solution of the advection dispersion equation and GIS tools guarantees intuitive spatial data management. Authorities may also benefit from the ease of use of such tools in order to set in place pollution prevention measures. This solution, because few parameters are needed, could hence be applied to data-scarce environments. Furthermore, using this tool values for longitudinal dispersivity and first-order decay rate coefficients may be derived. In our case study application, we estimated the longitudinal dispersion coefficient and the decay rate coefficient to be 35 m^2/s and $3 \cdot 10^{-5}$ s^{-1} for the pharmaceutical compound carbamazepine. Another potential application could be in the feasibility stage of the design of water-related green infrastructures for the improvement of water quality [46,52]. On the other hand, the increasing number of integrated geographical databases (including surface water bodies characteristics) along with the increasing availability of sensors gathering and distributing quasi real-time monitoring data (such as surface water heads) may allow for the combined use of monitoring and modelling to set up early warning systems to track pollution events [53–55]. To this aim, further research and pilot experimental sites are needed.

The open and free characteristics of the developed code guarantee reproducibility of the run analyses, and also use of the tool, one of the most important aspects of science, making free software an ideal framework for scientific work. In this context, the use of free software is consolidating in our societies in a gradual, but constant way [56]. Integration in the FREEWAT plugin [18] and inclusion within the list of the official plugin of QGIS will guarantee the dissemination and potentially the application of this research product.

In the present formulation, the ORGANICS tool does not allow users to simulate the transport of substances under conditions where flow may increase/decrease downstream not only by tributaries, but also by continuous groundwater drainage from the surrounding domain. Nor does it allow for the spatial or time variability of the degradation rate. The latter could be beneficial to differentiate, for example, the relative importance in time of biodegradation from photodegradation processes [50,57,58]. Future development may include integration of more complex analytical solutions, comprising also source terms.

The ORGANICS plugin is a first attempt to popularize the use of simple modelling tools. For more complex solutions and the inclusion of time-varying source/sink terms, a wide range of numerical tools exist [7,10]. We wish therefore to stress that the main element of this tool resides in its simplicity. Finally, the ORGANICS plugin provides GIS experts the ability to perform approximate, but fast simulations of the evolution of pollutants concentration in surface water bodies at selected targets.

Supplementary Materials: The following supporting information can be downloaded at: https://www.mdpi.com/article/10.3390/w14182850/s1. The Supplementary Material folder contains: folder organics: contains the plugin to be installed in QGIS v3.xx; folder example_problems: contains the files used to run the Example problem described in Section 3.2 of the paper.

Author Contributions: Conceptualization, R.R. and I.B.; methodology, R.R., I.B., L.S., L.E., A.C. and N.D.L.; software, I.B. and R.R.; validation, R.R., N.D.L. and L.S. and L.E.; formal analysis, R.R. and L.E.; resources, A.B. and R.R.; data curation, R.R. and I.B.; writing—original draft preparation, R.R. and I.B.; writing—review and editing, R.R., A.C., N.D.L., L.S. and L.E.; visualization, R.R. and I.B.; funding acquisition, A.B. and R.R. All authors have read and agreed to the published version of the manuscript.

Funding: This paper presents the ORGANICS plugin. The development of the ORGANICS plugin was financed through the research agreement between Distretto Idrografico Alpi Orientali and Istituto di Scienze della Vita—Scuola Superiore Sant'Anna: *CONVENZIONE DI RICERCA per lo sviluppo di un modello con metodi numerici integrati in software per la gestione dei dati spaziali (GIS) che descriva i meccanismi di trasporto, abbattimento e circolazione in acque superficiali di fitofarmaci e composti azotati* (2019–2021), co-funded within the GREVISLIN INTERREG Italia-Slovenia Project (Green infrastructures for the conservation and improvement of the condition of habitats and protected species along the rivers, CUP number: G76I18000080007) [https://www.ita-slo.eu/en/grevislin (accessed on 1 August 2022)].

Data Availability Statement: Data used in the section "Case study application" come from research run within the Italian-Israeli bilateral project PHARM-SWAP MED (co-funded by the Italian Ministero degli Affari Esteri e della Cooperazione Internazionale).

Acknowledgments: The authors wish to thank Sara Pasini and Matteo Bisaglia.

Conflicts of Interest: The authors declare no conflict of interest.

References

1. Walker, D.B.; Baumgartner, D.J.; Gerba, C.P.; Fitzsimmons, K. Chapter 16—Surface water pollution. In *Environmental and Pollution Science*, 3rd ed.; Mark, L.B., Ian, L.P., Charles, P.G., Eds.; Academic Press: Cambridge, MA, USA, 2019; pp. 261–292, ISBN 978-0-12-814719-1. [CrossRef]
2. Montuori, P.; De Rosa, E.; Di Duca, F.; Provvisiero, D.P.; Sarnacchiaro, P.; Nardone, A.; Triassi, M. Estimation of Polycyclic Aromatic Hydrocarbons Pollution in Mediterranean Sea from Volturno River, Southern Italy: Distribution, Risk Assessment and Loads. *Int. J. Environ. Res. Public Health* **2021**, *18*, 1383. [CrossRef] [PubMed]

3. Beermann, A.J.; Elbrecht, V.; Karnatz, S.; Ma, L.; Matthaei, C.D.; Piggott, J.J.; Leese, F. Multiple-stressor effects on stream macroinvertebrate communities: A mesocosm experiment manipulating salinity, fine sediment and flow velocity. *Sci. Total Environ.* **2018**, *610–611*, 961–971. [CrossRef] [PubMed]
4. Jackson, M.C.; Loewen, C.J.G.; Vinebrooke, R.D.; Chimimba, C.T. Net effects of multiple stressors in freshwater ecosystems: A meta-analysis. *Glob. Chang. Biol.* **2016**, *22*, 180–189. [CrossRef] [PubMed]
5. Bavumiragira, J.P.; Ge, J.; Yin, H. Fate and transport of pharmaceuticals in water systems: A processes review. *Sci. Total Environ.* **2022**, *823*, 153635. [CrossRef]
6. Pistocchi, C.; Silvestri, N.; Rossetto, R.; Sabbatini, T.; Guidi, M.; Baneschi, I.; Bonari, E.; Trevisan, D. A simple model to assess nitrogen and phosphorus contamination in ungauged surface drainage networks: Application to the Massaciuccoli Lake Catchment. Italy. *J. Environ. Qual.* **2012**, *41*, 544–553. [CrossRef]
7. Anderson, E.J.; Phanikumar, M.S. Surface storage dynamics in large rivers: Comparing three-dimensional particle transport, one-dimensional fractional derivative, and multirate transient storage models. *Water Resour. Res.* **2011**, *47*, W09511. [CrossRef]
8. De Wit, M.; Pebesma, E.J. Nutrient fluxes at the river basin scale. II: The balance between data availability and model complexity. *Hydrol. Processes* **2001**, *15*, 761–775. [CrossRef]
9. Runkel, R.L. *One-Dimensional Transport with Inflow and Storage (OTIS): A Solute Transport Model for Streams and Rivers*; USGS Water Resources Investigations Report 98-4018; U.S. Geological Survey: Denver, CO, USA, 1998. Available online: http://co.water.usgs.gov/otis/ (accessed on 1 July 2022).
10. Van Genuchten, M.T.; Leij, F.J.; Skaggs, T.H.; Toride, N.; Bradford, S.A.; Pontedeiro, E.M. Exact analytical solutions for contaminant transport in rivers 1. The equilibrium advection-dispersion equation. *J. Hydrol. Hydromech.* **2013**, *61*, 146–160. [CrossRef]
11. De Smedt, F.; Brevis, W.; Debels, P. Analytical solution for solute transport resulting from instantaneous injection in streams with transient storage. *J. Hydrol.* **2005**, *315*, 25–39. [CrossRef]
12. Šimůnek, J.; Van Genuchten, M.T.; Sejna, M.; Toride, N.; Leij, F.J. *The STANMOD Computer Software for Evaluating Solute Transport in Porous Media Using Analytical Solutions of Convection-Dispersion Equation*; Version 2.0, IGWMC-TPS-71; International Ground Water Modeling Center (IGWMC), Colorado School of Mines: Golden, CO, USA, 2000; p. 32. Available online: http://www.pc-progress.com/en/Default.aspx?stanmod (accessed on 1 July 2022).
13. Ince, D.C.; Hatton, L.; Graham-Cumming, J. The case for open computer programs. *Nature* **2012**, *482*, 485. [CrossRef]
14. Brovelli, M.A.; Minghini, M.; Moreno-Sanchez, R.; Oliveira, R. 2017. Free and open source software for geospatial applications (FOSS4G) to support Future Earth. *Int. J. Digit. Earth* **2017**, *10*, 386–404. [CrossRef]
15. *Communication to the Commission, Open Source Software Strategy 2020–2023*; Think Open: Brussels, Belgium, 2020.
16. Coetzee, S.; Ivánová, I.; Mitasova, H.; Brovelli, M.A. Open Geospatial Software and Data: A Review of the Current State and A Perspective into the Future. *ISPRS Int. J. Geo-Inf.* **2020**, *9*, 90. [CrossRef]
17. Kresic, N.; Mikszewski, A. *Hydrogeological Conceptual Site Models: Data Analysis and Visualization*; CRC Press: Boca Raton, FL, USA, 2012.
18. Rossetto, R.; De Filippis, G.; Triana, F.; Ghetta, M.; Borsi, I.; Schmid, W. Software tools for management of conjunctive use of surface-and ground-water in the rural environment: Integration of the Farm Process and the Crop Growth Module in the FREEWAT platform. *Agric. Water Manag.* **2019**, *223*, 105717. [CrossRef]
19. Foglia, L.; Borsi, I.; Mehl, S.; de Filippis, G.; Cannata, M.; Vasquez-Sune, E.; Criollo, R.; Rossetto, R. FREEWAT, a Free and Open Source, GIS-Integrated, Hydrological Modeling Platform. *Groundwater* **2018**, *56*, 521–523. [CrossRef] [PubMed]
20. Rosas-Chavoya, M.; Gallardo-Salazar, J.L.; López-Serrano, P.M.; Alcántara-Concepción, P.C.; León-Miranda, A.K. QGIS a constantly growing free and open-source geospatial software contributing to scientific development. *Cuad. Investig. Geográfica* **2022**, *48*, 197–213. [CrossRef]
21. Vieux, B.E. Geographic information systems and non-point source water quality and quantity modelling. *Hydrol. Processes* **1991**, *5*, 101–113. [CrossRef]
22. Olivera, F.; Valenzuela, M.; Srinivasan, R.; Choi, J.; Cho, H.; Koka, S.; Agrawal, A. ArcGIS-SWAT: A geodata model and GIS interface for SWAT. *J. Am. Water Resour. Assoc.* **2006**, *42*, 295–309. [CrossRef]
23. Neitsch, S.L.; Arnold, J.G.; Kiniry, J.R.; Williams, J.R.; King, K.W. *Soil and Water Assessment Tool (SWAT): Theoretical Documentation*; Version 2000; TWRI Report TR-191; TexasWater Resources Institute: College Station, TX, USA, 2002.
24. Becker, M.W.; Jiang, Z. Flux-based contaminant transport in a GIS environment. *J. Hydrol.* **2007**, *343*, 203–210. [CrossRef]
25. Akbar, T.A.; Lin, H.; DeGroote, J. Development and evaluation of GIS-based ArcPRZM-3 system for spatial modeling of groundwater vulnerability to pesticide contamination. *Comput. Geosci.* **2011**, *37*, 822–830. [CrossRef]
26. Rossetto, R.; Borsi, I.; Schifani, C.; Bonari, E.; Mogorovich, P.; Primicerio, M. SID&GRID: Integrating hydrological modeling in GIS environment. *Rend. Online Soc. Geol. Ital.* **2013**, *24*, 282–283.
27. Anguix, A.; Díaz, L. gvSIG: A GIS desktop solution for an open SDI. *J. Geogr. Reg. Plan.* **2008**, *1*, 41–48.
28. Oliveira, M.M.; Martins, T.N. A methodology for the preliminary characterization of the river boundary condition in finite difference groundwater flow numerical models. *Acque Sotter. Ital. J. Groundw.* **2019**, *8*, 21–27. [CrossRef]
29. Harbaugh, A.W. *MODFLOW-2005, the U.S. Geological Survey Modular Groundwater Model—The Ground-water Flow Process*; U.S. Geological Survey, Techniques and Methods 6–A16; U.S. Geological Survey: Reston, VA, USA, 2005.
30. Bittner, D.; Rychlik, A.; Klöffel, T.; Leuteritz, A.; Disse, M.; Chiogna, G. A GIS-based model for simulating the hydrological effects of land use changes on karst systems—The integration of the LuKARS model into FREEWAT. *Environ. Model. Softw.* **2020**, *127*, 104682. [CrossRef]

31. QGIS A Free and Open Source Geographic Information System. Open Source. Geospatial Foundation Project. 2009. Available online: http://qgis.osgeo.org (accessed on 1 July 2022).
32. Nielsen, A.; Bolding, K.; Hu, F.; Trolle, D. An open source QGIS-based workflow for model application and experimentation with aquatic ecosystems. *Environ. Model. Softw.* **2017**, *95*, 358–364. [CrossRef]
33. Ellsäßer, F.; Röll, A.; Stiegler, C.; Hendrayanto Hölscher, D. Introducing QWaterModel, a QGIS plugin for predicting evapotranspiration from land surface temperatures. *Environ. Model. Softw.* **2020**, *130*, 104739. [CrossRef]
34. Dile, Y.T.; Daggupati, P.; George, C.; Srinivasan, R.; Arnold, J. Introducing a new open source GIS user interface for the SWAT model. *Environ. Model. Softw.* **2016**, *85*, 129–138. [CrossRef]
35. Criollo, R.; Velasco, V.; Nardi, A.; De Vries, L.M.; Riera, C.; Scheiber, L.; Jurado, A.; Brouyère, S.; Pujades, E.; Rossetto, R.; et al. AkvaGIS: An open source tool for water quantity and quality management. *Comput. Geosci.* **2019**, *127*, 123–132. [CrossRef]
36. De Filippis, G.; Ercoli, L.; Rossetto, R. A Spatially Distributed, Physically-Based Modeling Approach for Estimating Agricultural Nitrate Leaching to Groundwater. *Hydrology* **2021**, *8*, 8. [CrossRef]
37. Benedini, M.; Tsakiris, G. Water quality modelling for rivers and streams. In *Water Science and Technology Library*; Springer: Berlin/Heidelberg, Germany, 2013; Volume 70.
38. Python Software Foundation. Python 3 Release. Available online: https://www.python.org/download/releases/3.0/ (accessed on 1 July 2022).
39. Qt5 Library. One framework. One codebase. Any platform. Available online: https://www.qt.io/ (accessed on 1 July 2022).
40. QGIS. QGIS Python API. Available online: https://qgis.org/pyqgis/3.10/ (accessed on 1 July 2022).
41. Pullar, D.; Springer, D. Towards integrating GIS and catchment models. *Environ. Model. Softw.* **2000**, *15*, 451–459. [CrossRef]
42. Wang, L.; Jackson, C.R.; Pachocka, M.; Kingdon, A. A seamlessly coupled GIS and distributed groundwater flow model. *Environ. Model. Softw.* **2016**, *82*, 1–6. [CrossRef]
43. Toride, N.; Leij, F.J.; van Genuchten, M.T. *The CXTFIT Code for Estimating Transport Parameters from Laboratory or Field Tracer Experiments*; Version 2.0. Research Report No. 137; U.S. Salinity Laboratory, USDA, ARS: Riverside, CA, USA, 1995.
44. Van Genuchten, M.T.; Alves, W.J. *Analytical Solutions of the One-Dimensional Convective Dispersive Solute Transport Equations*; Technical Bulletin No. 1661; U.S. Department of Agriculture: Washington, DC, USA, 1982.
45. Tixier, C.; Singer, H. Occurrence and fate of carbamazepine, clofibric acid, diclofenac, ibuprofen, ketoprofen, and naproxen in surface waters. *Environ. Sci. Technol.* **2003**, *37*, 1061–1068. [CrossRef] [PubMed]
46. Barbagli, A.; Jensen, B.N.; Raza, M.; Schueth, C.; Rossetto, R. Assessment of soil buffer capacity on nutrients and pharmaceuticals in nature-based solution applications. *Environ. Sci. Pollut. Res.* **2019**, *26*, 759–774. [CrossRef] [PubMed]
47. Zhang, Y.; Geißen, S.U.; Gal, C. Carbamazepine and diclofenac: Removal in wastewater treatment plants and occurrence in water bodies. *Chemosphere* **2008**, *73*, 1151–1161. [CrossRef]
48. Cardini, A.; Pellegrino, E.; Ercoli, L. Predicted and Measured Concentration of Pharmaceuticals in Surface Water of Areas with Increasing Anthropic Pressure: A Case Study in the Coastal Area of Central Italy. *Water* **2021**, *13*, 2807. [CrossRef]
49. Julínek, T.; Říha, J. Longitudinal dispersion in an open channel determined from a tracer study. *Environ. Earth Sci.* **2017**, *76*, 592. [CrossRef]
50. Liu, Y.; Zarfl, C.; Basu, N.B.; Cirpka, O.A. Modeling the fate of pharmaceuticals in a fourth-order river under competing assumptions of transient storage. *Water Resour. Res.* **2020**, *56*, e2019WR026100. [CrossRef]
51. Acuña, V.; von Schiller, D.; García-Galán, M.J.; Rodríguez-Mozaz, S.; Corominas, L.; Petrovic, M.; Poch, M.; Barceló, D.; Sabater, S. Occurrence and in-stream attenuation of wastewater-derived pharmaceuticals in Iberian rivers. *Sci. Total Environ.* **2015**, *503–504*, 133–141. [CrossRef]
52. Piacentini, S.M.; Rossetto, R. Attitude and Actual Behaviour towards Water-Related Green Infrastructures and Sustainable Drainage Systems in Four North-Western Mediterranean Regions of Italy and France. *Water* **2020**, *12*, 1474. [CrossRef]
53. Rossetto, R.; Barbagli, A.; Borsi, I.; Mazzanti, G.; Vienken, T.; Bonari, E. Site investigation and design of the monitoring system at the Sant'Alessio Induced RiverBank Filtration plant (Lucca, Italy). *Rend. Online Soc. Geol. Ital.* **2015**, *35*, 248–251.
54. Luo, L.; Lan, J.; Wang, Y.; Li, H.; Wu, Z.; McBridge, C.; Zhou, H.; Liu, F.; Zhang, R.; Gong, F.; et al. A Novel Early Warning System (EWS) for Water Quality, Integrating a High-Frequency Monitoring Database with Efficient Data Quality Control Technology at a Large and Deep Lake (Lake Qiandao), China. *Water* **2022**, *14*, 602. [CrossRef]
55. Liu, J.; Wang, P.; Jiang, D.; Nan, J.; Zhu, W. An integrated data-driven framework for surface water quality anomaly detection and early warning. *J. Clean. Prod.* **2020**, *251*, 119145. [CrossRef]
56. Robles, G.; Steinmacher, I.; Adams, P.; Treude, C. Twenty Years of Open Source Software: From Skepticism to Mainstream. *IEEE Softw.* **2019**, *36*, 12–15. [CrossRef]
57. Hanamoto, S.; Nakada, N.; Yamashita, N.; Tanaka, H. Modeling the photochemical attenuation of down-the-drain chemicals during river transport by stochastic methods and field measurements of pharmaceuticals and personal care products. *Environ. Sci. Technol.* **2013**, *47*, 13571–13577. [CrossRef]
58. Guillet, G.; Knapp, J.L.A.; Merel, S.; Cirpka, O.A.; Grathwohl, P.; Zwiener, C.; Schwientek, M. Fate of wastewater contaminants in rivers: Using conservative-tracer based transfer functions to assess reactive transport. *Sci. Total Environ.* **2019**, *656*, 1250–1260. [CrossRef] [PubMed]

Article

Simulation of Heat Flow in a Synthetic Watershed: Lags and Dampening across Multiple Pathways under a Climate-Forcing Scenario

Daniel T. Feinstein [1,*], Randall J. Hunt [2] and Eric D. Morway [3]

[1] U.S. Geological Survey Upper Midwest Water Science Center, 3209 North Maryland Avenue, Milwaukee, WI 53211, USA

[2] U.S. Geological Survey Upper Midwest Water Science Center, 1 Gifford Pinchot Drive, Madison, WI 53726, USA

[3] U.S. Geological Survey Nevada Water Science Center, 2730 N. Deer Run Road, Carson City, NV 89701, USA

* Correspondence: dtfeinst@usgs.gov

Abstract: Although there is widespread agreement that future climates tend toward warming, the response of aquatic ecosystems to that warming is not well understood. This work, a continuation of companion research, explores the role of distinct watershed pathways in lagging and dampening climate-change signals. It subjects a synthetic flow and transport model to a 30-year warming signal based on climate projections, quantifying the heat breakthrough on a monthly time step along connected pathways. The system corresponds to a temperate watershed roughly 27 km on a side and consists of (a) land-surface processes of overland flow, (b) infiltration through an unsaturated zone (UZ) above an unconfined sandy aquifer overlying impermeable bedrock, and (c) groundwater flow along shallow and deep pathlines that converge as discharge to a surface-water network. Numerical simulations show that about 40% of the warming applied to watershed infiltration arrives at the water table and that the UZ stores a large fraction of the upward-trending heat signal. Additionally, once groundwater reaches the surface-water network after traveling through the saturated zone, only about 10% of the original warm-up signal is returned to streams by discharge. However, increases in the simulated streamflow temperatures are of similar magnitude to increases at the water table, due to the addition of heat by storm runoff, which bypasses UZ and groundwater storage and counteracts subsurface dampening. The synthetic modeling method and tentative findings reported here provide a potential workflow for real-world applications of climate-change modeling at the full watershed scale.

Keywords: heat transport; watershed modeling; temperature; climate change

Citation: Feinstein, D.T.; Hunt, R.J.; Morway, E.D. Simulation of Heat Flow in a Synthetic Watershed: Lags and Dampening across Multiple Pathways under a Climate-Forcing Scenario. *Water* **2022**, *14*, 2810. https://doi.org/10.3390/w14182810

Academic Editor: Cristina Di Salvo

Received: 23 June 2022
Accepted: 29 August 2022
Published: 9 September 2022

1. Introduction and Objectives

As the climate warms, researchers are increasingly focused on characterizing the effects of atmospheric change on different parts of the natural environment, including surface and subsurface pathways within a watershed (Figure 1). The warming of a groundwater/surface-water system is conditioned by two primary factors. First, the top of the system is separated from the warming atmosphere by an unsaturated zone (UZ). The UZ transmits and stores water and heat as they move downward to the water table [1,2]. It acts as a low-pass filter on water and heat impulses integrated over time by lagging and dampening the thermal load after it leaves the bottom of the root zone. The influence of this filtering is influenced by the thickness of the UZ [3]. Second, in temperate regions, younger groundwater stored near the top of the saturated zone can have different temperatures than older groundwater from deeper parts of the aquifer. As these flow paths converge near stream, lake, and wetland discharge locations [4], the total amount of heat transmitted back to the surface-water system is determined by the combined effect of all the water pathways. Such factors influence the timing, magnitude, and distribution of the thermal energy that eventually discharges to the surface-water system through distinct watershed pathways. Transient effects can be difficult to evaluate using qualitative analyses of downstream receptors but are crucially important for realistically forecasting system response to future temperature increases. Such increases—even when modest—can form tipping points that transform surface-water ecology and habitats [5] and change subsurface nutrient cycling [6].

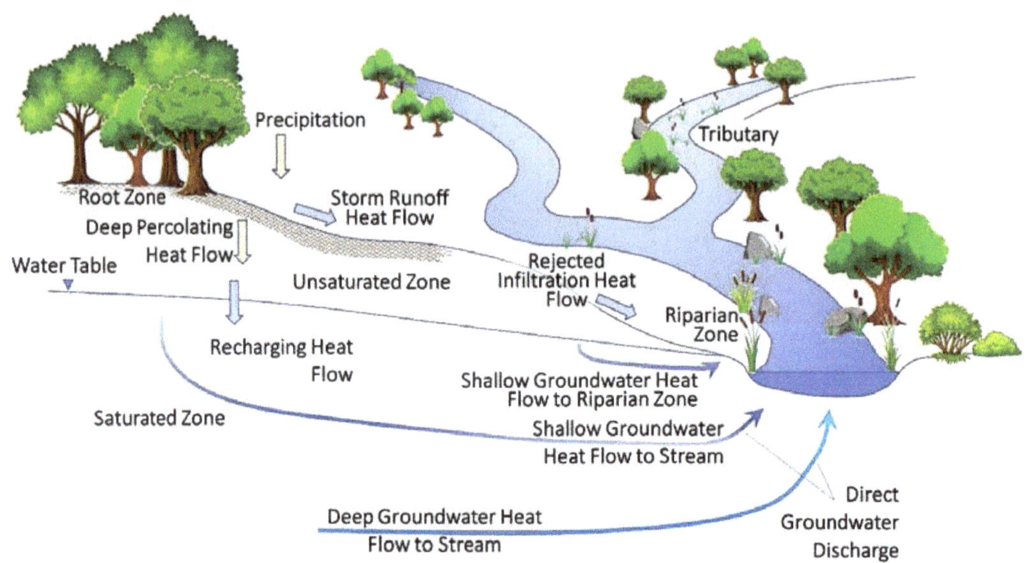

Figure 1. Schematic of heat-flow pathways through watershed.

Because downstream aquatic habitats typically integrate influences from upstream, understanding the potential effects of climate change on water resources requires studies and simulations at a watershed scale. The conditions that influence watershed heat transport pathways include the variable thickness of the UZ, variation of groundwater residence time with depth, and topographic/geographic factors such as the width of riparian areas and stream density of the surface-water system. In addition, quantitative studies must also account for different flow and transport processes along these system pathways, for example, the propagation of heat by convection, conduction, and dispersion.

The translation of atmospheric warming to aquatic resources has been a focus of previous work, including analytic [7], process-based watershed [8–10], regression-based [11–13], remote sensing [14], and field measurement [15] approaches. Our work extends this rich history through transient quantitative numerical simulation of salient watershed pathways that store and transport heat. Specifically, this study leverages recent advances in quantitative transient numerical methods detailed in Morway et al. [1] and builds upon initial testing of the synthetic watershed presented in Morway et al. [2]. Here the previous work is extended in three important ways:

1. Heat transport at monthly intervals is explicitly tracked at the watershed scale (1) between the top of the UZ and the water table, (2) between the water table and groundwater discharge zones, and (3) from upstream to downstream in the surface-water network fed by groundwater. The quantification of the heat-flow across various boundaries within a watershed, for example, the water table, enables a more detailed evaluation of the thermal response of a watershed to a changing climate, and in particular to warming infiltration. This type of analysis will be increasingly important as the thermal impacts of a changing climate affect, for example, cold-water fisheries.

2. Whereas Morway et al. [2] employed a heuristic synthetic heat inflow time series to isolate distinct warming effects, this study uses a climate forcing function based on the "mean model" Global Climate Model (GCM) high-emission scenarios [16]. The forcing combines the effects of, first, monthly temperature changes and, second, monthly changes in precipitation that are carried into changes in the infiltration rate through the root zone to the top of the UZ. Application of a specific climate scenario permits a more realistic lag [17] and dampening assessment, which in turn facilitates an extension of these methods to real-world decision support settings.
3. In most watersheds, stream flow is a combination of surface (e.g., overland runoff) and subsurface flows. The transfer of heat to the stream network is therefore dependent on both pathways. Variation and extremes observed in stream temperatures tend to be much greater than what is observed in ambient groundwater temperatures-even in baseflow-dominated systems [18,19]. The difference is attributable in large measure to the influence of quick-flowpath additions of storm runoff during warm, wet months that overprint slower/steadier rates of groundwater thermal discharge. Therefore, this analysis expands upon Morway et al. [2] by simulating and analyzing the heat load returned to the surface-water network from storm runoff in addition to the heat load returned by the groundwater system.

The simulations under study use the groundwater flow model MODFLOW-NWT [20] and a recently augmented version of the companion transport code MT3D-USGS [1,21]. Explicit simulation of heat transport through the UZ makes this work distinct from previous watershed-scale efforts [22–24]. The model output includes the two dependent variables head and temperature, as well as volumetric water and heat fluxes, all of which have utility for watershed-scale assessments.

The methods, results and discussion presented in this article are accompanied by information in a Supplementary Material Section [25]. It consists of three appendices giving additional detail (including Supplementary Figures and Tables) on subjects referenced below.

2. Methods

Heat flow travels through the watershed via linked thermal pathways (Figure 1). For example, subsurface heat loading often begins with heat inflows that originate as infiltration below the bottom of the root zone. The infiltrating heat next moves downward through the UZ to the water table and, upon recharging the aquifer, begins migrating toward a discharge location via shallow and deep groundwater flow paths. Heat associated with precipitation that is unable to infiltrate the subsurface (when the water table is at/above the land surface, or the precipitation rate is faster than the soil's ability to infiltrate) flows more quickly to surface-water features. In this effort, we simulate and analyze heat flow pathways in the synthetic model to illustrate the occurrence and magnitude of lags (changes in phase) and dampening (change in amplitude) of atmospheric warming applied at the top of the UZ as it travels through the watershed and associated surface water system.

The pathways shown in Figure 1 correspond to the following flow and storage terms simulated by the model:

- Groundwater runoff is defined as the sum of groundwater discharge to land surface and rejected infiltration from the land surface under conditions of Hortonian or Dunnian flow.
- Baseflow is defined as the sum of direct groundwater discharge to surface water plus groundwater runoff.
- Total streamflow is defined as the sum of baseflow and storm runoff.
- From a watershed flow system perspective, the following partitioning occurs:
- Precipitation is partitioned into infiltration, storm runoff and evapotranspiration (ET) from the root zone.
- Infiltration is partitioned into rejected infiltration, recharge and storage changes in the UZ (no ET is simulated from the UZ since the infiltration is considered to be what percolates below the root zone).

- Recharge is partitioned into storage changes in the groundwater system, direct discharge to surface water and discharge to land surface.

Note that some streamflow generation conceptualizations include an "interflow" subsurface pathway reflecting late-storm seepage that contributes to a recessional limb of a storm hydrograph [26]. Here, however, we focus on more time-integrated results from the watershed (monthly to multi-decadal) and consider such interflow contributions as included in the other subsurface components to streamflow. Moreover, although rejected infiltration is a form of runoff, for purposes of this analysis it is included in the baseflow term, leaving stormflow as the only remaining surface runoff component of total streamflow (because direct precipitation on the stream surface is not simulated).

Figure 1 illustrates this framework by showing the pathways from precipitation and infiltration through the various forms of runoff, recharge and discharge for both water and heat.

2.1. Construction of Spin-Up and Climate Change Forcing Function

The amount of heat that enters the top of the UZ is the product of the infiltration rate and its temperature. A brief description of how each time series was generated is offered below. A more detailed description is provided in the Supplementary Material Section S1.

Because the effects of a warming climate cannot be represented by steady state conditions, careful selection of the time discretization and initial conditions used within a transient model are important. Because the time-integrated effects of warming infiltration associated with climate change over a 30-year period of analysis was the focus of this study, monthly timesteps were deemed sufficient to represent the seasonal, random, and non-stationary aspects of the warming infiltration on temperatures in the subsurface. To establish initial condition by the start of the 30-year warming period, a 30 year spin-up period with annually-cyclic infiltration rates and temperatures (i.e., the same values were specified for all 30 Januarys, for example) was employed to ensure a dynamic equilibrium by the start of the warming period (see page 313 in Anderson et al. [27]). After spin-up, the infiltration rates and temperatures continue to vary monthly with a seasonal periodicity, but also have an underlying warming trend and a random noise component. As described in detail in the Supplementary Material Section S1, initial conditions used here differ from those used in the companion study [2]. In this work, both the monthly infiltration and its specified temperature time series varied during spin-up. In the companion paper, temperature varied monthly while the infiltration rate was held constant during its spin-up period at 0.2 m/yr. For this work, the average monthly infiltration rates and temperatures used during spin-up are based on a watershed located in southern Wisconsin, USA [19].

After spin-up, infiltration rates and its accompanying temperature are based on the high-emissions RCP-8.5 climate scenario [16] results for the Midwest United States, which generally reflect wetter and warmer conditions compared to the spin-up period. Downscaled regional results from southern Wisconsin, USA (Figure 2) were applied to the synthetic watershed, where simulated warming corresponds to the period 2022 through 2051. The warming trend applied to the infiltration temperature is imposed on the seasonal signal, which also includes random noise generated from a uniform distribution centered on 0 °C and a range of 4 °C. By the end of the 30-year warming period, the average annual temperature of the infiltration is approximately 2 °C warmer relative to the end of the spin-up period. The amount of heat inflow at the end of the warming period, which is the product of the infiltration rate and its temperature, is roughly 25% higher than the amount of heat inflow at the end of the spin-up period. For our simulations, this value effectively represents an upper limit of the expected climate change in terms of the heat added to the subsurface attributable to wetter and warmer conditions. Most of the increase in heat inflow is due to the trend applied to the infiltration temperature; only a small part is due to the trend applied to the infiltration rate (Supplementary Material Section S1). Because flow and heat transport are simulated separately, spin-up infiltration rates are specified in the flow model via the UZF1 package [28] while the temperatures assigned to the infiltration are specified in the UZT package of MT3D-USGS [21].

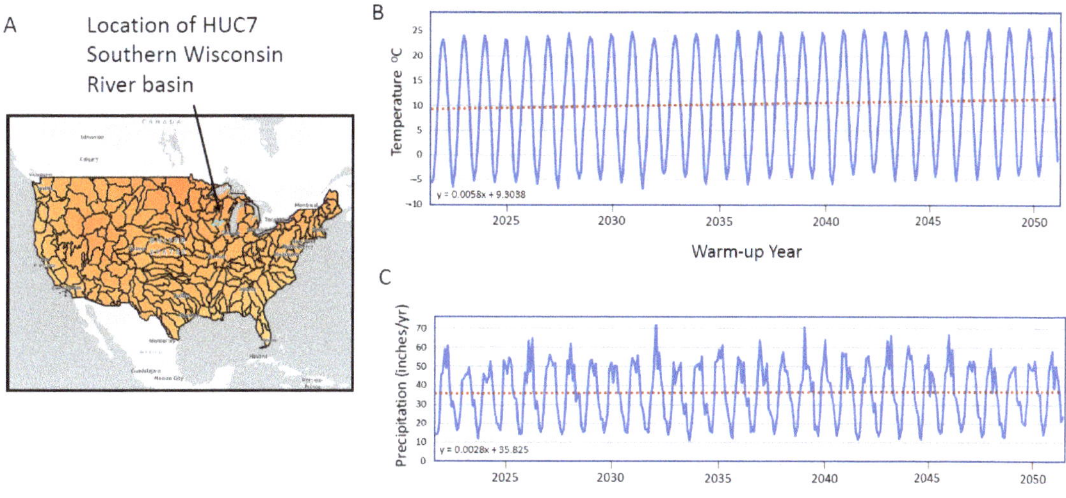

Figure 2. Climate forcing used in the synthetic model from high-emissions RCP8.5 scenario [16] downscaled to southern Wisconsin. (**A**) Location of southern Wisconsin basin [29] to which [16] data correspond, (**B**) the atmospheric temperature used to set the temperature of the infiltration (°C), and (**C**) precipitation and infiltration rate forcing (inch/year) where infiltration through the root zone is assumed to be one quarter the precipitation. Equations on plots correspond to dashed linear trend lines. RCP stands for Representative Concentration Pathway.

2.2. Model Construction

The synthetic model uses the same geometry, parameter values, and boundary conditions as described in Morway et al. [2]. Additional detail of the model construction is provided in Supplementary Material Section S2. In brief, the salient aspects of the model design are characterized by:

1. spin-up specification of temporally varying infiltration rates and temperatures that, when multiplied, result in a single time-dependent heat infiltration rate that represents a warming climate signal;
2. the climate forcing described in (1) is applied in a spatially uniform manner to the entire model domain; that is, infiltration rates and temperatures are temporally variable but spatially uniform;
3. aquifer/flow and transport parameters [e.g., the hydraulic conductivity (flow) and porosity (transport)], are spatially uniform across the model domain.

The model approximates a mid-sized watershed (about 290 square miles or 750 square km, falling into the HUC10-size category according to the U.S. watershed scheme) that includes streams, wetlands, and a lake (Figure 3). The surface-water system is strongly gaining ("baseflow dominated")-there is very little loss from streams to the aquifer. The subsurface system consists of a sandy aquifer separated from the land surface by an UZ and overlying effectively impermeable bedrock. No-flow boundaries are specified along the east, west, and bottom of the model. Regional groundwater flow gradients from north to south are generated by general head boundaries along the northern and southern model boundaries, but the flow system is strongly influenced by local groundwater divides which reflect the effects of topography and the surface-water network. Cells in layer 1 are typically unsaturated but may contain the water table in riparian zones adjacent to surface-water features. Cells in layers 2 through 4 can be either unsaturated, partially, or fully saturated. Layers 5 through 8 are fully saturated for the duration of the simulation. Parameter values for the flow and transport simulations are listed in the Supplementary Material Section S1. Monthly stress periods are used in both the flow and transport simulations.

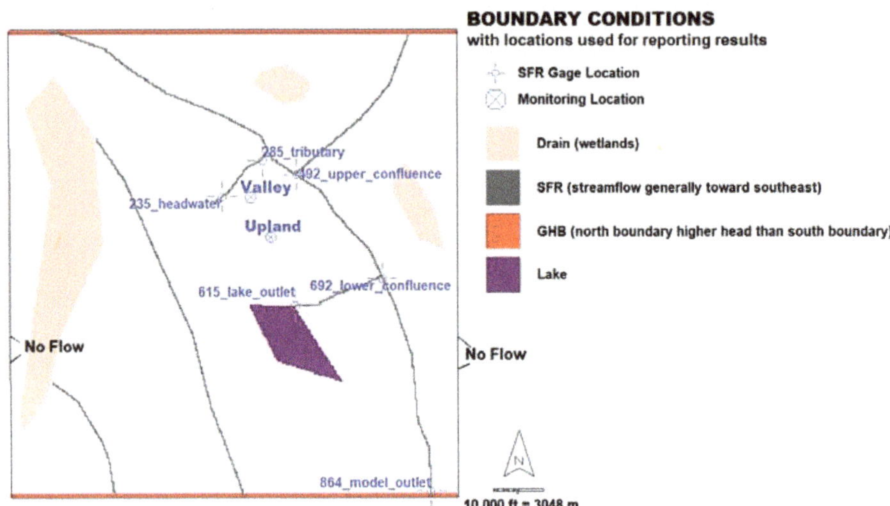

Figure 3. Synthetic model setup showing domain and boundary conditions, as well as locations for monitoring temperature results. Cross-sections are shown in Figure 4. [SFR: Stream Flow Routing; GHB: General Head Boundary].

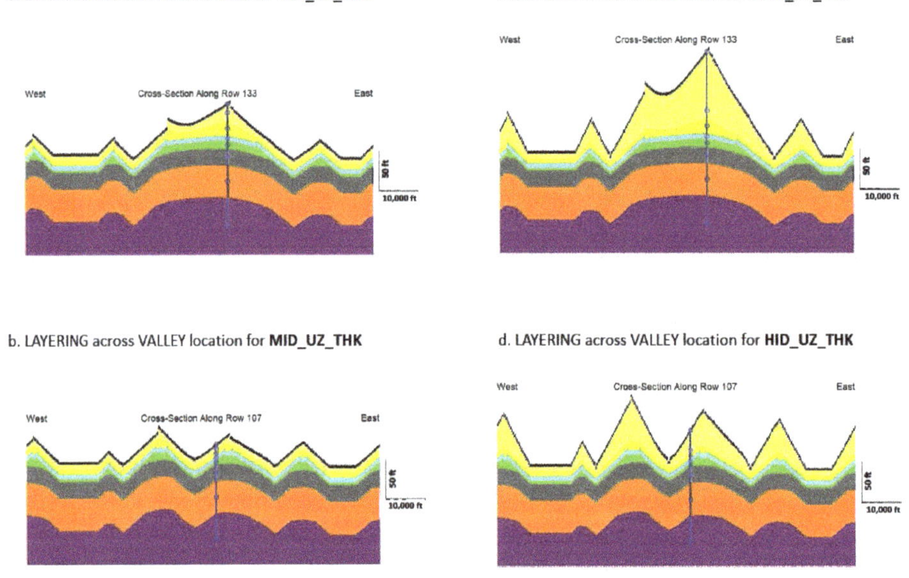

Figure 4. Layering through Upland and Valley locations; (**a**) MID_TRENDED model version, Upland cross section. (**b**) MID_TRENDED model version, Valley cross section. (**c**) HI_TRENDED model version, Upland cross section. (**d**) HI_TRENDED model version, Valley cross section. Cross section locations are shown in Figure 3. Top black layer is 3-ft thick receptor layer for receiving infiltration.

An important simplification incorporated in the proposed methodology is to equate the heat signal with infiltration that has *already* passed the root zone. Root zone processes in humid areas, which bear on both the movement of water and heat, include evaporation, transpiration, and conduction, leading to flow that can be both downward and upward. The key assumption in our methodology is that at a monthly transport time step, these root zone processes can be neglected and the warming signal at the top of the UZ can be equated with the average amount of water passing the root zone over the month and with the average monthly atmospheric temperature. This assumption is discussed in detail in Morway et al. [1].

To elucidate the importance of including the UZ in regional-scale heat transport simulations, two versions of the model were constructed for highlighting the effects of UZ thickness on heat transport:

1. MID_TRENDED model (Figure 4a,b): This version is designed to produce, on average, a moderate water table depth (i.e., moderate UZ thickness) that varies from 0 m in riparian areas (approximately 20% of the model domain) to about 15 m below land surface in the upland areas.
2. HI_TRENDED model (Figure 4c,d): using steeper topography, this version simulates a thicker UZ compared to the MID_TRENDED model. The water table depth varies from 0 m in riparian areas (approximately 4% of the model domain) to more than 30 m in the upland areas.

Thus, the main difference between the MID_ and HI_TRENDED models is the thickness of layers 1 through 4; the deeper groundwater system represented by layers 5 through 8 is the same in both versions. More information on model construction and model versions, including specification of model flow and transport parameters and selection of parameter values, is provided in Morway et al. [2] and Supplementary Material Section S2.

One of the key differences in the model setup in this analysis compared to that documented in Morway et al. [2] is that surface-water runoff is here explicitly simulated using options available in the UZF1 and SFR [30] packages (Supplementary Material Section S2). Because overland runoff is passed to MT3D-USGS via the linker file [31], it automatically accounts for the heat transported to streams and associated with runoff. Precipitation is assumed partitioned into storm runoff, infiltration, and evapotranspiration. In the synthetic model, storm runoff was set equal to 8.3% of the specified monthly precipitation rate, which is equivalent to 33% of the infiltration rate since the infiltration rate is set to 25% of the precipitation rate. The remaining precipitation is taken up by evapotranspiration rate, equal to 67% of the precipitation. These ratios are intended to represent a porous/sandy watershed where infiltration through the root zone is several times greater than storm runoff. Figure 5 shows the flow budget fluxes for the most upgradient eastern stream subbasin. Whereas the companion analysis [2] focused on understanding the impacts of warming infiltration on baseflow temperatures, this study considers the effects of all return flows on heat transport in the surface-water network, including runoff from the land surface associated with stormflow. The model does not, however, simulate precipitation or evaporation directly on or from the surface water, respectively.

The 75% of the total water and heat flux that enters the watershed over any year (net of evapotranspiration) as infiltration is the source of recharge to the water table. The recharge flux is divided among the following down-basin pathways: groundwater discharge to the stream channels and water bodies, groundwater discharge to land surface (that is, to riparian areas bordering surface water), and rejected infiltration from riparian areas. The portion of these three down-basin terms that terminate as water and heat flux to streams collectively sum to stream baseflow. Stormflow runoff contributes the remaining 25% of the water and heat that enters the watershed (net of evapotranspiration). It runs off instantaneously to the surface-water feature that is directly downslope in the form of a stream segment or lake (Supplementary Material Section S1). The infiltrating heat across the model domain is therefore subject to the low-pass filtering effects (phase and amplitude shifts) of heat transport through unsaturated and saturated flow pathways. Heat contributed to streams from stormflow runoff is not filtered by the subsurface pathways and is therefore left un-modified (i.e., no lags or dampening is associated with this particular heat transport pathway).

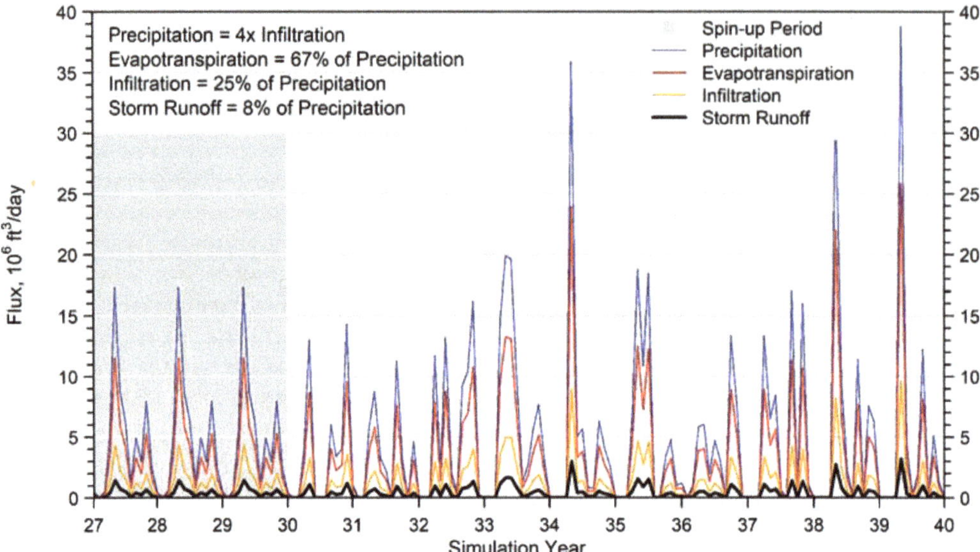

Figure 5. Flux terms for an example basin showing assumed relationships of precipitation, evapotranspiration, and storm runoff with infiltration rate at the end of the spin-up and warming periods. The absolute fluxes correspond to amounts for basin contributing runoff to an example stream segment (see Supplementary Material Figure S2-3a for the location of the basin comprising Segment 1 in the synthetic model). Warming begins in Simulation Year 30.

3. Results and Discussion

The process-based modeling approach used here produces time series of simulated temperatures throughout the model domain in response to climate forcing, including above the water table (i.e., the UZ), at the water table, in the deep groundwater system, and at various locations in the surface water network. Here, for both versions of the synthetic model, we focus on: (1) the temperature trends along the pathways shown in Figure 1, (2) the distribution, magnitude, and timing of heat transfers (fluxes and flows) within subbasins of the watershed, and (3) the lag and dampening effect of the UZ on the infiltrating heat signal at particular locations and across subbasins within the watershed as well as the lag and dampening effects induced by down-system pathways.

3.1. Temperature Trends along Pathways

Although the annual average temperature during the spin-up period is 8.55 °C, the flow-weighted (or infiltration-weighted) average temperature during spin-up is 9.97 °C. At the end of the spin-up period, the simulated stream and lake temperatures converge to about 10 °C for both the simulation with thinner and with thicker UZ thickness (the MID_ and HI_TRENDED simulations, respectively). After reaching dynamic equilibrium conditions by the end of the spin-up period, temperatures assigned to both the infiltration and storm runoff followed the same time-series scheme described above (see section titled "Construction of spin-up and climate forcing function") in the warming period. The average infiltration-weighted temperature over the 30-year warming period is 10.95 °C. Our analysis focuses on the final 10 years of warming, where the infiltration-weighted average temperature was 12.23 °C, a 2.26 °C rise compared to the last year of the spin-up period. If no lag or dampening occurred, the temperatures throughout the watershed would reflect this higher 12.23 °C temperature.

Warming in the subsurface was observed in the water table at two locations hereafter referred to as the Upland and Valley locations (Figure 3). The thinner UZ associated with the MID_TRENDED model contributes to a greater thermal response at the water table by the end of the simulation compared to the HI_TRENDED result (Figure 6A,B). In addition, a thinner UZ contributes to a flashier thermal response at the water table where the UZ thickness is even smaller (<15 ft) at the Valley location (Figure 6A). At the Upland location, the water-table temperature is smooth and muted in both models with a subtle temperature increase simulated at the water table for the first 22 years of the warming period. During the final 8 years of the simulation period, the temperature response at the water table to the overall warming trend is better defined with a clear rise in water table temperatures in year 52 of the simulation. Before that, limited sensitivity to year-by-year fluctuations in the temperature of the infiltration is shown in the MID_TRENDED simulation and no sensitivity is evident in the HI_TRENDED simulation results (Figure 6B). At the Valley location, where the water-table depth is approximately 3.3 m (11 ft), simulated temperatures in the MID_TRENDED simulation exhibit a much more responsive behavior in the last 10 years of the simulation compared to the HI_TRENDED simulation where the water-table depth averages 9.6 m (31 ft).

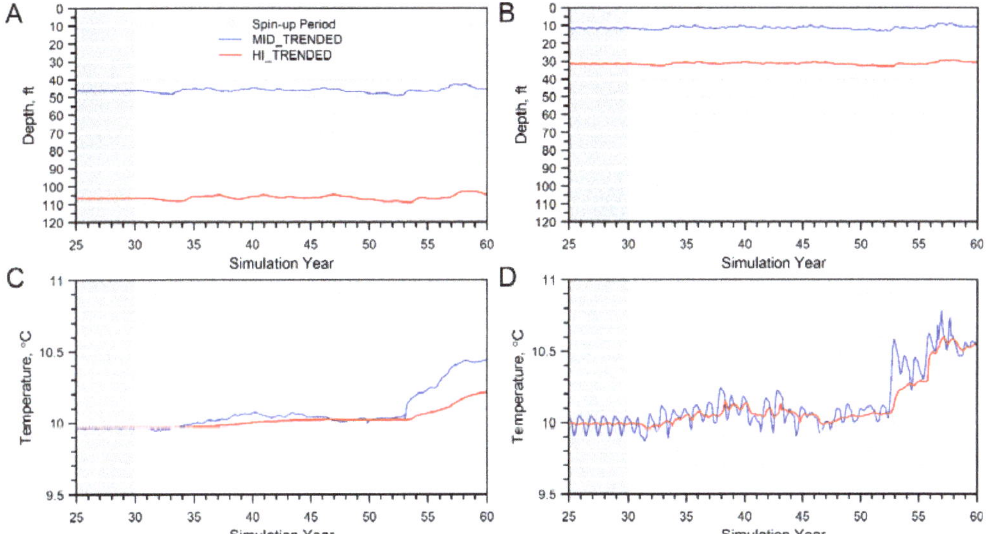

Figure 6. Time-series results for Synthetic Model at Upland and Valley Locations: (**A**,**B**) Depth to water table for MID_TRENDED and HI_TRENDED simulations; (**C**,**D**) Water table temperature for MID_TRENDED and HI_TRENDED simulations. Upland and Valley locations are shown in Figure 3. Warming period begins in Simulation Year 30.

Figure 7 shows the percent of the model domain with a water-table temperature at a given threshold (y-axis) over the warming period (x-axis). Red indicates the temperature corresponding to the warmest 20% of the domain, blue correspond to the coolest 20% of the domain. For example, at the beginning of the warming period in the MID_TRENDED simulation (year 0 on the x-axis; Figure 7A), the water table temperature across entire model domain is roughly 10 °C, but by the end of the warming period the water table temperature is at or below 10.5 °C. Note also that the MID_TRENDED simulation (Figure 7A) because of its thinner UZ consistently shows flashier water-table temperature responses for the warmest (reds) and coolest (blues) parts of the watershed compared to the HI_TRENDED simulation (Figure 7B). Direct comparison of the median watershed temperatures through time [represented by the dotted (MID_TRENDED) and solid (HI_TRENDED) contour lines in Figure 7B] indicates that a thicker UZ yields, on average, a more subdued water-table temperature response to warming infiltration, highlighting the ability of a thicker UZ to store and filter heat transport to the water table. As a reminder, the annual average temperature of the infiltration warmed by 2 °C during the 30-year warming period. In response, the shallow groundwater temperatures in the MID_ and HI_TRENDED simulations rose by more than 1 °C across about 20% of the model domain.

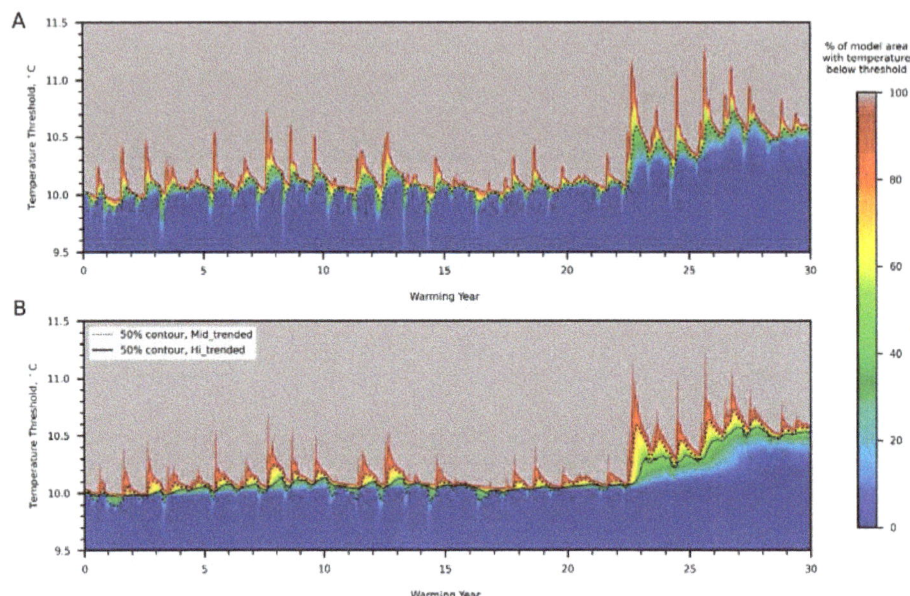

Figure 7. The percent water-table area of model domains that is simulated below increasing temperature thresholds (y-axis) over the time of the warming period (x-axis), for the (**A**) MID_TRENDED and (**B**) HI_TRENDED simulations. The contours indicate the temperature for the 50% threshold. The MID_TRENDED 50% contour displayed in (**A**) also is shown in (**B**) for comparison with the HI_TRENDED 50% contour. Surface-water cells are excluded from calculations.

3.2. Heat Fluxes and Heat Flows

It is often instructive to evaluate the response to warming in terms of heat movement instead of temperature change. Heat movement consists of three flux components–convection, conduction and dispersion [1]. A component of heat flux (measured in Watts) normalized by an area perpendicular to the flux direction yields the corresponding component of heat flow (for example, in units of Watts/m^2.) There are three main interfaces at the beginning or end of watershed pathways where flux or flow components can be calculated: across the top of the UZ (infiltration), across the water table (recharge) and across a streambed or lake bed (baseflow). They are discussed in turn:

- In this study the thermal infiltration is equated with the heat movement downward from the bottom of the root zone which occurs after runoff and evapotranspiration have rerouted some of the water and heat along the land surface or to the atmosphere. This net infiltrating heat flow is imposed as a purely downward convective process into the top of the UZ. For our purposes, the combined effect of heat conduction and dispersion at the root zone/UZ interface, either upward or downward, is considered to be unimportant in comparison to the surface and root zone processes that determine the average monthly rate of infiltrating heat flux.

- Our analysis of the relative weight of simulated heat transport components out of the UZ show that for the humid temperate conditions of the synthetic model, the convective heat flow dominates the conductive and dispersive flow at the water-table interface. This relation persists over time at the scale of individual model locations (Supplementary Material Section S3 Figure S3-10) and when averaged over the entire model domain (Figure 8). For the MID_TRENDED simulation, the absolute value of the conductive heat flows at the water table average about 11% of the convective heat flow, whereas the dispersive heat flow is only 0.05% of the convective heat flow. For the HI_TRENDED simulation, incorporating a generally thicker UZ, the corresponding ratios are 7% and 0.03%. It is worth noting that thermal dispersion is a negligible heat flow component owing to the relatively small longitudinal dispersivity specified (0.9 m) relative to the lateral grid spacing [91 m (300 ft)], a choice consistent with a homogeneous synthetic aquifer. These findings suggest that there is in general only minor loss of accuracy if the heat flow across the interface at the top of the groundwater system is approximated by considering the convective heat flux alone.
- The temperature gradient across the streambed between the stream water in the channel and the ambient groundwater could be incorporated in equations that yield convective and dispersive components of heat flow. Thermal conduction would occur whether the temperature gradient is in the same direction as baseflow or in the opposite direction away from the stream; dispersion would occur only when the gradient is in the same direction as the flow through the streambed. However, the MT3D-USGS code neglects these theoretical components of heat flux and only calculates the convective component, either as a function of groundwater temperature in the presence of baseflow or as a function streamflow temperature in the presence of stream.

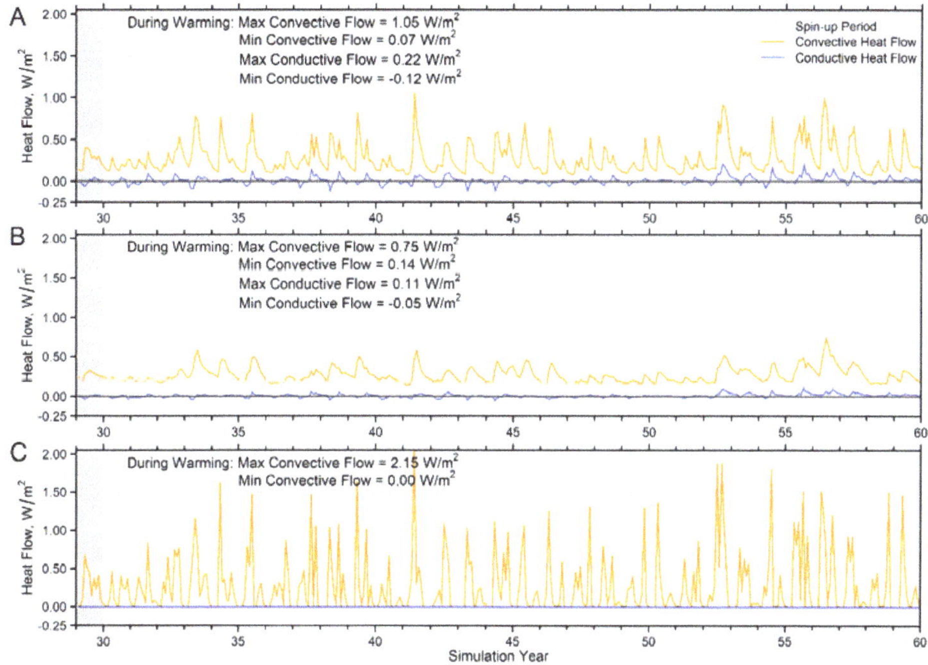

Figure 8. Recharge heat flow components (W/m^2) averaged monthly over the model domain for the (**A**) MID_TRENDED, (**B**) HI_TRENDED, and (**C**) MID_TRENDED (riparian area only) simulations. Heat flow components are shown for conduction and convection. Dispersive heat flow is negligible. Warming period begins in Simulation Year 30.

Given that heat flux across the major watershed interfaces is either imposed as convective flux, approximated by convective flux, or only calculated for convective flux, it is convenient in the pathway analysis of thermal lagging and dampening presented below to define heat movement strictly in relation to the magnitude and direction of water flow, neglecting the conductive and dispersive fluxes.

The convective flux of heat for any part of the model domain is calculated as the flux of water through the model cells constituting the given volume multiplied by the temperature of the water and by the density and heat capacity of fresh water. A convenient unit for the convective heat flux accumulated over one second is Watts (W, equivalent to 1 joule/sec). The rate of convective heat flow is the flux normalized by the area corresponding to the flow. For example, the heat flow in recharge, baseflow and runoff associated with the upstream areas of gages shown in Figure 3 is equal to the accumulated upstream heat flux divided by the areas reported in Table 1 (also see Supplementary Material Section S2, Figure S2-3b for map view of areas upstream of gages). A convenient unit for the rate of heat flow is Watts per square m (W/m^2). Interested readers are directed to Supplementary Material Section S3 for a more detailed discussion of the calculation of the quantities heat flux and heat flow.

Table 1. Topographic areas upstream of stream gages identified in Figure 3.

Stream Gage Number	Gage Description	Upstream Topographic Area	
		(equated with Gage Recharge, Baseflow and Runoff Areas)	
		$mile^2$	km^2
235	Headwater	2.07	5.35
285	Tributary	12.06	31.23
492	Upper Confluence	58.62	151.82
615	Lake Outlet	30.34	78.59
692	Lower Confluence	107.22	277.69
864	Model Outlet	134.38	348.04

Notes: Total area of model domain is 290.5 $mile^2$ = 752.5 km^2, taken to be extent of watershed. Upstream area associated with Gage 864 includes entire eastern basin of watershed including all upstream gages.

Convective heat flux diminishes in strength as it moves through the subsurface. The first reduction occurs in the UZ from where heat enters the simulation as infiltration to where it recharges the groundwater system (Figure 9). This loss of heat is largely due to changes in the amount of heat stored in the UZ. Additional losses to the total heat flux through the watershed occur in the saturated zone or as recharge makes its way to discharge locations, for example, as groundwater discharges directly to streams (Figure 9). In this case, as the shallow groundwater is warmed by the recharge associated with warmer infiltration, it mixes with cooler (and deeper) groundwater as it travels through the saturated zone. The effect of mixing is evident in the heat flux results along pathways. In Figure 9, Gage 492 represents integrated conditions over the upper basin of the eastern part of the stream network, and Gage 864 represents conditions for the entire eastern stream network (see Figure 3 for locations). The simulated heat flux in the upgradient stream network (above Gauge 492) is only a fraction of the flux integrated over the entire eastern basin (above Gage 864). However, it is striking that for both gage locations (in both the MID_ and HI_TRENDED versions of the model), the convective heat fluxes entering the subsurface as infiltration, subsequently converted to recharge, increases appreciably over the 30 years due to climate warming. By contrast, the simulated convective heat flux out of the subsurface (i.e., baseflow) increases by a comparatively small amount in response to that warming, pointing to the substantial dampening (from mixing as well as heat storage) that occurs in the saturated zone (Figure 9).

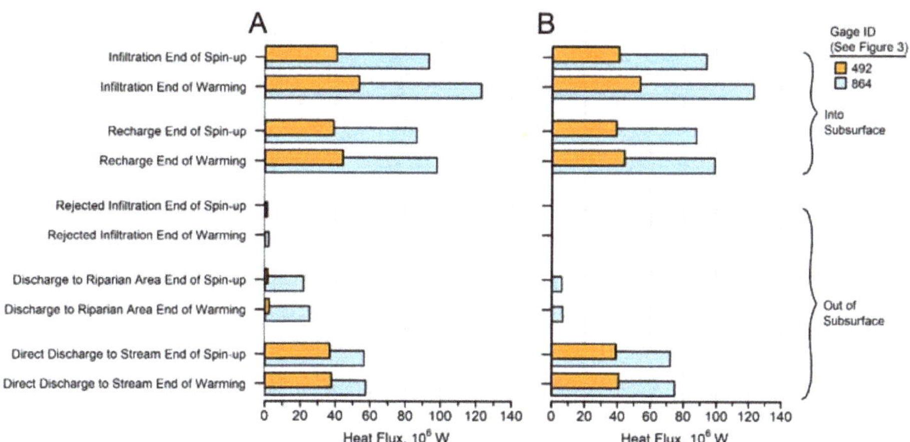

Figure 9. Convective heat flux (Watts) accumulated over contributing basins upgradient of two gage locations. (**A**) MID_TRENDED simulation. (**B**) HI_TRENDED simulation. Infiltration flux is compared to flux transmitted by down-system pathways. Convective flux for each pathway corresponds to the average for last year of spin-up ("end of spin-up") and to the average of last 10 years of warming ("end of warming").

Visualization and comparison of results are facilitated by extending the analysis of how simulated heat is propagated across pathways in terms of heat fluxes normalized by the area of model cells or by the area of watershed subbasins. In what follows recharge and discharge thermal transfers are analyzed in terms of heat flows. Heat transmission losses in the UZ (that is, from infiltration to recharge) are primarily the result of heat storage effects due to warming of water in the UZ. A secondary loss of heat can occur when cooler water enters the UZ behind warmer water, producing an upward thermal gradient which gives rise to upward thermal conduction from the deeper part of the system Recall that thermal gradients drive conductive and dispersive flows and can be upward and downward in the UZ whereas the convective flows, given the kinematic wave formulation in the UZF1 packages, only simulates downward flow [1,27,32].

Where the UZ is thin, for example, in riparian areas adjacent to the surface-water features, the infiltrating heat flow readily warms the water table since there is little opportunity to store additional heat in the UZ (Figure 10). During a cool month with moderate infiltration (e.g., March at 15.25 years), the heat flow to the water table is modest, i.e., less than 0.5 W/m^2 in both the MID_ and HI_TRENDED models (Figure 10A,B). This is not the case for a relatively warm and wet month (e.g., August at 25.67 years) when the heat flow generally exceeds 2.0 W/m^2 in the riparian areas adjacent the surface-water features (Figure 10C,D), though the riparian area is much narrower in the HI_TRENDED simulation. Thus, the spatial distribution of heat flow in the watershed is influenced by the watershed topography, which affects the UZ thickness, as well as by the lateral extent of the riparian area.

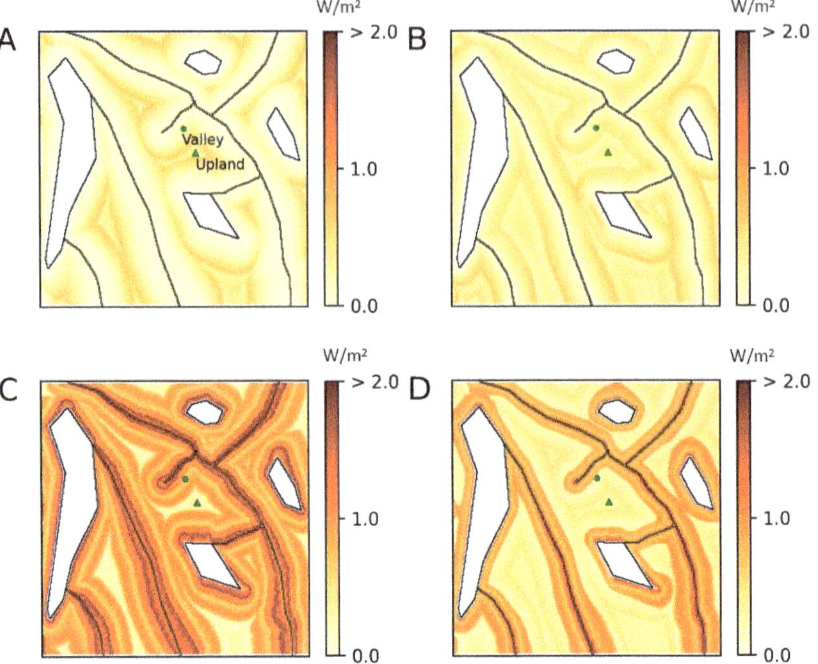

Figure 10. Maps of total recharging heat flow at the water table in watts per square meter (W/m^2) for the (**A**) MID_TRENDED simulation in the month of March after 15.25 years of warming, (**B**) HI_TRENDED simulation (also in March) after 15.25 years of warming, (**C**) MID_TRENDED simulation in the month of August after 25.67 years of warming, and (**D**) HI_TRENDED simulation (also in August) after 25.67 years of warming. The plotted total heat flow is the sum of the convective, conductive and dispersive heat flows.

3.3. *Lags and Dampening of Heat Signal*

To understand better the role different parts of the hydrologic system have on lagging convective heat transport in the subsurface (that is, changing the phase of the thermal impulse), a lag analysis was performed in terms of correlation coefficients computed at different monthly offsets. The time series of the (causal) infiltrating heat flow was paired with the simulated convective heat flow time series at different locations within the watershed, for example, at the water table, using a set of monthly lags (1, 2, 3, etc. monthly offsets). Correlation coefficients were calculated for each monthly lag and compared across months to yield a measure of the delay in heat transport through different parts of the subsurface system. A separate dampening analysis (that is, the change of amplitude along pathways with respect to the infiltrating signal) was performed by computing the ratio of the average convective heat flow (or temperature in the case of baseflow and streamflow) for the last 10 years of warming to the average convective heat flow (or temperature) value during the last year of the spin-up period. The lag and dampening ratios were calculated for the major pathways shown in Figure 1. Additional details on the calculation procedures are offered in in Supplementary Material Section S3.

The lags (phase) and dampening (amplitude) applied to the infiltrating heat flow prior to its recharging the aquifer is strongly influenced by the thickness of the UZ. The phase and amplitude shifts associated with distinct UZ thicknesses are evident in Figure 11 when comparing the convective heat flow arriving at the water table (red and blue lines) to the heat inflow at the top of the UZ (light blue bars). The heat flow associated with the infiltration at the top of UZ is identical for both runs. At the Upland location, for example, where a relatively thick UZ exists (Figure 11A), lags are longer with more significant muting compared to the Valley location where the UZ is relatively thin (Figure 11B). The effect of the UZ is further highlighted by contrasting the recharging heat flow only at the Upland location for the MID_ and HI_TRENDED simulations (Figure 11A). That is, the additional UZ thickness in the HI_TRENDED simulation adds months to the arrival time of the infiltrating heat flow at the water table and further subdues the magnitude of the heat flow (Figure 11A, note the peaks of the red line are lower than the peaks of the blue line). The dashed lines in Figure 11, corresponding to the temperature of the water table cell, reflect the effects of the recharging heat flow mixing with water table. At the Upland location, the simulated temperature at the water table is not as responsive (Figure 11A) as at the Valley location (Figure 11B).

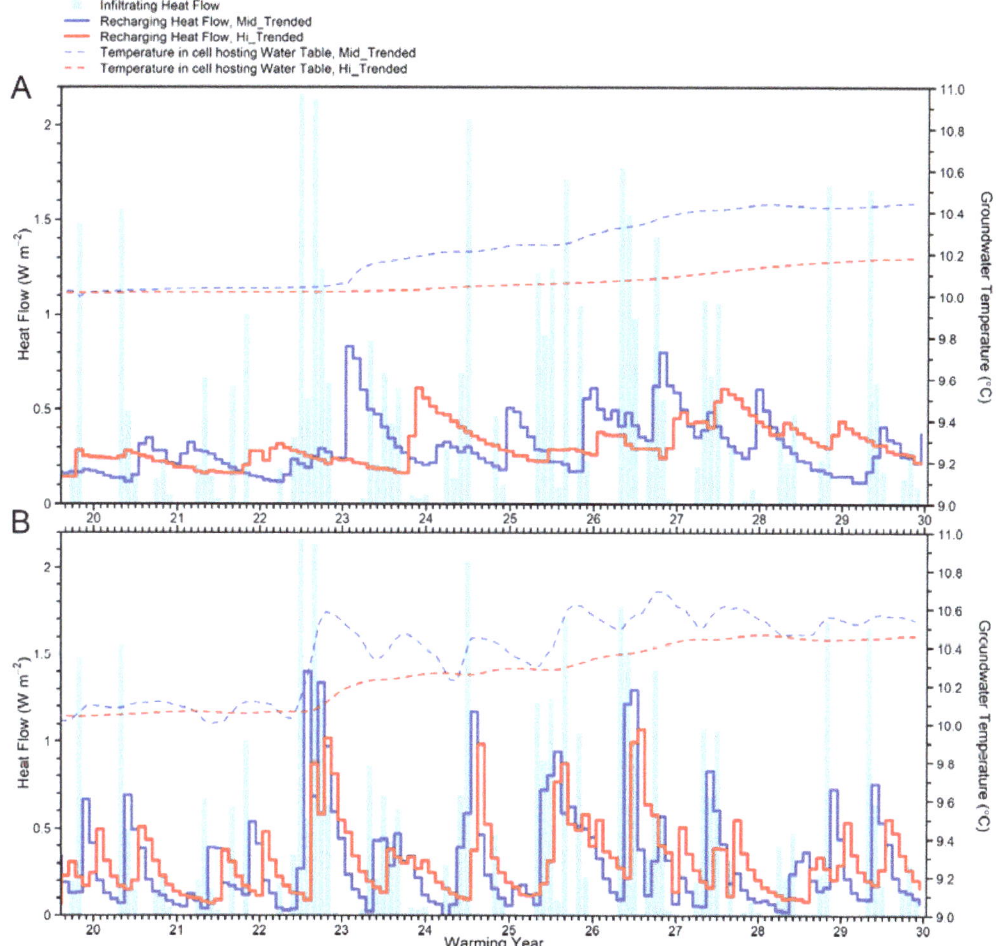

Figure 11. Convective heat flow (W/m^2) and temperature (°C) of the recharge at the (**A**) Upland and (**B**) Valley locations for the MID_TRENDED and HI_TRENDED simulations for the last 10 years of the warming period.

At the Upland location, the heat-flow lag in the MID_TRENDED simulation shown in Figure 11 is quite long, about 7–8 months. (See also Supplementary Material Table S3-2). In addition, the temperature response of the shallow groundwater in both the MID_ and HI_TRENDED simulations is noticeably subdued (Figure 11A), attributable to UZ thicknesses exceeding 40 ft (12 m; Figure 6A) and 100 ft (30 m; Figure 6B), respectively.

A clearer correspondence between the infiltrating and recharging heat flows is seen in the shallow groundwater at the Valley location where the UZ is relatively thin (Figure 11B). For example, the MID_TRENDED simulation, featuring relatively thin UZ thickness, shows strong heat-flow correlation to infiltration changes, peaking at a one-month lag, while the HI_TRENDED simulation, with relatively thick UZ thickness, shows a slightly less strong correlation peaking at a two-month lag (Supplementary Material Table S3-2). The water-table temperature time series at the Valley location, especially for the MID_TRENDED simulation, also shows a definite, if lagged, response to the infiltrating heat signal (Figure 11B). This responsiveness is due to reduced UZ thickness at this location, and, therefore, to reduced capacity for heat storage.

The local lagging and dampening of heat flow through the UZ evident in Figure 11 at the scale of a single water-table cell can be integrated over basins within the watershed by dividing the total heat flux recharging the basin by its area. In Figure 12A the heat-flow behavior over time in recharge is compared to the infiltration forcing at a small headwater basin upstream from Gage 235, 2 mile^2 (5 km^2) in extent (see Figure 3 and Table 1). The offset and attenuation of the infiltration forcing in the recharge time series over the last 10 years of warming, tends to be greater for the simulation with relatively thick UZ (HI_TRENDED) than the simulation with relatively thin UZ (MID_TRENDED), but both model versions show pronounced inertial effects due to the UZ. The lagging and dampening of heat flow in recharge corresponding to the entire stream and lake network on the east side of the domain (that is, to the area upstream of Gage 864, equal to 134 mile^2 (348 km^2)) is similar to that registered for the small headwater basin, although slightly more attenuated (i.e., Figure 12A versus Figure 12B). This similarity is a reflection of the spatially homogeneous conditions that obtain in the synthetic model.

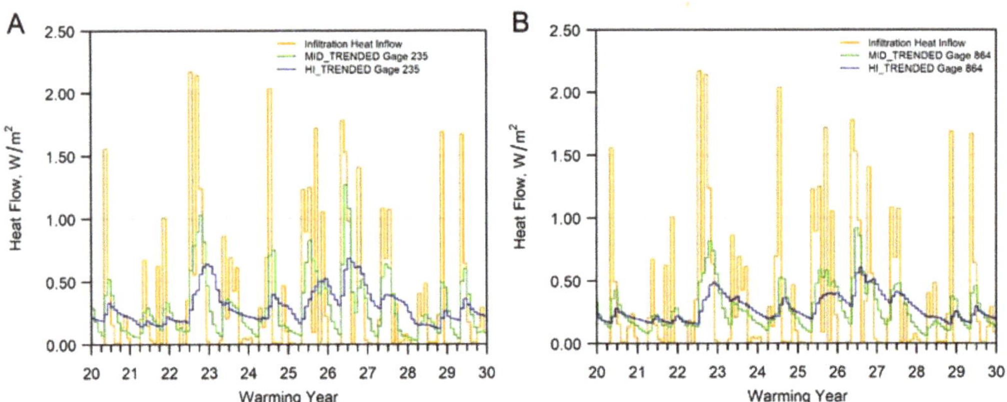

Figure 12. Heat flow response (Watts/m^2) along recharge pathway for (**A**) MID_TRENDED and (**B**) HI_TRENDED simulations. Graphs compare impulse heat flow in INFILTRATION to response heat flow in the pathway during last 10 years of warming for contributing basins corresponding to headwater gage (235) and model outlet gage (864). Heat Flow = Heat Flux over basin normalized by basin area.

Graphs similar to Figure 12 show the lagging and dampening behavior at the 235 and 864 gages for downgradient pathways associated with the stream interface (see Supplementary Information Figure S3-11b–f). For all pathways, the amount of lagging between the infiltration forcing and the downgradient heat flow or downgradient temperature response is quantified over a series of nested basins upstream of gage locations, according to the correlation method discussed in Supplementary Material Section S3. The tables accompanying the graphs contain an array of these calculations. The amount of dampening for the same nested basins is also tabulated. Dampening is quantified based on a direct comparison between the average heat flow or temperature amplitude over the last 10 years of warming and the average amplitude over the last year of spin-up before warming. These pathway results constitute the key hypothetical findings arising from the deployment of the synthetic model.

Whereas distinct lag correlations between the infiltrating and recharging heat flows are evident along the UZ pathway, there is less coherence when considering the longer groundwater pathways that that terminates as direct groundwater discharge to streams (Supplementary Material Table S3-3). This weakening is attributed to mixing of shallow and deep groundwater flow paths and, to an expected lesser extent, changes in heat stored in the aquifer matrix.

The total streamflow carries the heat contribution of both the baseflow components and the storm runoff component. The temperature and heat flow in total streamflow at different gage locations along the stream network show little lag with respect to the infiltration temperature and heat flow ("From Inflow to Total Streamflow" row in Supplementary Material Table S3-3). This result is expected given that storm runoff is the dominant contributor of heat to the stream; that is, storm runoff carries heat quickly overland to the streams with minimal lag and dampening along its path (Supplementary Material Figure S3-10).

Watershed-scale dampening is expressed when incoming heat flow of the infiltration is partially stored in the UZ first, with additional heat storage occurring later in the groundwater system (Supplementary Material Section S3). Table 2 (top) and Figure 13 summarize the pathway dampening that occurs between the last year of spin-up and the last 10 years of warming in terms of heat flow at the basin outlet Gage 864. The climate forcing imposes a 31.3% average increase in the infiltrating heat flow within the contributing area of the basin over the warming period. The recharge transmits a fraction of that signal, producing about a 13% increase in heat flow relative to the spin-up conditions for both the MID_TRENDED and HI_TRENDED simulations. That is, the recharge delivers roughly four tenths of the warm-up entering at the top of the UZ to the water table. For down-system pathways, there is a further and rather sharp reduction in heat propagation through increased dampening of the original infiltrating heat signal. For example, where groundwater discharges to the stream, only about one tenth of the original infiltrating warm-up signal is simulated. There is comparatively less dampening when considering the total baseflow discharge; only about two tenths of the infiltrating warm-up signal is transported to the surface-water system. The baseflow to the stream network carries more of the climate forcing than the direct discharge component because it also includes heat flow from groundwater runoff, which is not subject to dampening. However, total streamflow at the outlet gage, because it incorporates undampened storm runoff, shows less dampening altogether–similar to that shown by recharge (about four tenths of the heat impulse–top of Table 2 and Figure 13). Storm runoff is simulated to be a powerful driver of stream heat-flow conditions.

Table 2. Dampening of warming along watershed pathways: Attenuation of convective heat flow and temperature along pathways contributing to the 864 gage location for MID_TRENDED and HI_TRENDED simulations.

Heat Flow Pathway	Unit	Location	Model Version	Amplitude Percent Increase Due to Warming [1]
Infiltration	Heat flow (Watts/m^2)	Uniform across watershed	MID_Trended and HI_Trended	31.3%
Recharge to Water Table	Heat flow (Watts/m^2)	Contributing Basin for Model Outlet (864)	MID_Trended HI_Trended	13.3% 12.7%
Direct Discharge to Streams	Heat flow (Watts/m^2)	Contributing Basin for Model Outlet (864)	MID_Trended HI_Trended	1.6% 3.7%
Baseflow to Streams	Heat flow (Watts/m^2)	Contributing Basin for Model Outlet (864)	MID_Trended HI_Trended	6.8% 4.4%
Total Streamflow	Heat flow (Watts/m^2)	Contributing Basin for Model Outlet (864)	MID_Trended HI_Trended	12.7% 11.1%

Heat Flow Pathway	Unit	Location	Model Version	Amplitude Percent Increase Due to Warming [1]
Infiltration	Flux-weighted temperature (°C)	Uniform across watershed	MID_Trended and HI_Trended	22.7%
Direct Discharge to Streams	Flux-weighted temperature (°C)	Contributing Basin for Model Outlet (864)	MID_Trended HI_Trended	1.3% 1.3%
Total Streamflow	Flux-weighted temperature (°C)	Contributing Basin for Model Outlet (864)	MID_Trended HI_Trended	5.6% 5.4%

Note: [1] Calculated from ratio of average values of last 10 years of warming to last one year of spinup.

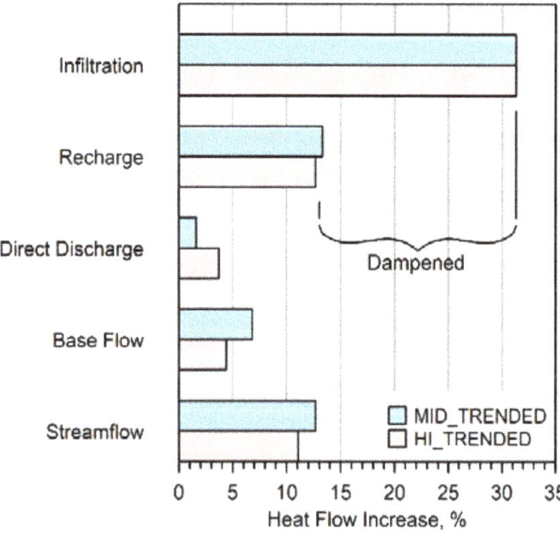

Figure 13. Changes to convective heat flow for different watershed compartments within the simulation are compared for the MID_TRENDED and HI_TRENDED model versions. An increase in heat flow is calculated as the average heat flow over the last 10 years of the of the warming period minus the average heat flow for last year of the spin-up period.

Dampening is also evident in the temperature response to the infiltrating heat flow impulse for down-system pathways (bottom of Table 2). Recall that the flux-weighted temperature in the last year of spin-up averages 9.97 °C and rises to an average of 12.23 °C for the last 10 years of warming, equivalent to a 23% increase (taking 0 °C as the reference point). For both simulations, the direct groundwater discharge temperature increases by only 1.3% of the spin-up average condition; the corresponding relative increase in total streamflow temperature is about 5.5%.

3.4. Implications for Modeling Watershed Heat Transport

Several findings are important for watershed-scale heat transport simulations. First, changes to temporal dynamics of the system in the form of heat storage, lags, and dampening, can all be considered aspects of thermal inertia along watershed pathways. The synthetic model shows how the inertial strength of the UZ and the full groundwater system acts on phase responses to heat fluctuations and on amplitude responses to heat trends originating at the top of the system. The mitigating effect is opposed by the quick heat flows associated with groundwater runoff and, especially, storm runoff. The distinct inertial strengths of watershed pathways combine to produce the complex down-system baseflow and total streamflow responses.

Second, watershed heat transport must consider all heat transport pathways together to accurately simulate the complexities of the down-system response to warming infiltration. Consider the difference in system baseflow thermal response to warming as compared to total streamflow. The total streamflow thermal response is dominated by the amount of heat added to it by storm runoff, which during warm, wet months is conveyed rapidly (in the model context, instantaneously) to stream segments. A baseflow-only characterization would show much less effect of warming. Put otherwise: in the synthetic model, storm runoff constitutes only one-quarter of total streamflow in these simulations, but its thermal effect is disproportionally large due to the absence of any lag or dampening effects on its contributed heat load. Thus, the thermal load contributed by storm runoff overwhelms cooler thermal flows from direct groundwater discharge to the streams. It is worth noting, however, that there are periods of the year (often ecologically important) when storm runoff is largely absent and baseflow dominates total streamflow (and by extension its thermal regime).

Third, this modeling exercise was limited to evaluating the response to a high-emission climate scenario over 30 years. If the warming were to extend over a longer period, there is an expectation that the ability of the UZ and groundwater system to store heat would diminish over time and provide less dampening of the infiltrating heat flow before it reaches a down-system discharge location. In addition, watersheds with lower storage capacity, higher thermal conduction, lower thermal sorption, and higher UZ vertical hydraulic conductivity are expected to produce less lagging and dampening. Thus, the transferability of the results presented here should focus on the heat transport relationships between watershed components (e.g., the UZ, or the saturated zone) rather than on the specific percentages or absolute relative differences resulting from the use of the synthetic model.

3.5. Limitations and Suggestions for Future Work

The thrust of this study is to demonstrate that groundwater/surface-water models can be combined with climate scenarios to simulate water and heat flow at the watershed scale in ways that facilitate science-based forecasting of global warming effects on resources such as stream habitats. There are several limitations and lessons from our hypothetical study that may apply to future applications:

- The climate scenarios, appropriately downscaled, are a promising basis for forecasting effects of climate change on watersheds. For the synthetic model, we imposed a linear temperature rise in line with the high-emissions scenario for an area in the Upper Midwest, USA. No effort was made here to partition the expected heat inflow increase among the seasons or months–but this kind of refinement over and above simple linear infiltration trends might be warranted in a real-world application based on different statistical moments of the GCM results for the region under study.

- Similarly, the present study imposed a linear trend on future precipitation and infiltration: expected changes in precipitation intensity and seasonal distribution patterns were neglected but may be important for applications to real-world watersheds.
- A key expansion of the method presented in the companion paper [2] is the inclusion of storm runoff as part of the watershed flow and heat budgets. However, certain thermal mechanisms are still omitted, such as the effect of heat-bearing precipitation and solar radiation on streams as well as the latent heat effects due to evaporation from surface water bodies. Future developments could add these processes to the MT3D-USGS code if sufficiently important for calibration and forecasting.
- This study presents a lagging and dampening analysis of heat flows performed strictly in terms of the convective component. In a real-world application, this approximation, justified for the synthetic model, might not always prove adequate because of the particular importance of conductive and/or dispersive components at watershed interfaces. In such cases, it might be necessary to expand the heat-flow analysis to include all heat transport components, including possibly conduction and dispersion across streambeds. However, it is worth noting that the lagging and dampening analysis in terms of simulated temperature is not an approximation but reflects all heat transport components.
- In this study, the temperature of infiltration at the bottom of the root zone is set equal to the time series of the atmospheric temperature. The assumption may have its validity reduced with time steps shorter than a month or for seasons subject to high rates of evapotranspiration. Additional studies may be needed to determine if and at what time scale temperatures at the top of the UZ can be reliably equated with atmospheric conditions.
- The specific findings presented here regarding lags and dampening correspond to assumed uniform sandy subsurface conditions. In a heterogeneous setting with finer deposits and preferential flow, the phase and amplitude patterns might appreciably change (consider, for example, the effect of confining beds in the unsaturated and/or saturated systems).
- For the synthetic model under study, it was not necessary to impose a calibration period between spin-up and warming periods: only two periods, in this case both set to 30 years, were sufficient for demonstration purposes. However, a real-world application would likely include calibration to historical observations of heads, flows and temperatures. The length of the calibration period would depend on the available data but would need to be long enough to represent properly the transition from the dynamic equilibrium of the pre-calibration spin-up period to the more variable forcing during the calibration and prediction phases.
- If the model setup were modified to extend the warming trends incorporated in the heat inflow forcing function beyond 30 years, the amplitude of the energy and temperature effects over time would of course be magnified. Any applications of the method to real-world settings would likely simulate forecasts into the second half of the 21st century. Given the large degree of uncertainty around future thermal forcing, it is reasonable for applications of the proposed method to real-world watersheds that a range of GCM emission scenarios be considered (as is done in Hunt et al. [18,19]) to treat simulated heat flow findings in a more statistical fashion.
- Monthly time steps may not be sufficiently refined for some forecasts arising from climate warming (for example, fish vulnerability to short-term stream temperature fluctuations). In such cases, simulations with time discretization finer than a month might be warranted, though practitioners may consider restricting temporal refinement of the model to only those stress periods where it is needed.
- The surface-water network in the synthetic model is baseflow-dominated. Losing streams might be more common in a given real-world watershed, but both the MODFLOW and MT3D-USGS codes can handle any combination of gaining and losing conditions with respect to both water and heat flow.

- There are other simplifications adopted in this hypothetical approach that might require more attention in a real-world application. We assumed a no-flow boundary between an unconfined aquifer and underlying bedrock–in real-world settings, the possibility of water and heat loss below an unconfined aquifer might require explicit modeling of deeper units. We equated groundwater contributing area to stretches of stream completely within their topographic basins (Supplementary Material Figure S2-6)–in a real-world study the researcher might want to use the model to delineate groundwater divides more precisely based on the simulated flow system. We also assumed a small dispersivity value of about 0.91 m (3 ft) relative to a grid spacing of 91 m (300 ft). Higher values of dispersivity or identification of preferential pathways could have a strong influence on convection processes. Finally, an important limitation arises from the primitive lake physics in current versions of MODFLOW and MT3D-USGS, where mechanisms such as lake stratification, ice formation, and latent heat transfers during evaporation from surface water are neglected. Such lake processes continue to be subjects of active research that could lead to more sophisticated treatment of water bodies within the watershed thermal regime.

4. Conclusions

The objectives of this research were as follows:

(a) to forge a robust approach for applying numerical models to study the hydrologic effects of long-term climate change at the full watershed scale and at a monthly time interval, as deemed appropriate for taking account of how a warming trend imposed on background seasonal and random variability propagates through space from the top of the unsaturated zone downward;

(b) to use a synthetic model of a temperate watershed to not only develop the method but also to draw tentative conclusions about the degree of lagging and damping that a future climate forcing would undergo along distinct surface and subsurface pathways, resulting in predictable changes to the warming signal at unsaturated/saturated and groundwater/surface-water interfaces.

The first phase of this work demonstrated the utility of recent model enhancements for simulating how a climate signal is modified as water moves through the UZ and the groundwater system, as well as over the land surface, on its way to a surface-water network. The synthetic model was used to demonstrate the power of the widely used MODFLOW and MT3D-USGS software to track the watershed response to warming. The method yielded quantitative results for the transient distribution of heat flow conditions in the water table, as determined by the propagation of convective and conductive energy components, where it was shown that convection is more important than conduction for the simulated system. The method also allowed us to perform detailed impulse-response analyses of the convective heat signal integrated over time and its transient effect on the groundwater/surface-water system. The dominant effect of UZ thickness, highlighted in Morway et al. (2022b) [2], was confirmed when two model versions with different water-table depths at the watershed scale were applied to the study of heat-flow lags and dampening. The potential importance of the riparian zone was also evident when comparing the direct groundwater discharge response to the more integrated total baseflow response.

The time delays identified by the modeling exercise represented thermal inertia processes resulting from travel through the UZ and the presence of long flow pathlines in groundwater, as opposed to the quick flow resulting from groundwater discharge in riparian areas and storm runoff components. Lags in integrated response time for convective heat flows and for temperature of the streamflow were very short due to the large heat load carried rapidly to streams in warm wet months by undampened storm runoff. The imposed increase in the heat impulse at the top of the UZ was appreciably dampened along unsaturated, saturated, and surface-water pathways, but in complex ways. When the average convective heat flow in the last 10 years of a 30-year warming period was compared to the dynamic equilibrium conditions at the onset of warming, the heat inflow signal was reduced at the water table to 40% the original signal. The presence of both short and long groundwater flow paths and variable path depths further reduced the strength of the thermal loading such that at the stream interface, it was a small fraction of the warming signal. However, when other components of the total baseflow to streams were considered (stormflow, rejected infiltration and groundwater discharge to riparian areas), the model simulated more efficient heat propagation, and the reductions in the warming trend relative to the initial impulse were similar to what was seen at the water table. The simulated dampening response in the streamflow itself could be evaluated in terms of both convective heat flow (diminished by roughly half at the watershed scale with respect to the initial warming impulse) and temperature (registering about one quarter the strength of the assumed near-surface rise).

Because not all parts of a watershed are equal from an ecological standpoint, future modeling studies will need to be tuned to simulate the lag and dampening effects of the subsurface system at interfaces of biological importance. For example, the thermal dynamics at the groundwater/surface-water interface will be of particular importance for a portion of the life cycle of some benthic invertebrates. A holistic watershed representation, i.e., one that includes the UZ, will likely prove useful for capturing complex water and heat flow interactions along the various watershed pathways and through interfaces of special importance.

Supplementary Materials: The following supporting information can be downloaded at: https://www.mdpi.com/article/10.3390/w14182810/s1.

Author Contributions: D.T.F., R.J.H. and E.D.M. shared equally in the conceptualization of the material. D.T.F. did most of the writing with help from R.J.H., while E.D.M. and Feinstein collaborated on the figures. D.T.F. took the lead on the response to reviewers with contributions from coauthors. All authors have read and agreed to the published version of the manuscript.

Funding: The authors would like to acknowledge support from the U.S. Geological Survey Land Change Science/Climate Research & Development for this work.

Institutional Review Board Statement: Not applicable.

Informed Consent Statement: Not applicable.

Data Availability Statement: The model executables, input and output files are available in an online model archive [33].

Acknowledgments: The authors acknowledge the assistance of Alden Provost (U.S. Geological Survey) and Vivek Bedekar (S.S. Papadopulos) in helping us to formulate the modeling approach. We remain especially grateful for the counsel and friendship of our recently departed colleague Richard Healy-he will be greatly missed by everyone who knew him. Any use of trade, firm, or product names is for descriptive purposes only and does not imply endorsement by the U.S. Government.

Conflicts of Interest: The authors have no conflict of interest.

References

1. Morway, E.D.; Feinstein, D.T.; Hunt, R.J.; Healy, R.W. New capabilities in MT3D-USGS for Simulating Unsaturated-Zone Heat Transport. *Groundwater* **2022**, *in press*.
2. Morway, E.D.; Feinstein, D.T.; Hunt, R.J. Simulation of heat flow in a synthetic watershed: The role of the unsaturated zone. *Water* **2022**, *in press*.
3. Hunt, R.J.; Prudic, D.E.; Walker, J.F.; Anderson, M.P. Importance of unsaturated zone flow for simulating recharge in a humid climate. *Groundwater* **2008**, *46*, 551–560. [CrossRef] [PubMed]
4. Marçais, J.; Derry, L.A.; Guillaumot, L.; Aquilina, L.; de Dreuzy, J.R. Dynamic contributions of stratified groundwater to streams controls seasonal variations of streamwater transit times. *Water Resour. Res.* **2022**, *58*, e2021WR029659. [CrossRef]
5. Hunt, R.J.; Wilcox, D.A. Ecohydrology—Why hydrologists should care. *Groundwater* **2003**, *41*, 289–290. [CrossRef] [PubMed]
6. Cogswell, C.; Heiss, J.W. Climate and Seasonal Temperature Controls on Biogeochemical Transformations in Unconfined Coastal Aquifers. *J. Geophys. Res. Biogeosci.* **2021**, *126*, e2021JG006605. [CrossRef]

7. Burns, E.R.; Zhu, Y.; Zhan, H.; Manga, M.; Williams, C.F.; Ingebritsen, S.E.; Dunham, J.B. Thermal effect of climate change on groundwater-fed ecosystems. *Water Resour. Res.* **2017**, *53*, 3341–3351. [CrossRef]
8. Suzuki, H.; Nakatsugawa, M.; Ishiyama, N. Climate Change Impacts on Stream Water Temperatures in a Snowy Cold Region According to Geological Conditions. *Water* **2022**, *14*, 2166. [CrossRef]
9. Yao, Y.; Tian, H.; Kalin, L.; Pan, S.; Friedrichs, M.A.; Wang, J.; Li, Y. Contrasting stream water temperature responses to global change in the Mid-Atlantic Region of the United States: A process-based modeling study. *J. Hydrol.* **2021**, *601*, 126633. [CrossRef]
10. Mancewicz, L.K.; Davisson, L.; Wheelock, S.J.; Burns, E.; Poulson, S.R.; Tyler, S.W. Impacts of Climate Change on Groundwater Availability and Spring Flows: Observations from the Highly Productive Medicine Lake Highlands/Fall River Springs Aquifer System. *J. Am. Water Resour. Assoc.* **2021**, *57*, 1021–1036. [CrossRef]
11. Caldwell, T.G.; Wolaver, B.D.; Bongiovanni, T.; Pierre, J.P.; Robertson, S.; Abolt, C.; Scanlon, B.R. Spring discharge and thermal regime of a groundwater dependent ecosystem in an arid karst environment. *J. Hydrol.* **2020**, *587*, 124947. [CrossRef]
12. Hemmerle, H.; Bayer, P. Climate change yields groundwater warming in Bavaria, Germany. *Front. Earth Sci.* **2020**, *8*, 523. [CrossRef]
13. Benz, S.A.; Bayer, P.; Winkler, G.; Blum, P. Recent trends of groundwater temperatures in Austria. *Hydrol. Earth Syst. Sci.* **2018**, *22*, 3143–3154. [CrossRef]
14. Cartwright, J.; Johnson, H.M. Springs as hydrologic refugia in a changing climate? A remote-sensing approach. *Ecosphere* **2018**, *9*, e02155. [CrossRef]
15. Blum, P.; Menberg, K.; Koch, F.; Benz, S.A.; Tissen, C.; Hemmerle, H.; Bayer, P. Is thermal use of groundwater a pollution? *J. Contam. Hydrol.* **2021**, *239*, 103791. [CrossRef] [PubMed]
16. IPCC (Ed.) *Climate Change 2022: Impacts, Adaptation, and Vulnerability, Technical Summary*; Cambridge University Press: Cambridge, UK, 2022.
17. Markle, J.M.; Schincariol, R.A. Thermal plume transport from sand and gravel pits–Potential thermal impacts on cool water streams. *J. Hydrol.* **2007**, *338*, 174–195. [CrossRef]
18. Hunt, R.J.; Walker, J.F.; Selbig, W.R.; Westenbroek, S.M.; Regan, R.S. *Simulation of Climate-change Effects on Streamflow, Lake Water Budgets, and Stream Temperature using GSFLOW and SNTEMP, Trout Lake Watershed, Wisconsin*; U.S. Geological Survey Scientific Investigations Report 2013-5159; U.S. Geological Survey: Reston, VA, USA, 2013; 118p. [CrossRef]
19. Hunt, R.J.; Westenbroek, S.M.; Walker, J.F.; Selbig, W.R.; Regan, R.S.; Leaf, A.T.; Saad, D.A. *Simulation of Climate Change Effects on Streamflow, Groundwater, and Stream Temperature Using GSFLOW and SNTEMP in the Black Earth Creek Watershed, Wisconsin*; 2328-0328; U.S. Geological Survey, Scientific Investigations Report 2016-5091; U.S. Geological Survey: Reston, VA, USA, 2016; pp. 1–117. [CrossRef]
20. Niswonger, R.G.; Panday, S.; Ibaraki, M. *MODFLOW-NWT, a Newton Formulation for MODFLOW-2005*; U.S. Geological Survey: Reston, VA, USA, 2011; p. 44. [CrossRef]
21. Bedekar, V.; Morway, E.D.; Langevin, C.D.; Tonkin, M.J. *MT3D-USGS Version 1: A U.S. Geological Survey Release of MT3DMS Updated with New and Expanded Transport Capabilities for Use with MODFLOW*; 2328-7055; U.S. Geological Survey: Reston, VA, USA, 2016. [CrossRef]
22. Gunawardhana, L.N.; Kazama, S.J.J.O.H. Using subsurface temperatures to derive the spatial extent of the land use change effect. *J. Hydrol.* **2012**, *460*, 40–51. [CrossRef]
23. Loinaz, M.C.; Davidsen, H.K.; Butts, M.; Bauer-Gottwein, P. Integrated flow and temperature modeling at the catchment scale. *J. Hydrol.* **2013**, *495*, 238–251. [CrossRef]
24. Loinaz, M.C.; Gross, D.; Unnasch, R.; Butts, M.; Bauer-Gottwein, P. Modeling ecohydrological impacts of land management and water use in the Silver Creek basin, Idaho. *J. Geophys. Res. Biogeosci.* **2014**, *119*, 487–507. [CrossRef]
25. Feinstein, D.T.; Hunt, R.J.; Morway, E.D. Supplementary Material to accompany present article: Simulation of heat flow in a synthetic watershed: Lags and dampening across multiple pathways under a climate-forcing scenario. *Water* **2022**, *in press*.
26. Beven, K. Interflow. In *Unsaturated Flow in Hydrologic Modeling*; Morel-Seytoux, H.J., Ed.; NATO ASI Series; Springer: Dordrecht, The Netherlands, 1989; Volume 275, pp. 191–219. ISBN 978-94-009-2352-2. [CrossRef]
27. Anderson, M.P.; Woessner, W.W.; Hunt, R.J. *Applied Groundwater Modeling: Simulation of Flow and Advective Transport*; 0080916384; Academic Press: London, UK, 2015. [CrossRef]
28. Niswonger, R.G.; Prudic, D.E.; Regan, R.S. *Documentation of the Unsaturated-Zone Flow (UZF1) Package for Modeling Unsaturated Flow between the Land Surface and the Water Table with MODFLOW-2005*; U.S. Geological Survey Techniques and Methods Report 6-A19; U.S. Geological Survey: Reston, VA, USA, 2006; 62p. [CrossRef]
29. USGS. National Climate Change Viewer. Available online: https://www2.usgs.gov/landresources/lcs/nccv/maca2/maca2_watersheds.html (accessed on 31 January 2022).
30. Niswonger, R.G.; Prudic, D.E. *Documentation of the Streamflow-Routing (SFR2) Package to Include Unsaturated Flow Beneath Streams-A Modification to SFR1*; U.S. Geological Survey Techniques and Methods Report 6-A13; U.S. Geological Survey: Reston, VA, USA, 2005; 51p. [CrossRef]
31. Zheng, C.; Hill, M.C.; Hsieh, P.A. *MODFLOW-2000, the US Geological Survey Modular Ground-Water Model: User Guide to the LMT6 Package, the Linkage with MT3DMS for Multi-Species Mass Transport Modeling*; U.S. Geological Survey Open-File Report 2001-82; U.S. Geological Survey: Reston, VA, USA, 2001. [CrossRef]

32. Morway, E.D.; Niswonger, R.G.; Langevin, C.D.; Bailey, R.T.; Healy, R.W. Modeling variably saturated subsurface solute transport with MODFLOW-UZF and MT3DMS. *Groundwater* **2013**, *51*, 237–251. [CrossRef]
33. Morway, E.D.; Feinstein, D.T.; Hunt, R.J. MODFLOW-NWT and MT3D-USGS Models for Evaluating Heat Flows, Lags, and Dampening under High Emission Climate Forcing for Unsaturated/Saturated Transport in a Synthetic Watershed. *U.S. Geol. Surv. Model Arch.* **2022**, *in press*. [CrossRef]

Article

A Stepwise Modelling Approach to Identifying Structural Features That Control Groundwater Flow in a Folded Carbonate Aquifer System

Elisabetta Preziosi [1], Nicolas Guyennon [1], Anna Bruna Petrangeli [1], Emanuele Romano [1] and Cristina Di Salvo [2,*]

[1] CNR-Water Research Institute, Area della Ricerca di Roma 1, Via Salaria Km 29,300-C.P. 10, Monterotondo Stazione, 00015 Rome, Italy
[2] CNR-Institute of Environmental Geology and Geoengineering, Area della Ricerca di Roma 1, Via Salaria Km 29,300-C.P. 10, Monterotondo Stazione, 00015 Rome, Italy
* Correspondence: cristina.disalvo@igag.cnr.it

Abstract: This paper concerns a stepwise modelling procedure for groundwater flow simulation in a folded and faulted, multilayer carbonate aquifer, which constitutes a source of good quality water for human consumption in the Apennine Range in Central Italy. A perennial river acts as the main natural drain for groundwater while sustaining valuable water-related ecosystems. The spatial distribution of recharge was estimated using the Thornthwaite–Mather method on 60 years of climate data. The system was conceptualized as three main aquifers separated by two locally discontinuous aquitards. Three numerical models were implemented by gradually adding complexity to the model grid: single layer (2D), three layers (quasi-3D) and five layers (fully 3D), using an equivalent porous medium approach, in order to find the best solution with a parsimonious model setting. To overcome dry-cell problems in the fully 3D model, the Newton–Raphson formulation for MODFLOW-2005 was invoked. The calibration results show that a fully 3D model was required to match the observed distribution of aquifer outflow to the river baseflow. The numerical model demonstrated the major impact of folded and faulted geological structures on controlling the flow dynamics in terms of flow direction, water heads and the spatial distribution of the outflows to the river and springs.

Keywords: carbonate aquifer; faults and folds; groundwater modelling; multilayer aquifer; MODFLOW-NWT formulation; Central Italy

Citation: Preziosi, E.; Guyennon, N.; Petrangeli, A.B.; Romano, E.; Di Salvo, C. A Stepwise Modelling Approach to Identifying Structural Features That Control Groundwater Flow in a Folded Carbonate Aquifer System. *Water* **2022**, *14*, 2475. https://doi.org/10.3390/w14162475

Academic Editor: Adriana Bruggeman

Received: 5 July 2022
Accepted: 9 August 2022
Published: 11 August 2022

1. Introduction

Carbonate aquifers are important groundwater resources worldwide due to their high permeability and rapid groundwater velocities. Ford and Williams [1] estimate that 20% of the world population largely depends on groundwater from carbonate aquifers. These are often defined as karst aquifers due to their "self-organized, high permeability channel networks formed by positive feedback between dissolution and flow" [2]. An important feature of the carbonate aquifers, especially when diffuse flow prevails, is their capability to store large quantities of groundwater during humid periods and gradually release them during dry periods. Hence, they are fundamental to sustain both human uses and groundwater-related ecosystems in many parts of the world. Water quality in carbonate aquifers is often excellent; hence, they are regarded as strategic both for human consumption as well as to sustain environmental uses. The high permeability often results in thick unsaturated zones, so exploitation of groundwater is often from low-elevation springs, especially in mountainous areas [3]. Pumping wells located near the springs are used to overcome spring discharge shortage in dry seasons [4].

These resources are very often exploited to supply large urban areas, and the evaluation of the possible negative effects of climate changes on their discharge is challenging. The decrease in annual precipitation and increase in temperature and evapotranspiration due to climate change in the Mediterranean area [5–9] is leading to a general decline both in surface water discharge [10] and in groundwater recharge [11]. Preziosi and Romano (2013) [12] highlighted that the spring discharge of large carbonate aquifers in Central Italy has decreased by 6% to 42% in the period 1938–2011, the highest percentages estimated for the smallest springs [13].

Recent advances in groundwater flow and level prediction include the application of diverse methods including wavelet analysis, Gaussian process regression [14], machine learning modes [15] and integrated numerical modeling [16].

The development of numerical models is considered a fundamental step for the adoption of water management plans aiming to preserve groundwater resources and the related ecosystems [17]. Many authors have focused on the numerical modelling of karst systems to assess the risk of spring discharge shortage due to climate change [18,19] or to evaluate the effects of withdrawals [4]. However, numerical modeling of carbonate aquifers in folded and faulted terrains is a challenge due to the complexity of the hydrogeological systems, and excessive simplification may lead to an unsatisfactory predictive capability of the model. Carbonate aquifers are often characterized by highly conductive conduit flow paths embedded in a less conductive fissured and fractured matrix [20]. In spite of these strong permeability contrasts, the equivalent porous medium approach (EPM) can be applied to karstified aquifers with some limitations; specifically, it may not be suitable for deterministically modeling flow along faults, and it often fails to predict flow direction and velocity. However, it can correctly approximate flow and spring discharge at the regional scale [21]. Further, Abusaada and Sauter (2013) [22] affirm that EPM models can simulate flow in karst aquifers as long as the simulated saturated volume is large enough to average out the local influence of karst conduits. The significant influence of the geological structure (especially folding and lithology) and the karst system on the location of the springs and their flow regime has been addressed by [23,24]. Structural folds may divert groundwater flow from the general hydraulic gradient. The presence of marl layers may sustain perched sub-aquifers above the regional aquifer, and karstification may locally increase the hydraulic conductivity by several orders of magnitude. Moreover, the complex geometry of model layers can result in different thicknesses of saturated portions, which can imply the drying and rewetting of cells during model iterations, leading to numerical instabilities, preventing convergence and increasing numerical error [25]. Including all these characteristics in the conceptual and numerical model requires the adequate definition of layer thickness, dipping and hydraulic properties. Recent advances in groundwater numerical modeling include the development of solvers which facilitate achieving convergence and/or reducing computational errors due to model nonlinearities, as well as packages tailored for solving specific problems [26]. This allows the development of fully 3D numerical models which are able to reproduce complex settings with stable solutions. However, by increasing model structure complexity, the number of input parameters increases as well as the related uncertainty. A model with too many parameters is susceptible to over-fit the data [27–29]. The higher the complexity, the more accurate the calibration procedure should be, requiring an adequate number of calibration targets. Simulating complexity not supported by the data can be useless and misleading.

The aim of this research was to develop and test a modelling procedure for the simulation of groundwater flow in a complex karst, folded, multilayer aquifer using the EPM approach. In this framework, three steady-state numerical models of a carbonate aquifer in Central Italy (Monte Coscerno) were developed with increasing complexity as warranted by the inability of the simpler model to adequately reproduce observations [30]. A stepwise procedure was developed for testing the ability of the models to reproduce observations. Seeking parsimony, we compare the results from a simple one-layer model and more complex quasi-3D and fully 3D models with a different number of layers and spatial variability of parameters.

2. Conceptual Model

2.1. Geological and Hydrogeological Setting

The case study aquifer (Monte Coscerno, 230 km^2) is located in the Apennine Range in Central Italy (Figure 1). It can be described as a box-fold anticline with a nearly meridian axis and a vertical-to-overturned forelimb [31] belonging to the frontal thrust ramp of a regionally curved thrust known as the Mount Coscerno–Rivodutri thrust [32,33]. The above mentioned thrust on the east and the Valnerina thrust [34] on the west, with an approximately meridian direction, bound nearly all the aquifer and act as a no-flow boundary (black solid line in Figure 1). The general groundwater flow is mainly in a south to north direction, parallel to the prevailing tectonic lines [35–37]. The areas where the recharge is most effective are the plateaus of Monte Coscerno, Monte Aspra and other summits that exceed 1000 m a.s.l., reaching 1685 m a.s.l. at Monte Coscerno (Figure 2). A sequence of Meso-Cenozoic calcareous formations interbedded with marl layers results in a multilayer aquifer system, with three main sub-aquifers (from bottom to top: Calcare Massiccio-Corniola, Maiolica, Scaglia limestone units) separated by two aquitards [36]. The aquitards are locally discontinuous due to depositional, erosional and tectonic effects, favoring vertical leakage between the three aquifers [37,38]. The base of the aquifer is represented by the top of the Triassic dolomitic evaporitic complex ("Marne a Rhaetavicula contorta" marlstones and "Anidriti di Burano" anhydrites, [39,40], mostly dipping westward. Groundwater is expected to flow according to the bedding attitudes in the direction of the steepest structural descent. This may result in large unsaturated portions in the eastern part of the aquifer (Figure 3); moving from east to west, the sub-aquifers 2 and 3 become confined, feeding the Scaglia sub-aquifer through vertical leakance upward. The sub-aquifers 1 and 2 do not extend through all the model area due to erosional processes which affected the anticline eastside where the more ancient formations outcrop (Figures 1 and 3). For this reason, rainfall infiltrates through the uppermost outcropping aquifer and flows westward and northward according to the dip of the layers.

Hourly hydrometric data and periodic discharge measurements of the Nera River in two stream gauging stations (Vallo di Nera and Torre Orsina, Figure 1) have been made available by the Umbria Region since 2006. The monthly discharge of the Lupa spring is available for the period 1985–1997. Moreover, the Lupa spring daily discharge (since 1998) and piezometric heads measured in the Scheggino well (since 2001) are available online from the local Regional Environmental Agency [41]. Head data are very scarce, except for the already mentioned Scheggino well and the very recently installed Renari di Capriglia well (Figure 1). The Nera River is incised into the carbonate aquifer, increasing its discharge from north to south. Several discharge measurements performed in the years 1991–1993 at the 6 gauging stations in Figure 1 [42] indicate a conspicuous discharge increment, nearly constant throughout the year, revealing gaining stream conditions between Vallo di Nera and Umbriano (reaches R1 to R4, Figure 1). The river baseflow was estimated between 3.2 and 3.4 m^3/s in the period 1991–1993 by Boni and Preziosi (1993) [35]. In addition, the aquifer feeds several point springs (Scheggino, Lupa and Pacce) and the Precetto stream (Figure 1, Table 1).

The aquifer discharge to the Nera River was estimated as the discharge increment between two gauges (Vallo di Nera and Torre Orsina gauges, Figure 1) using spot measurements in the years 1997–2012 provided by the Regional Environmental Agency [41]. The average of the 83 spot measurements is 3.28 m^3/s, ranging from 0.9 to 7.46 m^3/s (Figure 4). The total aquifer discharge was estimated at about 3.4–3.6 m^3/s, with an extremely regular seasonal regimen uncommon in karst areas [42]. However, there is no evidence in the area of developed karst conduits despite the presence of likely fractured limestones. Consequently, Monte Coscerno can be classified as a "diffuse flow aquifer" [44] i.e., a carbonate aquifer system with dispersive circulation due to a micro-fractured interconnected network with extremely reduced or even inexistent karstification without preferential drainage paths [45]. There is no concentration of flow towards localized springs, with the exception of the Lupa, Pacce and Scheggino springs. The Nera River flows parallel to the Valnerina thrust, as shown in Figure 1, that represents the no-flow boundary of the aquifer on the eastern side at the lowest topographic elevation, explaining the location of the springs between Vallo di Nera and Umbriano. The depth of the incision of the NE and NW-trending valleys is schematically shown in Figure 2 (left upper panel). Some of these valleys are very deep and profoundly incised. Nevertheless, they do not intercept the water table.

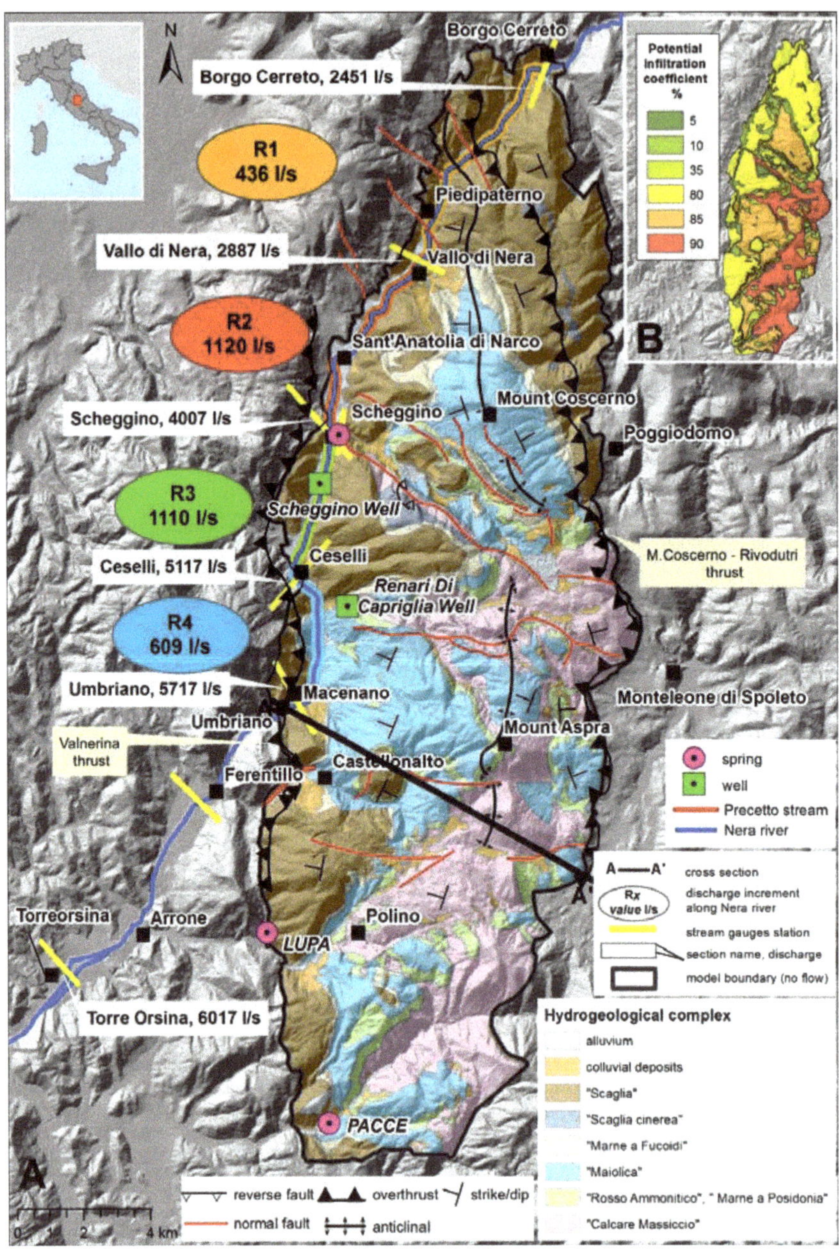

Figure 1. (**A**): Hydrogeological setting, conceptual model. (**B**): distribution map of the potential infiltration coefficients. Tectonic elements (faults, thrust, anticlinal axis) from [40].

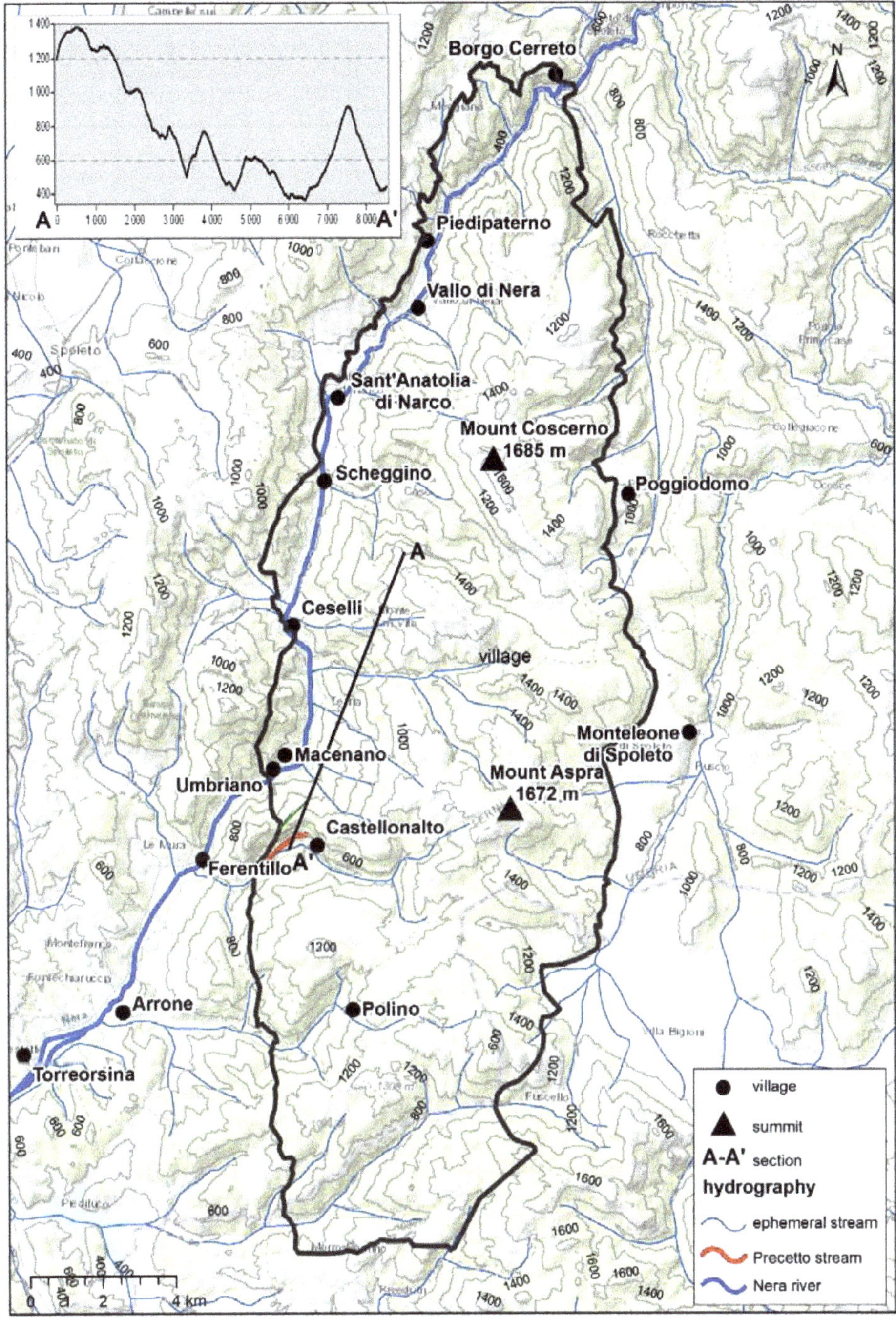

Figure 2. Topographic map of the study area. Elevation in meters above sea level.

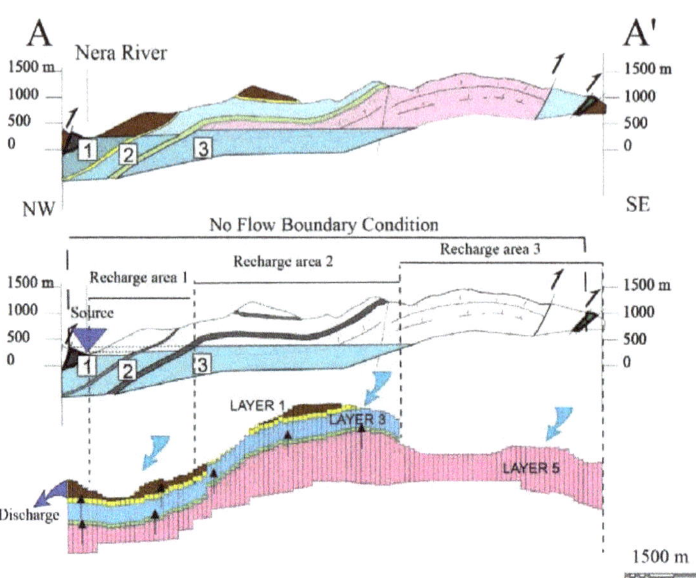

Figure 3. Schematic cross section and translation into model grid for the fully 3D model. Trace of cross section and legend are in Figure 1. (**Top panel**): schematic hydrogeologic cross section. The blue filling at the top and central panels represents the saturated zone; numbers 1, 2 and 3 refer to the sub-aquifers. Dotted lines in central panel represent the potentiometric levels of confined sub-aquifers. (**Bottom panel**): model grid for the fully 3D model; light blue arrows: recharge; blue arrow: discharge to the RIV cells.

Table 1. Gauging stations and springs in the study area.

Site Name	Altitude (m a.s.l.)	Average Discharge (L/s)	Reference Period	Reference
Borgo Cerreto	345	2451	1991–1993	[36]
Vallo di Nera	293	2887	1991–1993	[36]
Scheggino	275	4007	1991–1993	[36]
Ceselli	265	5117	1991–1993	[36]
Umbriano	242	5717	1991–1993	[36]
T.Orsina	211	6017	1991–1993	[36]
Scheggino spring	276	190	1991–1993	(accounted for in the Ceselli gauging site) [36]
Lupa spring	366	120	1997–2012	[41]
Pacce spring	480	60	2000–2001	Well field excluded. [43]
Precetto stream	325	120	1991–1993	[36]
Borgo Cerreto	345	2451	1991–1993	[36]
Vallo di Nera	293	2887	1991–1993	[36]

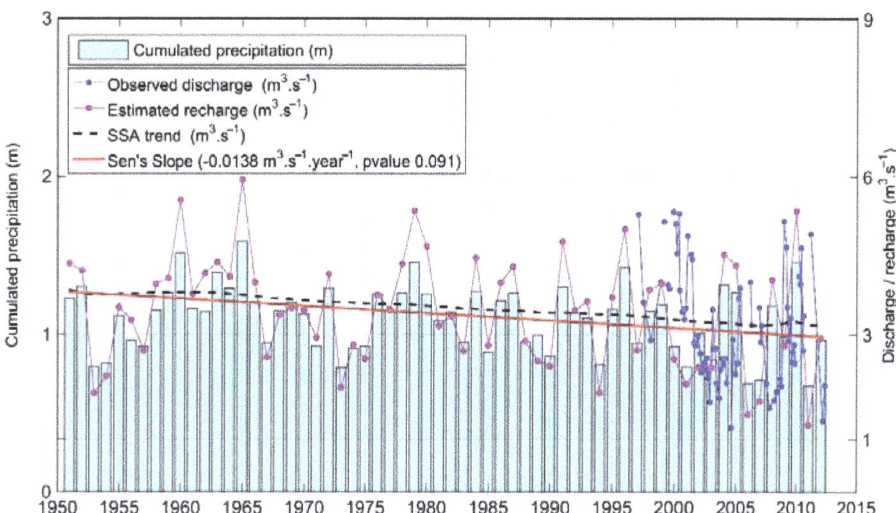

Figure 4. Estimated recharge to the aquifer (purple graph), with respect to the cumulated precipitation (histogram) and the observed aquifer discharge (blue graph). Trend (black dashed line) and Sen's slope refer to the calculated recharge.

2.2. Recharge Estimation

Daily precipitation and mean temperature (period: 1950–2013) were collected over the study area at the station locations of 16 and 7, respectively. Data were interpolated over a 1 km^2 regular grid through an ordinary kriging. Temperature data were previously detrended from the local lapse rate. Observed semivariograms were fitted through a spherical model at a monthly time step. The recharge to the aquifer was estimated, at a daily time step, over a 1 km^2 square grid, using the Thornthwaite–Mather model [46,47]. Recharge was assumed to be a fraction of water surplus when soil moisture exceeds the field capacity. Soil moisture was estimated as the difference between precipitation and actual evapotranspiration, the latter computed as a fraction of the potential evapotranspiration when the soil was partially saturated. Field capacity was set to 100 mm [36]. The potential infiltration coefficients have been ascribed to each hydrostratigraphic unit on the grounds of existing technical reports on the study area [48] and on an expert judgement basis, assuming that limestone formations have higher potential infiltration coefficients than marl formations and range from 5 to 90% of the water surplus (Figure 1B). Both field capacity and potential infiltration coefficients have been calibrated to match the observed aquifer contribution to the river and springs (3.4–3.6 m^3/s). The difference between water surplus and infiltration was ascribed to runoff. The validation was based on the comparison of the calculated annual infiltration with the observed aquifer discharge. The average estimated infiltration rate after calibration is 480 mm/y, corresponding to an average discharge of 3.5 m^3/s. Nonstationarity in the estimated recharge was assessed through a singular spectrum analysis (SSA) for the past 60 years (1950–2012) indicating an approximately linear reduction in the recharge without evident break points. The negative trend of the recharge is 1.8 mm/y, which is significant at 90% (Figure 4). In Figure 4, the aquifer discharge from 1997 to 2012, as estimated in Section 2.1, is also shown.

3. Numerical Model Description

3.1. Layers Discretization, Boundary Conditions and Codes

The steady-state models were implemented by gradually increasing the number of layers, from a 1-layer (2D) to a 3-layer (quasi-3D) and then a 5-layer model (fully 3D) with a uniform grid spacing of 100 × 100 m (Figure 5). The top and bottom of the layers were built using the available geological data (boreholes stratigraphy, geological maps, cross sections), by means of an ordinary kriging algorithm, with a spherical semivariogram. No-flow boundary conditions were assigned to the external boundary of the models. As the upper layers do not cover the whole extent in the quasi-3D and fully 3D models, no-flow boundary conditions (inactive cells) were assigned in the areas where the anticline is eroded (eastern portion), leading to the lowermost layers to outcrop. This allows the partitioning of the recharge into different layers according to the geological setting, assigning the calculated recharge rates to each outcropping cell. Drain and river packages (DRN and RIV) were used to simulate head-dependent flux boundary conditions for the springs, the Nera River and the Precetto stream, respectively. Initial hydraulic conductivity and transmissivity values were set based on the available on-site tests and on the results of previous numerical simulations [36,48]. The 2D model features one layer only. Simulating a unique aquifer, it neglects the behavior of the two aquitards. The groundwater flow is only in the x–y direction. The hydraulic conductivity values (K_h) were set for each cell by considering the average aquifer transmissivity divided by the cell thickness. The quasi-3D model consists of three layers, representing the sub-aquifers. Aquitards are not explicitly represented; the groundwater flow in each layer is in the x–y direction and exchanges between layers are regulated through vertical conductance known as the term vertical leakage. Vertical leakage is computed by the preprocessor Groundwater Vistas (Environmental Simulations International®) based on the saturated thickness and vertical K assigned to the implicit aquitard layer [49]. Finally, the fully 3D approach allows for vertical flow in aquifers and aquitards through the explicit representation of the aquitards. The fully 3D model was set up with five layers representing the three sub-aquifers and the two aquitards. Initially, the hydraulic conductivity of the explicit aquitards was assumed to be 1/100 of the hydraulic conductivity values set to each overlaying sub-aquifer. The mean estimated recharge (3.5 m^3/s, see Section 2.2) was uniformly assigned as the input recharge to the active cells of the models. Two-dimensional and quasi-3D models were run using MODFLOW 88/98 with a preconjugated gradient 2 solver (PCG2). In order to overcome dry-cell problems and reduce the model error in the fully 3D model, the Newton–Raphson formulation for MODFLOW-2005 (MODFLOW2005-NWT, [26]) was invoked. MODFLOW2005-NWT uses an alternative formulation of the GW-flow equation: the upstream weighting package (UPW) treats nonlinearities of cell drying and rewetting by using a continuous function of hydraulic head, instead of the discrete approach applied by the block-centered flow and layer property flow packages in the previous MODFLOW versions. Application of MODFLOW-NWT overcomes numerical problems by smoothing the transition from wet to dry cells and keeps all cells active [50]. MODFLOW-NWT keeps all cells active that are active at the start of the simulation. It assigns a head to unconfined cells even when the head falls below the cell bottom, allowing vertical flow in the form of recharge. However, dry cells no longer participate in horizontal aquifer flow. Use of the MODFLOW2005-NWT avoids solver instability in the presence of dry cells and diminishes the sensitivity of the solution to initial conditions.

In order to allow for a robust comparison of the 2D and quasi-3D simulations to the fully 3D model, an equivalent transmissivity was calculated for each cell for the quasi-3D and 2D models, taking account of the reduced number of layers:

$$\begin{aligned} T_1 + T_2 + T_3 + T_4 + T_5 = T_{1'} + T_{2'} + T_{3'} = T_{2D} = K_1 e_1 + K_2 e_2 + K_3 e_3 + K_4 e_4 + K_5 e_5 \\ = K_{1'} e_{1'} + K_{2'} e_{2'} + K_{3'} e_{3'} = K_{2D} e_{2D} \end{aligned} \quad (1)$$

- T_1, T_2, T_3, T_4, T_5 = transmissivity of the fully 3D model (subscript indicates the layer)
- $T_{1'}$, $T_{2'}$, $T_{3'}$ = transmissivity of the quasi-3D model (subscript indicates the layer)
- T_{2D} = transmissivity of the 2D model
- K_n = permeability (subscript indicates the layer)
- e_n = thickness (subscript indicates the layer)

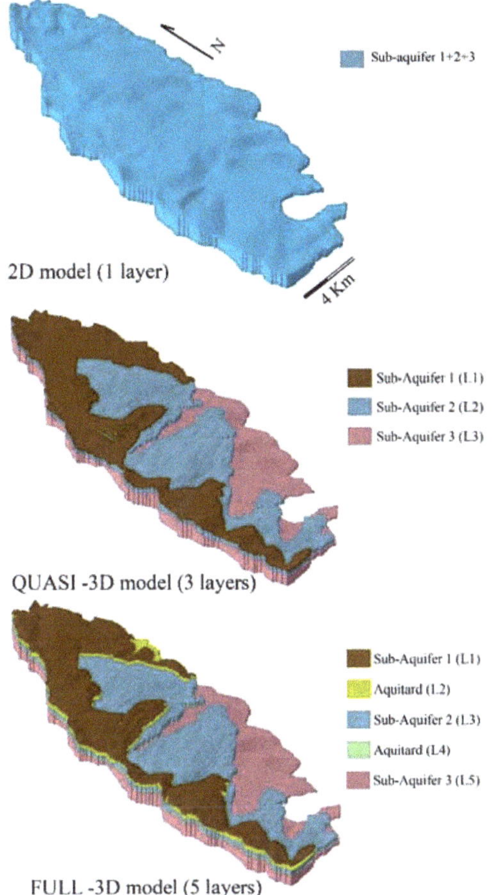

Figure 5. Sketch of the three steady-state models implemented with 2D, quasi-3D and fully 3D settings. Only active cells are shown.

3.2. Model Calibration

Models were calibrated in steady state through a manual trial-and-error procedure, using the available data, which comprise:

- the discharge increment in Nera River reaches R1 to R4;
- the discharge of Precetto stream;
- the discharge of Lupa, Scheggino and Pacce springs (Table 1).
- the head measured in two wells (Renari di Capriglia and Scheggino, Figure 1)

Flux and head targets used for model calibration are listed in Table 2. The topography elevation, especially in deep gorges, where the geological formations of the lower sub-aquifers are outcropping provided additional head constraints. The calibrated parameters were the horizontal and vertical hydraulic conductivities of each layer and river-bed hydraulic conductivity and thickness (Tables 3 and 4). The streambed conductance, which enters in the computation of the groundwater–surface water interaction, is computed by the software as the ratio of riverbed hydraulic conductivity and thickness.

Table 2. Flux and head targets used for model calibration.

Layer	Head Target	Flux Target
1 (sub-aquifer 1) Scaglia unit	n.a.	Nera River reach R1, Lupa spring, Scheggino spring, Precetto stream
3 (sub-aquifer 2), Maiolica unit	Scheggino well	Nera River reach R2, R3 and R4, Pacce spring
5 (sub-aquifer 3), Calcare massiccio and Corniola	Renari di Capriglia well	n.a.

Table 3. Range of calibrated Kh and Kz values.

Model Setting	N of Layers	N of Active Cells	N of Calibrated K Zones	Range of Kh Values, m/s	Range of Kz Values, m/s
2D	1	23,248	3	$1.1 \times 10^{-5}/2.8 \times 10^{-5}$	$1.15 \times 10^{-6}/2.8 \times 10^{-6}$
Quasi-3D	3	50,959	6	$0.1 \times 10^{-6}/4.6 \times 10^{-5}$	$1.1 \times 10^{-7}/2.8 \times 10^{-6}$
Fully 3D	5	77,212	12	$1.1 \times 10^{-6}/8.1 \times 10^{-5}$ (high K strip: 5.7×10^{-4} m/s)	$1.1 \times 10^{-7}/8.1 \times 10^{-5}$ (high K strip: 6.3×10^{-4} m/s)

Table 4. Initial and calibrated values of riverbed hydraulic conductivity and thickness.

River Reach	Riverbed Kv (m/s)		Riverbed Thickness (m)	
	Initial	Calibrated	Initial	Calibrated
1–4 (Nera River)	5.7×10^{-5}	2.3×10^{-3}	1	3.0
5 (Precetto stream)	5.7×10^{-5}	2.3×10^{-3}	1	3.0

The calibration phase involved an iterative refining of both aquifer K values and distribution and also riverbed Kv and thickness values, on the basis of a comparison between simulation results and observations. The calibration of hydraulic conductivity in each layer was performed by gradually modifying the K values by "zones", with each zone representing a homogeneous area of the layer. At first, the values from [36] were assigned to the sub-aquifers of the fully 3D model. During the calibration phase, the number of hydraulic conductivity zones was increased in order to refine the K distribution until the simulated values were close to the observations. After each K value variation, the model was rerun to compare results to the previous setting. Then, the pattern of K values was progressively refined, and many different K zones were added. The most difficult task was to reproduce the discharge of the river and maintain a sufficient discharge in the upper reaches. In order to reproduce this discharge distribution, a high conductivity strip was added, oriented N–S in layer 5 of the fully 3D model, which simulates faults bounding Calcare Massiccio Fm, acting as a preferential flow zone. The transmissivity distribution resulting in the fully 3D model after this calibration was transposed to the quasi-3D and 2D models following the Equation (1). Two-dimensional and quasi-3D models were further calibrated after this step in order to seek the best match using observations. In the final setting, the numbers of the calibrated conductivity zones was 3, 6 and 12 for the 2D, quasi-3D and fully 3D models, respectively (Table 3 and Figure 6).

The Nera River can be conceptualized as a bedrock stream with a limited thickness of loose alluvial sediments, up to a few meters of gravels and sands with very high permeability. In the calculation of the groundwater–surface water interaction, what really matters is the ratio between the riverbed Kv/thickness. During calibration, in order to obtain the observed high outflow to the river, this ratio was increased from 5.7×10^{-5} m/s to 7.7×10^{-4} m/s to obtain a minimal resistance to the groundwater flow from the aquifer to the river, assuming 3 m of riverbed thickness and 2.3×10^{-3} m/s for Kv (Table 4), which is consistent with the high permeability of the gravels and the presence of a few meters of riverbed above the bedrock.

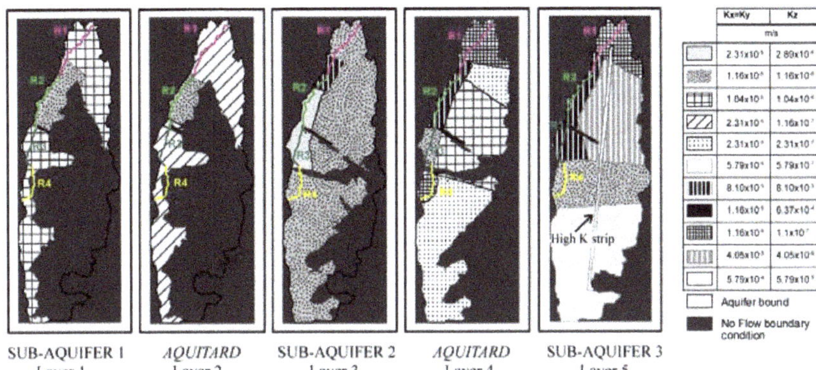

Figure 6. Distribution of K zones for fully 3D calibrated model. R1, R2, R3 and R4 indicates the river reaches (see also Figure 1).

Time of calculation ranges from a few seconds for the 2D model to several minutes for the fully 3D.

4. Results and Discussion

All the calibrated models give a good match between simulated and observed heads in the two only available wells (Figure 7 and Table 5). However, the 2D model was never able to match the observed distribution of river baseflow along each reach of the Nera River. In fact, assuming a single aquifer, this model is not able to feed the upstream cells of the river, and the groundwater converges toward the most downstream RIV cells (Figure 8A), which does not correspond to the observed groundwater–surface water interaction. The upstream reach R1 (Figure 1) loses water to the aquifer instead of gaining, while the downstream reach R4 gains the double of the observed discharge (Table 5).

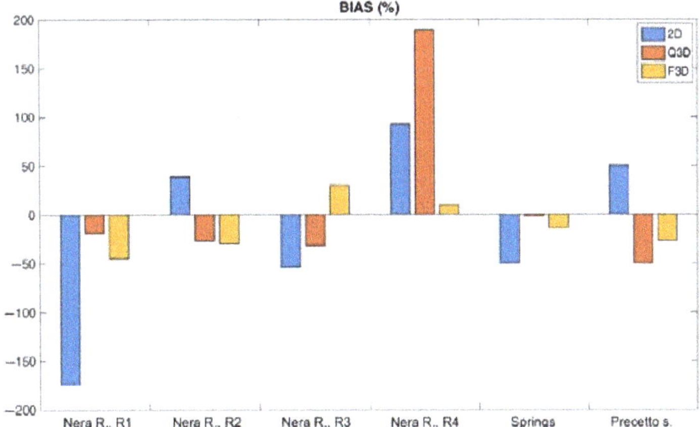

Figure 7. Target bias on the discharge (%).

Table 5. Results of the three models and comparison with the observed values. The bias % is shown.

Element	Observed	2D	Quasi-3D	Fully 3D
Well Scheggino, m a.s.l.	273	271.51	275.57	273.26
Well Renari, m a.s.l.	301	269.88	312.52	305.00
Nera R., R1, l/s	436	−322.00	351.92	238.56
Nera R., R2, l/s	1120	1557.82	812.62	778.68
Nera R., R3, l/s	1110	507.18	751.17	1445.15
Nera R., R4, l/s	609	1173.95	1760.49	669.35
springs, l/s	390	587.99	195.44	286.21
Precetto s., l/s	120	60.40	117.71	103.29

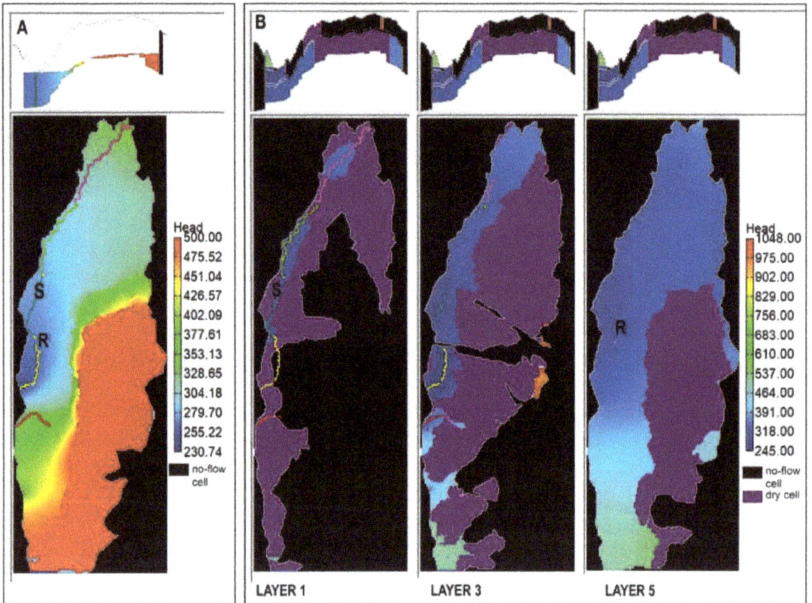

Figure 8. Simulated head for the 2D model (**A**) and for the layers 1, 3 and 5 in the fully 3D model (**B**). Model cross sections at the top of figures (row 162, column 23). Approximate locations of Scheggino well (S) and Renari well (R) are shown.

In order to reproduce the discharge correctly partitioned along the reaches, it was necessary to construct multilayer models with consideration of system aquitards.

An important advantage of quasi-3D and fully 3D models with respect to the 2D model is that the slope and dipping of single layers can be correctly reproduced, allowing flow to be reliably simulated along the layers. In addition, the quasi-3D and fully 3D models are able to represent localized areas of preferential upward flow with a nonuniform vertical leakage array or by adding high vertical permeability zones to simulate discontinuities in the aquitards, respectively.

Further, the quasi-3D and fully 3D models allow partitioning of the recharge into different numerical layers according to the geological setting. For example, where the anticline is eroded, the recharge is assigned to the outcropping layers; thus, each of the sub-aquifers directly receive a share of the total recharge (Figure 5). In the quasi-3D and fully 3D models, the simulations show that a large part of the anticline hosts dry cells due to the high elevation of the aquifer bottom with respect to the calculated heads. When using the PCG2 solver (quasi-3D model), the dry cells are excluded from the head calculations and recharge is passed to an underlying active cell. Conversely, in the NWT upstream weighting package (fully 3D model), dry cells remain active with a calculated head value that falls below the bottom of the cell; that head is used to set up the gradient to pass recharge to an underlying cell hosting the water table or toward adjacent cells in the case of the bottom layer. In particular, the quasi-3D model performs better than the 2D, but the simulated discharge of the reach R4 is still not acceptable (Table 5). The fully 3D model shows the best performance in terms of bias (Figure 7). The hydraulic heads calculated for the three aquifers in the fully 3D model (Figure 8B) show the important role of the basal sub-aquifer (layer five) to direct the high infiltration from the mountainous recharge areas to the northern, downstream, sectors of the structure. Due to the complex geometry, sub-aquifers one and two show extended dry cells areas, and the head potentials are strongly influenced by the RIV cells.

Indeed, the use of five layers allows for a very detailed setting of the distribution of K zones which, once calibrated, should reflect a possible pattern of permeability values, making the model results consistent with the observed discharge rates and heads. Previous runs, performed with relatively high K strips in the sub-aquifers three and five, e.g., the NW–SE faults cutting the structure near Scheggino (see Figures 1 and 6), were unable to drive the fluxes toward the northern part of the structure. Finally, the insertion of the high conductivity strip in the sub-aquifer three (Figure 6) ameliorated by far the calibration of the fully 3D model. Thus, the high hydraulic conductivity zones in a generalized and simplified view of the geological setting are likely to reflect fractures or faults zones. Tectonics also drive the vertical exchanges among the sub-aquifers through the aquitards. A better match was achieved considering zones of relatively high vertical permeability within the aquitards, simulating highly tectonized zones or stratigraphic gaps, for example in correspondence of reach R2 (Figure 6), which allows the groundwater to flow vertically upward from layer five to the upper layers.

However, high hydraulic conductivity zones alone were not sufficient to maintain the hydraulic gradients steep enough to feed the most upgradient reach R1. The groundwater mostly flows in the sub-aquifer three. Despite a prevalent diffuse circulation (which justifies the EPM approach), the high hydraulic conductivity strip along the anticline axis that facilitates the groundwater to flow from south to north is consistent with a discrete groundwater circulation pattern. This was confirmed by observation during the perforation of the Renari well for the Calcare Massiccio aquifer, which showed a prevalent circulation through fractures. Table 5 lists the results of the three models after calibration. Figure 7 shows the bias %.

Ultimately, the calibration process points to the presence of structural elements in the flow system that are not readily observable by other means. In this sense the stepwise process is not meant to produce a unique representation of the subsurface, but rather points to the existence of subsurface features that seem to be controlling flow to the river according to the fully 3D model, but which are otherwise very difficult to characterize by field work, i.e., the examination of outcrops or geophysical prospections. Such a model can be considered as an interpretive model [44] in the sense of a screening model that helps the modeler to develop an initial understanding of a groundwater system and/or test hypotheses about the system. Calibration through manual trial-and-error proved that a simple 2D model is not able to match field observations and thus is not an acceptable model for the site. Further, it provided insights into how parameter changes in different areas of the model could correspond to unknown subsurface features [51]. For example, the high permeability strip in the deepest layer in the fully 3D model could suggest the existence of a still unknown, developed karstic system. This manual process, although seemingly unable to find a unique solution, might be the only way forward given the impossibility of quantifying the "true uncertainty" in natural systems and the error inevitably associated with a complex structural model with few data supporting observations [52]. As stated by Fienen at al. (2009) [53], parsimonious models should avoid unnecessary or unsupported complexity while accurately delineating flow paths given our state of knowledge about the field setting.

5. Conclusions

Modelling a complexly folded and faulted hydrogeological system requires a stepwise procedure to reach the best solution with a parsimonious model setting. The inherent complexity of the conceptual model was reproduced in the numerical model by progressively adding elements to the model grid such as new layers from the 2D to the fully 3D model and hydraulic conductivity zones, and by calibrating the steady-state solution by manual trial and error. Furthermore, gradually increasing the model complexity can be a useful approach for simulating groundwater flow when few head data are available as it is common in karst aquifers. Due to the prevalent diffuse circulation, an EPM approach was used; however, the calibrated setting of hydraulic conductivity zones suggests a discrete groundwater circulation pattern, which was successfully simulated by adding a high permeability longitudinal strip. The calibrated pattern of K zones both for sub-aquifers and aquitards is likely to reflect the structural and stratigraphic setting. The quasi-3D and the fully 3D models both allow for recharge partitioning into the sub-aquifers. The vertical exchanges among the sub-aquifers are regulated by leakage coefficient or aquitards parametrization, respectively. The higher number of layers compared to the 2D model allows the 3D simulations to drive the groundwater flow towards the different parts of the river. However, the fully 3D model best matches the observed flow distribution at the different reaches along the river, simulating reliable flow paths and recharge partitioning into layers. The Newton–Raphson formulation of MODFLOW2005 is required to achieve convergence and reduce model error mainly due to cells drying and rewetting and producing numerical instabilities. The numerical model demonstrated the major impact of folded and faulted geological structures on controlling the flow dynamics in terms of flow direction, water heads and spatial distribution of the outflows to the river and springs. The stepwise process of model construction and calibration, even with a limited number of head and flux targets, points to the presence of structural elements in the subsurface that otherwise can escape observation in field studies of the terrain.

Author Contributions: Conceptualization, E.P. and C.D.S.; methodology, E.P., C.D.S., N.G. and E.R.; software, E.P., C.D.S., A.B.P. and N.G.; validation, E.P., C.D.S. and E.R.; formal analysis, E.P., C.D.S. and E.R.; investigation, E.P. and C.D.S.; resources, E.P.; data curation, E.P. and C.D.S.; writing—original draft preparation, E.P. and C.D.S.; writing—review and editing, E.P., C.D.S. and E.R.; visualization, C.D.S., A.B.P., E.R. and N.G.; supervision, E.P. All authors have read and agreed to the published version of the manuscript.

Funding: This research received no external funding.

Data Availability Statement: Data of springs discharge have been provided by the Regional Agency for Environmental Protection (ARPA Umbria). Precipitation, temperature and river discharge data have been provided by the Hydrographic Service of Umbria (Ufficio Idrografico della Regione Umbria).

Acknowledgments: The authors are deeply grateful to Daniel Feinstein and Rundy Hunt (USGS) for their useful suggestions and review of the manuscript. We also wish to thank Reviewer 1 for his/her interesting comments and constructive criticisms.

Conflicts of Interest: The authors declare no conflict of interest.

References

1. Ford, D.C.; Williams, P.W. *Karst Hydrogeology and Geomorphology*; John Wiley & Sons: Chichester, UK; Hoboken, NJ, USA, 2007.
2. Worthington, S.; Ford, D. Self-Organized Permeability in Carbonate Aquifers. *Ground Water* **2009**, *47*, 326–336. [CrossRef] [PubMed]
3. Worthington, S.R.H. Management of Carbonate Aquifers. In *Karst Management*; Van Beynen, P.E., Ed.; Department Geography, University of South Florida: Tampa, FL, USA, 2011; Volume XII, pp. 243–261. [CrossRef]
4. Dragoni, W.; Mottola, A.; Cambi, C. Modeling the effects of pumping wells in spring management: The case of Scirca spring (central Apennines, Italy). *J. Hydrol.* **2013**, *493*, 115–123. [CrossRef]
5. Fiorillo, F.; Guadagno, F.M. Karst Spring Discharges Analysis in Relation to Drought Periods, Using the SPI. *Water Resour. Manag.* **2010**, *24*, 1867–1884. [CrossRef]

6. Romano, E.; Preziosi, E. Precipitation pattern analysis in the Tiber River basin (central Italy) using standardized indices. *Int. J. Climatol.* **2013**, *33*, 1781–1792. [CrossRef]
7. Citrini, A.; Camera, C.; Beretta, G.P. Nossana Spring (Northern Italy) under Climate Change: Projections of Future Discharge Rates and Water Availability. *Water* **2020**, *12*, 387. [CrossRef]
8. Leone, G.; Pagnozzi, M.; Catani, V.; Ventafridda, G.; Esposito, L.; Fiorillo, F. A hundred years of Caposele spring discharge measurements: Trends and statistics for understanding water resource availability under climate change. *Stoch. Hydrol. Hydraul.* **2020**, *35*, 345–370. [CrossRef]
9. Romano, E.; Petrangeli, A.B.; Salerno, F.; Guyennon, N. Do recent meteorological drought events in central Italy result from long-term trend or increasing variability? *Int. J. Clim.* **2021**, *42*, 4111–4128. [CrossRef]
10. Romano, E.; Petrangeli, A.B.; Preziosi, E. Spati al and Time Analysis of Rainfall in the Tiber River Basin (Central Italy) in relation to Discharge Measurements (1920-2010). *Procedia Environ. Sci.* **2011**, *7*, 258–263. [CrossRef]
11. *IPCC 2022—Climate Change 2022: Impacts, Adaptation, and Vulnerability. Contribution of Working Group II to the Sixth Assessment Report of the Intergovernmental Panel on Climate Change*; Pörtner, H.-O.; Roberts, D.C.; Tignor, M.; Poloczanska, E.S.; Mintenbeck, K.; Alegría, A.; Craig, M.; Langsdorf, S.; Löschke, S.; Möller, V.; et al. (Eds.) Cambridge University Press: Cambridge, UK, 2022.
12. Preziosi, E.; Romano, E. Are large karstic springs good indicators for Climate Change effects on groundwater? In Proceedings of the European Geosciences Union General Assembly 2013, Vienna, Austria, 7–12 April 2013. *Geophys. Res. Abstr.* **2013**, *15*, 9367.
13. Romano, E.; Del Bon, A.; Petrangeli, A.; Preziosi, E. Generating synthetic time series of springs discharge in relation to standardized precipitation indices. Case study in Central Italy. *J. Hydrol.* **2013**, *507*, 86–99. [CrossRef]
14. Band, S.S.; Heggy, E.; Bateni, S.M.; Karami, H.; Rabiee, M.; Samadianfard, S.; Chau, K.-W.; Mosavi, A. Groundwater level prediction in arid areas using wavelet analysis and Gaussian process regression. *Eng. Appl. Comput. Fluid Mech.* **2021**, *15*, 1147–1158. [CrossRef]
15. Mallick, J.; Almesfer, M.K.; Alsubih, M.; Talukdar, S.; Ahmed, M.; Kahla, N.B. Developing a new method for future groundwater potentiality mapping under climate change in Bisha watershed, Saudi Arabia. *Geocarto Int.* **2022**, *1*, 1–33. [CrossRef]
16. Snoussi, M.; Jerbi, H.; Tarhouni, J. Integrated Groundwater Flow Modeling for Managing a Complex Alluvial Aquifer Case of Study Mio-Plio-Quaternary Plain of Kairouan (Central Tunisia). *Water* **2022**, *14*, 668. [CrossRef]
17. European Community. Directive 2000/60/EC of the European Parliament and of the Council of 23 October 2000 Establishing a Framework for Community Action in the Field of Water Policy. 2000. Available online: https://eur-lex.europa.eu/LexUriServ/LexUriServ.do?uri=CELEX:32000L0060:EN:HTML (accessed on 5 July 2022).
18. Gattinoni, P.; Francani, V. Depletion risk assessment of the Nossana Spring (Bergamo, Italy) based on the stochastic modeling of recharge. *Appl. Hydrogeol.* **2010**, *18*, 325–337. [CrossRef]
19. Levison, J.; Larocque, M.; Ouellet, M. Modeling low-flow bedrock springs providing ecological habitats with climate change scenarios. *J. Hydrol.* **2014**, *515*, 16–28. [CrossRef]
20. Reimann, T.; Giese, M.; Geyer, T.; Liedl, R.; Maréchal, J.C.; Shoemaker, W.B. Representation of water abstraction from a karst conduit with numerical discrete-continuum models. *Hydrol. Earth Syst. Sci.* **2014**, *18*, 227–241. [CrossRef]
21. Scanlon, B.R.; Mace, R.E.; Barrett, M.E.; Smith, B. Can we simulate regional groundwater flow in a karst system using equivalent porous media models? Case study, Barton Springs Edwards aquifer, USA. *J. Hydrol.* **2003**, *276*, 137–158. [CrossRef]
22. Abusaada, M.; Sauter, M. Studying the Flow Dynamics of a Karst Aquifer System with an Equivalent Porous Medium Model. *Ground Water* **2012**, *51*, 641–650. [CrossRef]
23. Ben-Itzhak, L.; Gvirtzman, H. Groundwater flow along and across structural folding: An example from the Judean Desert, Israel. *J. Hydrol.* **2005**, *312*, 51–69. [CrossRef]
24. Dafny, E.; Burg, A.; Gvirtzman, H. Effects of Karst and geological structure on groundwater flow: The case of Yarqon-Taninim Aquifer, Israel. *J. Hydrol.* **2010**, *389*, 260–275. [CrossRef]
25. Valota, G.; Giudici, M.; Parravicini, G.; Ponzini, G.; Romano, E. Is the forward problem of ground water hydrology always well posed? *Ground Water* **2002**, *40*, 500–508. [CrossRef] [PubMed]
26. Niswonger, R.G.; Panday, S.; Ibaraki, M. MODFLOW-NWT, A Newton formulation for MODFLOW-2005. *US Geol. Surv. Tech. Methods* **2011**, *6*, 44.
27. Mantoglou, A. On optimal model complexity in inverse modeling of heterogeneous aquifers. *J. Hydraul. Res.* **2005**, *43*, 574–583. [CrossRef]
28. Hill, M.C. The practical use of simplicity in developing groundwater models. *Ground Water* **2006**, *44*, 775–781. [CrossRef] [PubMed]
29. Moore, C.; Wöhling, T.; Doherty, J. Efficient regularization and uncertainty analysis using a global optimization methodology. *Water Resour. Res.* **2010**, *46*, W08527. [CrossRef]
30. Hill, M.C. Methods and Guidelines for Effective Model Calibration. In *U.S. Geological Survey, Water-Resources Investigations Report 98–4005*; U.S. Geological Survey: Denver, CO, USA, 1998; p. 98. Available online: https://water.usgs.gov/nrp/gwsoftware/modflow2000/WRIR98-4005.pdf (accessed on 10 August 2022).
31. Barchi, M.R.; Lemmi, M. Geology of the area of Monte Coscerno-Monte di Civitella (South-eastern Umbria). *Boll. Soc. Geol. Ital.* **1996**, *115*, 601–624.
32. Tavarnelli, E. Analisi geometrica e cinematica dei sovrascorrimenti compresi fra la Valnerina e la conca di Rieti (Appennino Umbro-Marchigiano-Sabino). *Boll. Soc. Geol. Ital.* **1996**, *113*, 249–259.

33. Pace, P.; Calamita, F.; Tavarnelli, E. Brittle-Ductile Shear Zones along Inversion-Related Frontal and Oblique Thrust Ramps: Insights from the Central-Northern Apennines Curved Thrust System (Italy). In *Ductile Shear Zones: From Micro- to Macro-Scales*; Mukherjee, S., Mulchrone, K.F., Eds.; John Wiley & Sons: Chichester, UK, 2015; Chapter 8; pp. 111–127.
34. Calamita, F.; Deiana, G. The arcuate shape of the Umbria-Marche-Sabina Apennines (central Italy). *Tectonophysics* **1988**, *146*, 139–147. [CrossRef]
35. Boni, C.; Preziosi, E. Una possibile simulazione numerica dell'acquifero basale di M.Coscerno—M.Aspra (bacino del fiume Nera). *Geol. Appl. E Idrogeol.* **1993**, *28*, 131–140.
36. Preziosi, E. Simulazioni Numeriche di Acquiferi Carbonatici in Aree Corrugate: Applicazioni al Sistema Idrogeologico Della Valnerina (Italia Centrale). In *Quad IRSA-CNR*; Istituto di Ricerca Sulle Acque-CNR: Rome, Italy, 2007; Volume 125, p. 225. ISSN 0390-6329.
37. Mastrorillo, L.; Baldoni, T.; Banzato, F.; Boscherini, A.; Cascone, D.; Checcucci, R.; Petitta, M.; Boni, C. Quantitative hydrogeological analysis of the carbonate domain of the Umbria Region (Central Italy). *Ital. J. Eng. Geol. Environ.* **2009**, *1*, 137–155. [CrossRef]
38. Mastrorillo, L.; Petitta, M. Hydrogeological conceptual model of the upper Chienti River Basin aquifers (Umbria-Marche Apennines). *Ital. J. Geosci.* **2014**, *133*, 396–408. [CrossRef]
39. Di Matteo, L.; Dragoni, W.; Ulderico, V.; Latini, M.; Spinsanti, R. Risorse idriche sotterranee e loro gestione: Il caso dell'ATO2 Umbria (Umbria meridionale). *Acque Sotter.* **2005**, *96*, 9–21.
40. Damiani, A.V.; Baldanza, A.; Barchi, M.R.; Boscherini, A.; Checcucci, R.; Decandia, F.A.; Felicioni, G.; Lemmi, M.; Luchetti, L.; Motti, A.; et al. *Geological Map of Italy. Scale 1:50,000. Scheet 563 "Spoleto"*; ISPRA-Servizio Geologico Nazionale: Roma, Italy, 2008.
41. ARPA Umbria. Available online: https://apps.arpa.umbria.it/acqua/contenuto/monitoraggio-in-continuo-acque-sotterranee (accessed on 13 June 2022).
42. Preziosi, E.; Romano, E. From hydrostructural analysis to mathematical modelling of regional aquifers. *Ital. J. Eng. Geol. Environ.* **2009**, *1*, 183–198. [CrossRef]
43. ARPA Umbria. Il Sistema Informativo di Arpa Umbria e la Conoscenza Ambientale, a cura di Mauro Emiliano, Quaderni Arpa Umbria. 2004. Available online: https://www.arpa.umbria.it/pagine/il-sistema-informativo-di-arpa-umbria-e-la-conosce (accessed on 5 July 2022).
44. White, W.B. *Geomorphology and Hydrology of Karst Terrains*; Oxford Universuty Press: New York, NY, USA, 1988; p. 464.
45. Vigna, B.; Banzato, C. The hydrogeology of high-mountain carbonate areas: An example of some Alpine systems in southern Piedmont (Italy). *Environ. Earth Sci.* **2015**, *74*, 267–280. [CrossRef]
46. Thornthwaite, C.W.; Mather, J.R. The water balance. *Climatology* **1955**, *8*, 5–86.
47. Thornthwaite, C.W.; Mather, J.R. Instructions and Tables for Computing Potential Evapotranspiration and the Water Balance. *Centerton* **1957**, *10*, 185–310.
48. Umbriadue Servizi Idrici S.C.A R.L.; Technologies for Water Services S.p.A. Acquedotto Scheggino-Pentima. In *Progetto Definitivo, Technical Report*; Servizio Idrico Integrato: Terni, Italy, 2006.
49. Rambaugh, J.O.; Rambaugh, D.B. Guide to Using Groundwater Vistas, Version 7, Environmental Simulation. 2017. Available online: https://d3pcsg2wjq9izr.cloudfront.net/files/4134/download/730323/4-gv7manual.pdf (accessed on 5 July 2022).
50. Feinstein, D.T.; Fienen, M.N.; Kennedy, J.L. Development and Application of a Groundwater/Surface-Water Flow Model Using MODFLOW-NWT for the Upper Fox River Basin, Southeastern Wisconsin. In *U.S. Geological Survey Scientific Investigations Report*; CreateSpace Independent Publishing Platform: Scotts Valley, CA, USA, 2012.
51. Anderson, M.P.; Woessner, W.W.; Hunt, R.J. *Applied Groundwater Modeling: Simulation of Flow and Advective Transport*, 2nd ed.; Academic Press, Inc.: Cambridge, MA, USA, 2015; p. 564. ISBN 9780120581030.
52. Hunt, R.J. Applied Uncertainty. *Groundwater* **2017**, *55*, 771–772. [CrossRef]
53. Fienen, M.F.; Hunt, R.J.; Krabbenhoft, D.P.; Clemo, T. Obtaining parsimonious hydraulic conductivity fields using head and transport observations: A Bayesian Geostatistical parameter estimation approach. *Water Resour. Res.* **2009**, *45*, W08405. [CrossRef]

 water

MDPI

Article

Groundwater Potential Assessment Using GIS and Remote Sensing Techniques: Case Study of West Arsi Zone, Ethiopia

Julla Kabeto [1,*], Dereje Adeba [1], Motuma Shiferaw Regasa [2] and Megersa Kebede Leta [3,*]

1. Department of Hydraulic and Water Resources Engineering, Institute of Engineering and Technology, Wollega University, Nekemte 395, Oromia Region, Ethiopia; mo_dereje2018@yahoo.com
2. Department of Hydrology and Hydrodynamics, Institute of Geophysics, Polish Academy of Sciences, 01-452 Warsaw, Poland; mregasa@igf.edu.pl
3. Faculty of Agriculture and Environmental Sciences, University of Rostock, Satower Str. 48, 18051 Rostock, Germany

* Correspondence: nafyomk1023@gmail.com (J.K.); megersa.kebede@uni-rostock.de (M.K.L.)

Abstract: Groundwater is a crucial source of water supply due to its continuous availability, reasonable natural quality, and being easily diverted directly to the poor community more cheaply and quickly. The West Arsi Zone residents remain surface water dependent due to traditional exploration of groundwater, which is a tedious approach in terms of resources and time. This study uses remote sensing data and geographic information system techniques to evaluate the groundwater potential of the study area. This technique is a fast, accurate, and feasible technique. Groundwater potential and recharge zone influencing parameters were derived from Operational Land Imager 8, digital elevation models, soil data, lithological data, and rainfall data. Borehole data were used for results validation. With spatial analysis tools, the parameters affecting groundwater potential (LULC, soil, lithology, rainfall, drainage density, lineament density, slope, and elevation) were mapped and organized. The weight of the parameters according to percent of influence on groundwater potential and recharge was determined by Analytical Hierarchy Process according to their relative influence. For weights allocated to each parameter, the consistency ratio obtained was 0.033, which is less than 0.1, showing the weight allocated to each parameter is acceptable. In the weighted overlay analysis, from a percent influence point of view, slope, land use/cover, and lithology are equally important and account for 24% each, while the soil group has the lowest percent of influence, which accounts only 2% according to this study. The generated groundwater potential map has four ranks, 2, 3, 4, and 5, in which its classes are Low, Moderate, High, and Very High, respectively, based on its groundwater potential availability rank and class. The area coverage is 9825.84 ha (0.79%), 440,726.49 ha (35.46%), 761,438.61 ha (61.27%), and 30,748.68 ha (2.47%) of the study area, respectively. Accordingly, the western part of district is expected to have very high groundwater potential. High groundwater potential is concentrated in the central and western parts whereas moderate groundwater potential distribution is dominant in the eastern part of the area. The validation result of 87.61% confirms the very good agreement among the groundwater record data and groundwater potential classes delineated.

Keywords: GIS; remote sensing; groundwater potential assessment; analytical hierarchy processes; weight overlay analysis; West Arsi Zone

Citation: Kabeto, J.; Adeba, D.; Regasa, M.S.; Leta, M.K. Groundwater Potential Assessment Using GIS and Remote Sensing Techniques: Case Study of West Arsi Zone, Ethiopia. *Water* **2022**, *14*, 1838. https://doi.org/10.3390/w14121838

Academic Editor: Cristina Di Salvo

Received: 6 May 2022
Accepted: 1 June 2022
Published: 7 June 2022

1. Introduction

Water is the most significant natural resource supporting human health, economic development, and ecological diversity. Groundwater is part of the water cycle, and which is stored in the saturated zones underneath the land surface and moves slowly through geologic formations called aquifers. Water could remain in an aquifer for hundreds or thousands of years. The existence and flow of groundwater is controlled by factors such as geological formations, soil type, lineament density, slope, drainage density, rainfall form, morphology, land-use/land-cover characteristics, and the interrelation between them [1,2]. Most groundwater originates from precipitation that percolate through the rock strata. Groundwater replenished, or recharged, by rain and snow melt that seep down in to the cracks and crevices beneath the land's surface. In different areas of the world, people face solemn water scarcities because groundwater discharge for use is faster than its natural replenishment (recharge). Due to the reason that groundwater is incessantly accessed and its reasonable natural quality, it becomes a vital source of water supply, both in urban and rural areas of any country. Indeed, due to groundwater being easily diverted directly to poor communities far more cheaply and quickly than surface water, it helps in poverty mitigation and reduction.

Lately, preparing a groundwater potential district map is crucial to delineate the location of a new abstraction well to fit the increasing demand of water. Groundwater resource thematic developing helps in the optimum use and appropriate safeguarding of groundwater resources [3–5]. The usual method of preparing a groundwater potential thematic depends on land surveying. Currently, GIS and RS techniques made groundwater resource potential detection easier, accurate, and faster [6]. Demarking of groundwater existence locations using RS data and GIS depends on indirect investigation of the directly visible factors mentioned above. A blend of these methods had been considered to be an effective instrument in locating and mapping groundwater potential [7,8]. The Geographic Information System is a very helpful and influential instrument in demarcation of groundwater potential and scarcity zones, analyzing and quantifying multivariate features of groundwater incidence [9]. It has the power of developing information in different thematic layers and integrating them with adequate accuracy within a short period of time. Satellite imageries are progressively used in groundwater investigation due to their usefulness in categorizing various ground topographies, which may help as either direct or indirect pointers of presence of groundwater [10–12]. Geospatial techniques help in generating and analyzing thematic layers such as geology, topography, soil, and land use for generating groundwater potential regions map [5,13,14].

Analytic Hierarchy Process (AHP) is a concept of measurement by pairwise comparisons, which depend on judgements of professionals to originate precedence scales. The judgements are made by method of rank of absolute judgements which denotes how the elements control each other with respect to an attribute given [15]. The AHP approach is a very flexible for the reason that it produces an easy way to discover the relationship among criteria and alternatives. AHP can be made in several ways, one of which is to use proficient choice software, by which its operation and calculation stages are done automatically. In the current study, Saaty's Analytic Hierarchy Process (AHP), which is a broadly used Multi-Criteria Decision Analysis (MCDA), was used to evaluate groundwater potential districts of regions. Decision making comprises various criteria and sub-criteria used to rank the alternatives of a decision. The criteria may be impalpable and have no measurements to help as a guide to rank the alternatives. Creating priorities for the criteria themselves in order to weigh the primacies of the options and add over all the criteria to gain the desired overall ranks of the alternatives is a thought-provoking task. In groundwater potential and recharge zone determination and groundwater potential zonation, the percent weight of factors will be calculated and determined by the AHP Excel calculator in which experts' experiences are deemed highly important and implemented. GIS techniques were used for the weighted overlay analysis and integrated with multi-criteria analysis.

High relief and steep slopes impart higher runoff, while topographical depressions and flat areas increase infiltration. Areas with a high drainage density increase surface runoff more than areas with a low drainage density. A high lineament density produces good groundwater potential and a low lineament density produces low groundwater potential. Areas covered with forest, other vegetation, and agriculture provide cracks and loosen the soil, so infiltration will be more and runoff will be less, whereas in urban areas and bare land the rate of infiltration may decrease. Loam, silty-loam, silt, sandy-loam, sand, and loamy-sand soil textures have high permeability whereas in clay, sandy-clay, silty-clay,

clay-loam, sandy-clay-loam, and silty-clay-loam the permeability is poor; in coarse granule loam the permeability is high and the flow is rapid. Areas having more rainfall have a good groundwater prospect and the areas with low rainfall have poor groundwater prospects. Geological formations that are deeply fractured, with cracks, fissures, folds, and discontinuities, are porous, which indicates high groundwater recharge prospects.

The purpose of groundwater potential modeling in this study is to identify groundwater-accessible locations throughout the study region in an easy and simple way. This will increase the accuracy and efficiency, save time, and the economy during groundwater resources management, planning, and developing by governmental and non-governmental organizations. The study was aimed to conduct (i) the demarcation of groundwater potential districts and isolation of appropriate sites for groundwater development using weighted overlay analysis techniques by means of AHP and ArcGIS; (ii) prepare thematic layers (LULC, soil, geology/lithology, rainfall, drainage density, lineament density, elevation, and slope) and reclassifying them for multicriteria overlay analysis; (iii) perform weighted overlay analysis in ArcGIS to decide on a suitable site for groundwater potential districts in the West Arsi Zone; and (iv) prepare well inventory data maps and in Excel to confirm the groundwater potential districts layers generated from the weight overlay analysis.

2. Material and Methods

2.1. Study Area Description

This study was conducted in West-Arsi zone; one of 20 zones of the Oromia regional government located in the central part of Ethiopia. It is situated between the 06°00′ and 08°00′ N latitudes and 38°00′ and 39°50′ E longitudes in the central part of Ethiopia (Figure 1). It covers an approximate total area of about 1,246,851.15 hectares or 12,468.51 square kilometers.

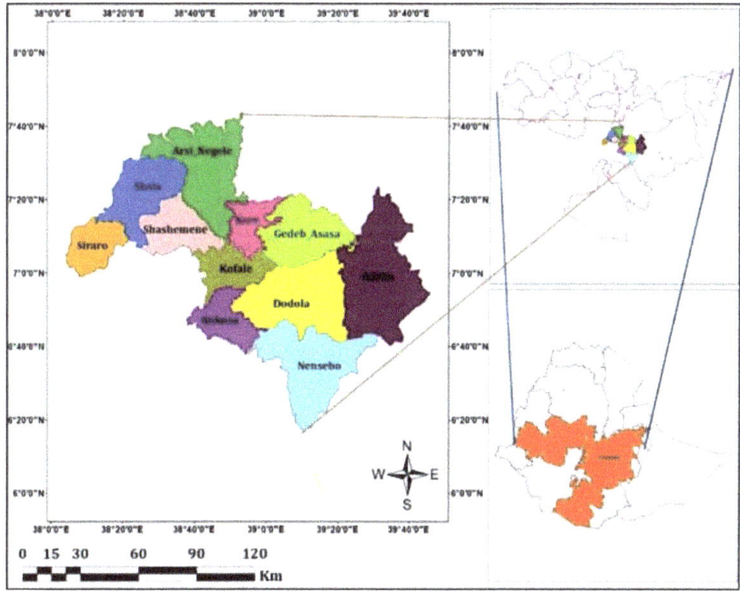

Figure 1. Location map of West Arsi Zone.

It is sub-divided into 11 districts (woredas), namely, Adaba, Arsi-Negele, Dodola, Gadab Asasa, Kokosa, Kofale, Kore, Shala, Siraro, Shashamane, and Nensebo.

The current (2019) population of West Arsi Zone is expected to be 2,761,464, as set by the world population prospect. The area has monthly average temperatures that vary between 20 and 24 °C and annual average temperatures between 13 and 28 °C. An average altitude of an area ranges between 1464 and 4171 m. a.s.l. The yearly average annual rainfall varies between 700.5 mm and 1976.7 mm. The average sunshine hour of an area varies between 3.8 and 8.8 h/day. Annual average relative humidity and wind speed of the study area are about 71.4% and 1.4 m/s, respectively. The area is drained by three river basins; the Rift Valley Lake basin, Wabe Shebele, and Genale Dewa (Figure 2).

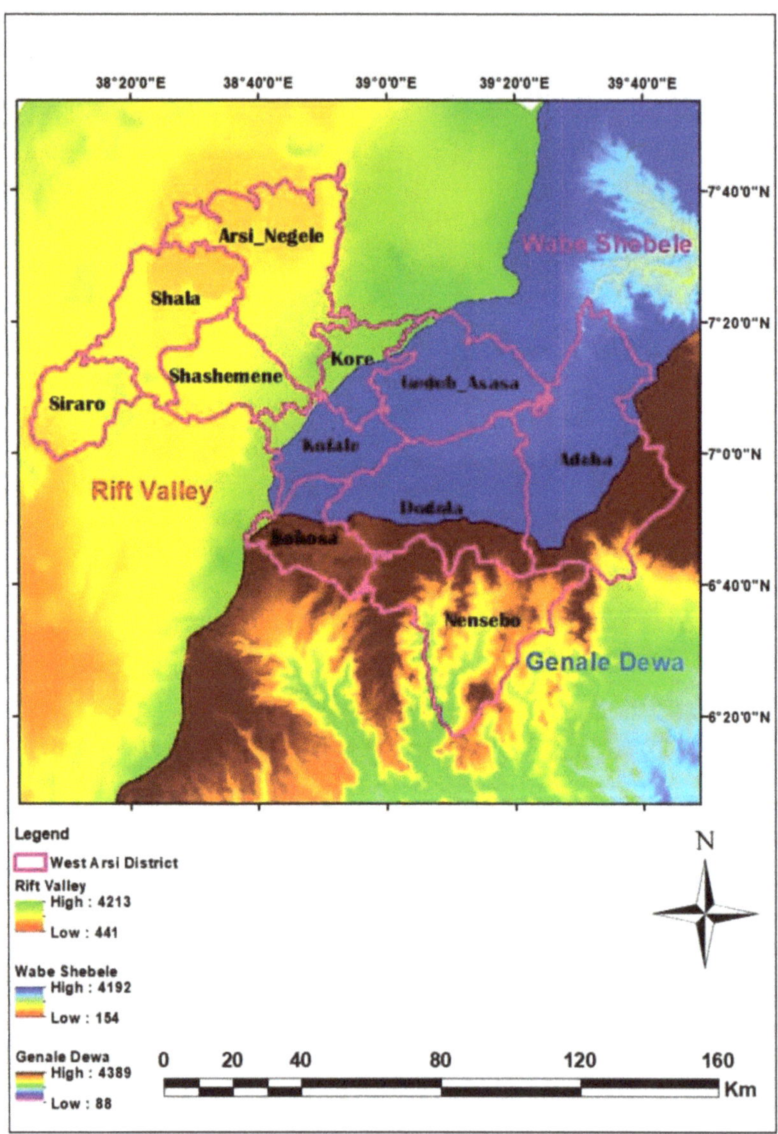

Figure 2. Three river basins drain the West Arsi Zone and West Arsi districts.

Hydrologically the area has a number of perennial and intermittent rivers as well as seasonal and non-seasonal springs. Lakes such as Shala, Abijata, and Langano are found in this study area. The greater part of the area is farming land with some bare land that is covered by sparsely populated natural plants (bushes; shrubs; thick, short, and long gasses; etc.) and eucalyptus plants; wheat, barley, potato, maize, teff, pea, and bean are the main product of the farming activities. The soil group of the study area is classified in to soil texture of clay, loam, sandy-loam, loamy-sand, sandy-clay, silty-clay-loam, and silt. The rock formation porosities are secondary porosities that have been developed due to weathering and tectonic fracture, which are suitable for groundwater storage and movement. The water supply source of the West Arsi Zone is rivers, springs, and hand dugs.

2.2. Methodology

2.2.1. Data Collection and Use

Data used for this study were collected from one-of-a-kind sectors, businesses, and extraordinary internet site sources. For this study, eight major surface and sub-surface groundwater potential-influencing criteria were separated and set for groundwater potential assessment. These criteria were selected due to being commonly used in previous literature [16–24] and advised by a number of experts to be used. Procedure followed to arrive at about the main objective of this study is as shown in (Figure 3).

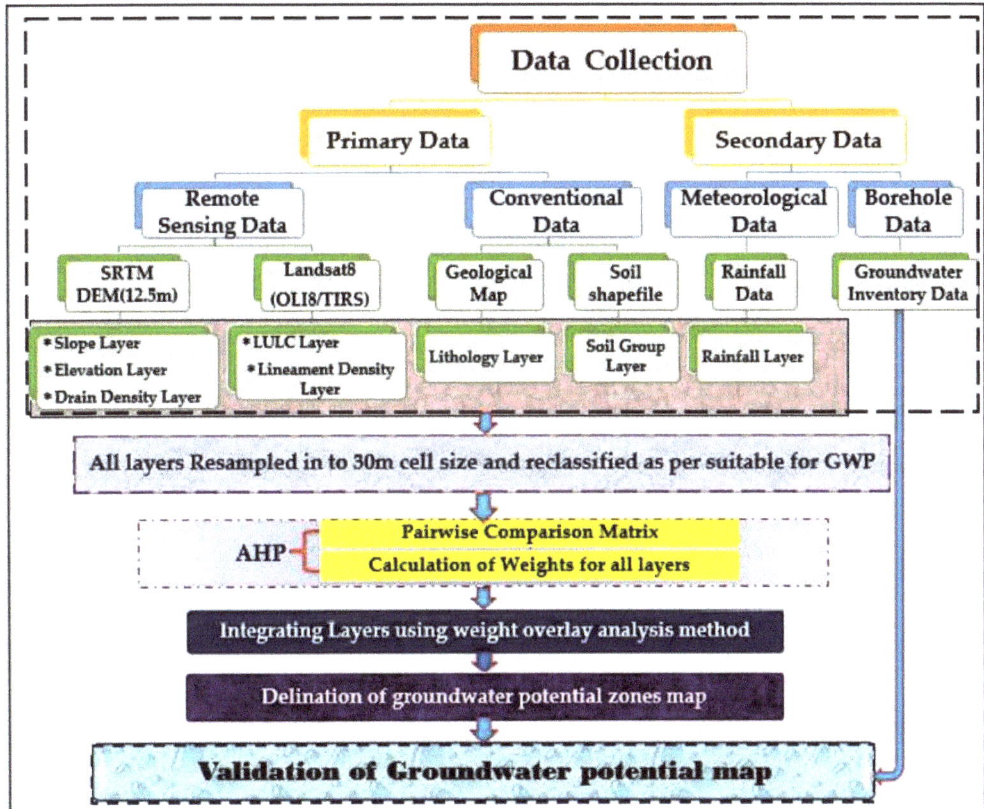

Figure 3. Research methodology flowchart.

Accordingly, based on accessible records and a literature review, eight groundwater-controlling factors were identified as proxy data, namely, slope, elevation, drainage density, lineament density, LULC, soil, rainfall, and lithology/geology. Groundwater stock records for validation purpose were additionally used. A DEM file of 30 m spatial resolution was downloaded from Image, courtesy of the USGS Earth Explorer webiste (https://earthexplorer.usgs.gov/ (accessed on 2 January 2020)) in the structure of Shuttle Radar Topographic Mission (SRTM) 1Arcsecond Global.

Operational Land Imager (OLI8)/Thermal Infrared Sensor (TIRS) with path and row of 168/055 scenes, received on 1 February 2019, and with a spatial resolution of 30 m for the seen to infrared and 15 m for the panchromatic, were downloaded from Image, courtesy of the USGS Earth Explorer (https://earthexplorer.usgs.gov/ (accessed on 22 November 2019)) website. From this source of data, slope, elevation, drain density, LULC, and lineament density layers were generated. Rainfall records were gathered from National Meteorological Agency of Ethiopia. The geology map used, at the scale of 1:1,000,000, was collected from the Geological Survey of Ethiopia and NB-37-2, 3, 6, and 7, which are the Dodola, Hosana, Asela and Dila hydrogeological maps, with notes downloaded from website https://gis.gse.gov.et/hg-maps/ (accessed on 15 April 2020). Groundwater inventory data (borehole, spring, and well data) of West Arsi Zone were from West Arsi Water, Mineral and Energy Bureau (WAWMEB). Ethiopia Soil records were gathered in the form of a shape file from the Food and Agricultural Organization [16] and Ministry of Water, Irrigation and Energy (MoWIE). The shape file to learn about the region was bought from West Arsi Zone Administrative Bureau and Oromia Administrative Bureau, used for extraction of the groundwater potential-influencing thematic layers.

2.2.2. Developing Groundwater Potential-Influencing Thematic Layers and Reclassifying

Land Use Land Cover Thematic Layer: Landsat8 OLI/TIRS has a path 168 and row 055 with cloud cover of land 0.01, Roll Angle of –0.001, Sun Azimuth of 130.39029597, Sun Elevation of 52.16321853, and spatial resolution/Cell Size of 30 m. Image composite using the process tool box from bands 1, 2, 3, 4, 5, 6, and 7 was done and the West Arsi Zone-representing image was extracted using the shapefile of study area with the help of the extraction tool of the spatial analysis tools. The LULC classification was done with help of the training sample manager tool, which was used to select the representative classes of the LULC, and the base map was used. This classification was a supervised classification because sample training was used.

The LULC classification accuracy was checked by a hundred random points (Table 1) edited on LULC-generated maps (Figure 4a) and opened on Google Earth Professional (Figure 4b). prediction accuracy, truth accuracy and overall accuracy were computed (Table 2).

Table 1. Random points taken for LULC accuracy-checking purposes.

Water Body	Built-Up Area	Barren Landscape	Forest	Vegetation Cover	Agriculture
10	10	6	19	24	31

Table 2. LULC accuracy-checking pivot table generated in Excel.

Sum of Value	Column Labels						
Row Labels	207	312	381	543	544	545	Grand Total
207	17					2	19
312	1	4				1	6
381			8		1	1	10
543				10			10
544		3			28		31
545	1				3	20	24
Grand Total	19	7	8	10	32	24	100

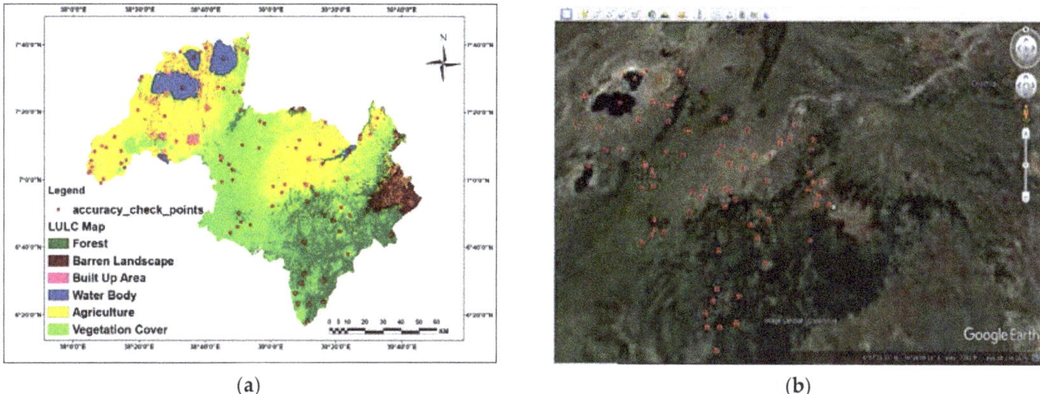

(**a**) (**b**)

Figure 4. (**a**) Random points plot on the LULC map. (**b**) Random points plotted on Google Earth Pro in kml format.

The kappa coefficient was used as a degree of agreement between the model predictions and reality [25] or to see if the values contained in a slip matrix represent a result considerably better than random [26]. The supported rating criteria for the kappa coefficient statistics, with the kappa coefficient ranging between 0.61 and 0.80 in strength agreement, are substantial, and the 0.81–1.00 strength agreement is almost perfect [27] (Table 3).

Table 3. Kappa coefficient rating and strength of agreement.

Sr. No	Kappa Coefficient	Strength of Agreement
1	<0.00	Poor
2	0.00–0.20	Slight
3	0.21–0.40	Fair
4	0.41–0.60	Moderate
5	0.61–0.80	Substantial
6	0.81–1.00	Almost perfect

Where 207 is Forest, 312 is Barren Landscape, 381 is Built Up, 543 is Water Body, 544 is Agriculture, and 545 is Vegetation Cover. Finally, the Land-Use/Land-Cover (LULC) layer of a district was ready and reclassified in line with the suitability of the parameters for groundwater potential availability (Figure 5a,b).

$$\text{OAA} = \left(\frac{\text{Total Properly classified pixels}}{\text{Total number of reference pixels}}\right) 100\% \tag{1}$$

where OAA is over all Accuracy

$$\text{PA} = \left[\frac{\text{Correctly classified pixels in each category}}{\text{Corresponding column total}}\right] 100\% \tag{2}$$

where PA is prediction Accuracy

$$\text{TA} = \left[\frac{\text{correctly classified pixels in every class}}{\text{Corresponding row total}}\right] 100\% \tag{3}$$

where TA is Truth Accuracy

$$\text{Kappa coefficient(K)} = \left[\frac{(\text{TCS} * \text{TS}) - \sum(\text{Ct} * \text{Rt})}{\text{TS}^2 - \sum(\text{Ct} * \text{Rt})}\right] 100\% \tag{4}$$

where Ct is column total, Rt is row total, TCS is total correct sample (87), and TS is total sample (100).

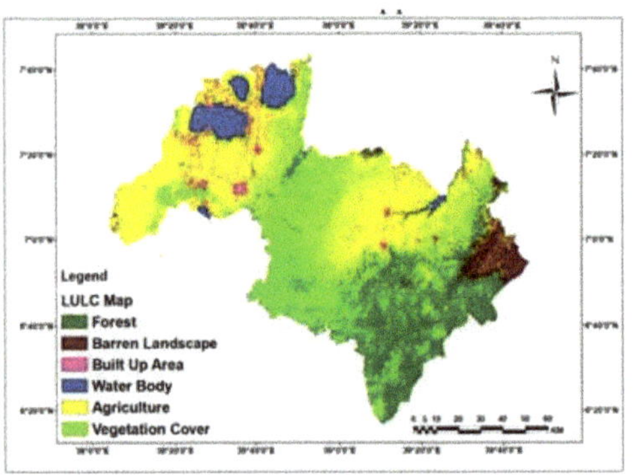

(a) Land Use Land Cover Map

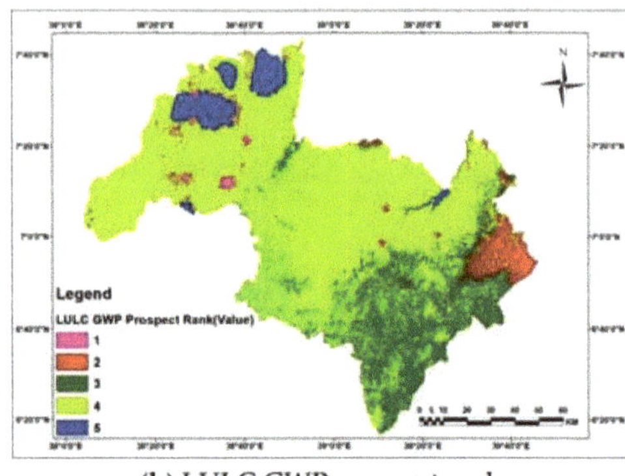

(b) LULC GWP prospect rank

Figure 5. LULC and groundwater potential prospect rank of the LULC map.

Rainfall Thematic Layer: Annual rain information of twenty-six years (1993–2019) from thirty-four stations in and from neighbors of the study space were obtained from National Earth Science Agency of the Federal Democratic Republic of Ethiopia. Annual point rain measures were regenerated to surface rain information employing a geo process tool of ArcGIS that interpolates a surface from points and rain map generated (Figure 6a). This rain map categories were reclassified into 5 category values in line with its rank as per the quality of the groundwater potential and the recharge victimization sort tool in the spatial analyst tools (Figure 6b).

Slope Thematic Layer: Closely spaced contours represent vessel slopes and distributed contours exhibit a light slope. The slope values area unit was calculated either in percentage or degrees in each vector and formation forms. The study space DEM was extracted applying the extraction tool of the spatial analyst tools from DEM file downloaded and mosaicked to a single DEM. The slope layer of the study space was generated applying 3D analyst tools of ArcGIS from the DEM (Figure 7a). The slope tool calculates elevation change at a degree applying elevations of the encircling [28]. The slope map classification was created applying natural breaks and therefore the slope degree of the West Arsi Zone ranges from 0° to 75.9°. These slope map categories were reclassified into 5 category values, in keeping with its rank as per the suitability for groundwater potential creation by means of the class tool in the spatial analyst tools (Figure 7b).

Elevation Thematic Layer: The elevation layer of the study space was generated from a DEM. Thus, the elevation information is required to be included in groundwater potential studies. The elevation layer of the West Arsi district was assessed given the 5 categories per its contribution to groundwater potential and recharge of the study space (Figure 8a). These elevation layer categories were reclassified into 5 category values per its rank as per the appropriateness for groundwater potential and recharge, applying the classify tool of the spatial analysis tools (Figure 8b).

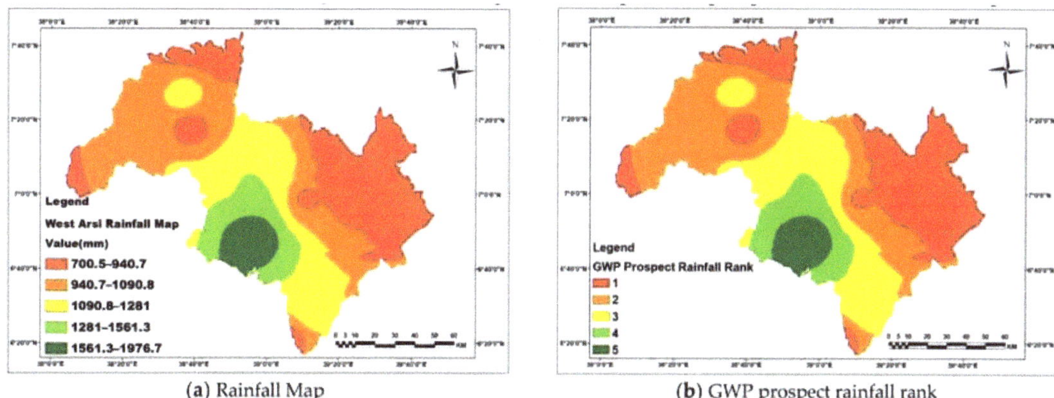

(**a**) Rainfall Map (**b**) GWP prospect rainfall rank

Figure 6. Rainfall and groundwater potential prospect rank of the rainfall map.

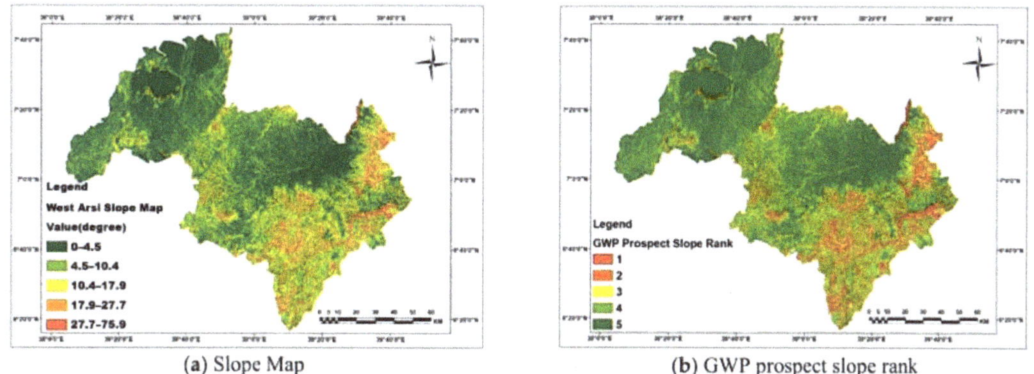

(**a**) Slope Map (**b**) GWP prospect slope rank

Figure 7. Slope and groundwater potential prospect rank of the slope map.

Drainage Density Thematic Layer: The drain density (km/km^2) expresses the nearness of space of waterway conduits, so providing a quantitative measuring of the typical span of waterway conduits of the entire basin [29]. To come up with a drain density map of a region, a filling sink was performed initially to get rid of the highest elevation and lowest elevation that lure the water applying the DEM manipulation tool of the terrain-preprocessing tool. A flow direction map was generated from the fill sink applying the flow direction tool of the land preprocessing tools. A flow accumulation map was generated from the flow direction applying the flow accumulation tools of the land preprocessing in Arc Hydro tools. The stream definition map was made from the flow accumulation data applying the raster calculator tool of the map algebra tool in the spatial analysis tools. A sink may be a cluster of 1 or a lot of cells that have lower elevations than all the encompassing cells whereas a peak may be a cluster of 1 or a lot of cells that have higher elevations than all the encompassing cells [30]. The drain density layer of the region has been created from a dissolved stream network applying density tool in ArcGIS spatial analyst tools (Figure 9a). The drain density layer of the study space was made applying the line density tool of the spatial analysis tools in ArcGIS software. The line density tool calculates drain density by dividing the span of the drain line by the encompassing watershed space, cells upstream of the cell, for every cell within the input flow direction grid. These drain density map categories were reclassified into 5 category values per its rank as per the suitability for groundwater potential and recharge, applying the separate tool of the spatial analysis tools (Figure 9b). Slope, elevation, and drain density maps of the space were extracted, processed, and generated from a Shuttle Radar Topographic Mission DEM of 12.5 m by 12.5 m spatial resolution, downloaded from USGS Earth explorer.

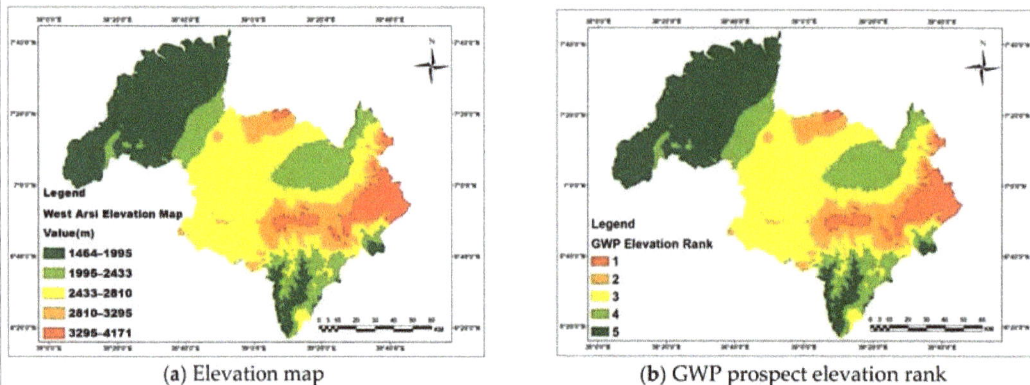

(a) Elevation map (b) GWP prospect elevation rank

Figure 8. Elevation and groundwater potential prospect rank of the elevation map.

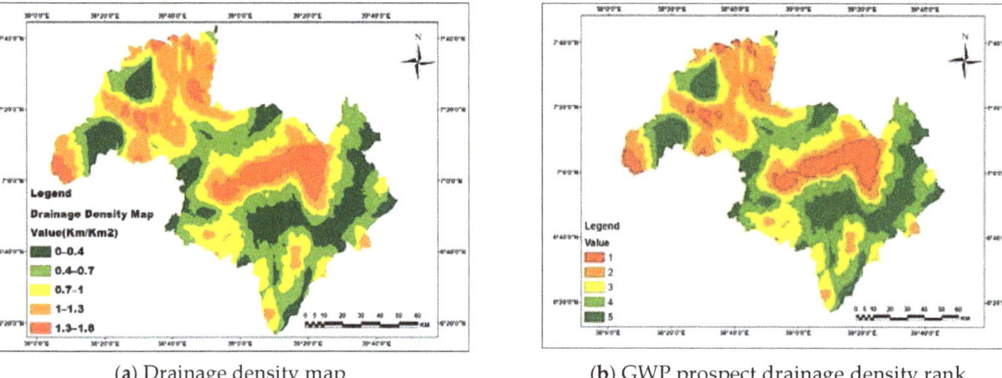

(a) Drainage density map (b) GWP prospect drainage density rank

Figure 9. Drainage density and groundwater potential prospect rank of the drainage density map.

Lineament Density Thematic Layer: Lineaments are unit straight linear parts visible at the surface as a major "line of landscape" [31]. These are units primarily being a mirrored image of the discontinuities on the Earth's surface caused by geologic or geomorphic processes [32]. Band 8 (0.50–0.68 μm), which is a panchromatic of the OLI8/TIRS image, was downloaded from USGS Earth explorer (https://earthexplorer.usgs.gov/ (accessed on 22 November 2019)) website and had a spatial resolution of 15 m extracted applying the West Arsi Zone shape file and exported in word format of the stretched sort. Lineament of a picture was extracted mechanically from images exported in .tiff format applying PCI Geomatica Banff applying the line tool in the algorithmic librarian tool saved as file sort Arc read. Line split, line split at vertices, and lineament density maps were generated applying the editor tool, feature tool of data management tool, and density tool of the spatial analysis tool operation. Principal Component Image (PCI) carry most data and is appropriate for lineament extraction functions. Band 8 of Landsat 8 was chosen and used because of its ability to identify linear and curvilineal features and having higher spatial resolution of 15 m and it is panchromatic mirrored band. Finally, the lineament and reclassified lineament density layer was produced from the band 8 OLI8/TIRS image (Figure 10a). These lineament density map categories were reclassified in to 5 category values in line with its rank as per suitable for groundwater potential and recharge zone delineation, applying the reclassify tool of the spatial analysis tools (Figure 10b).

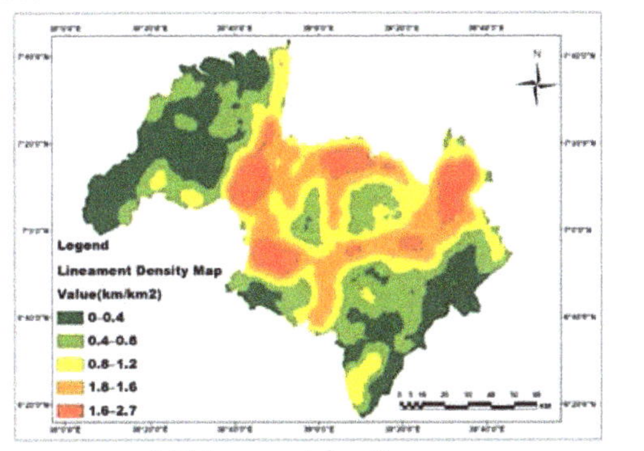

(a) Lineament density map

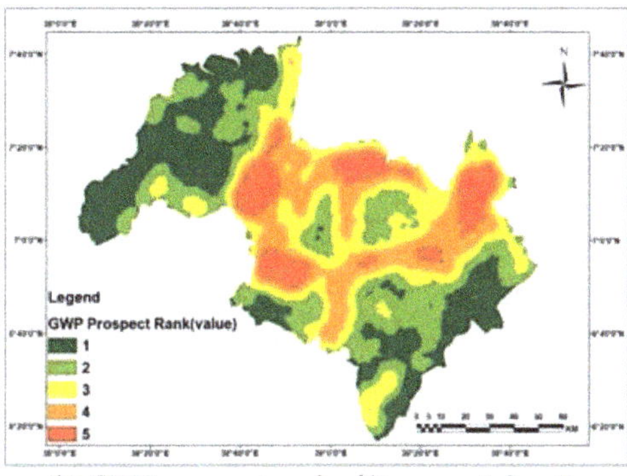

(b) GWP prospect rank of lineament density

Figure 10. Lineament density and groundwater potential prospect rank of the lineament density map.

Soil Group Thematic Layer: Soil group map (Figure 11a) of West Arsi Zone was generated from the dissolved shapefile of the study area clipped from the Ethiopia soil group shapefile using the clip tool of the analysis tools and converted into a raster using the polygon-to-raster tool of the conversion tools. This soil group map was regrouped into different six soil group texture and permeability. These soil map classes were reclassified in to five class values according to its rank as per the suitability for groundwater potential formation using the reclassify tool of the spatial analyst tools and a new soil group map was generated (Figure 11b).

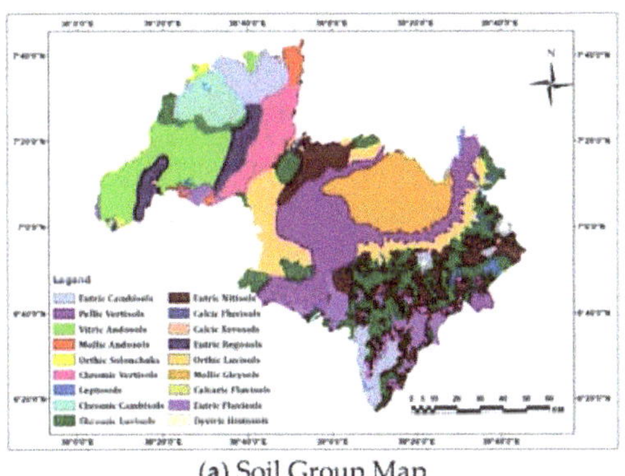

(a) Soil Group Map

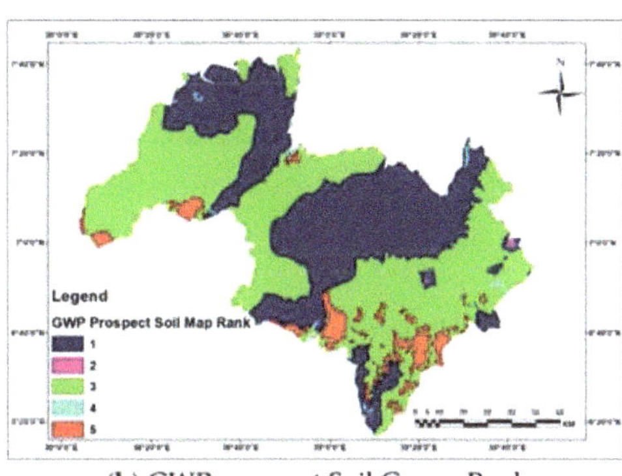

(b) GWP prospect Soil Group Rank

Figure 11. Soil group and groundwater potential prospect rank of the soil group map.

Lithological Map Preparation: The lithology layer of the region was generated by geo-referencing, digitizing, extracting of a region formation from the geological layer of the Oromia 1:1,000,000 scale [33] obtained from the Geological Survey of Ethiopian (GSE). The geology layer of Oromia was georeferenced applying geo-referencing tool of ArcGIS and corrected, and projected to the WGS1984 UTM Zone 37, applying projection and transformation tools of the data management tools. The West Arsi Zone geological formation image was clipped applying the study space shapefile with the assistance of clip tool of the analysis tools of ArcGIS. Study space lithology layer was generated from dissolved geology shape file reborn to raster applying conversion using the polygon-to-raster tool of the conversion tools (Figure 12a). The lithology layer categories were reclassified into 5 category values per its rank as per the suitableness for groundwater potential formation applying the reclassify tool of the spatial analysis tools, and a new lithology layer (Figure 12b) was generated.

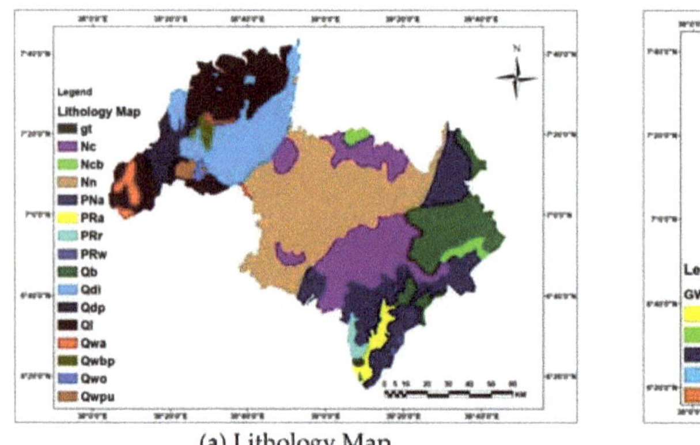

(a) Lithology Map

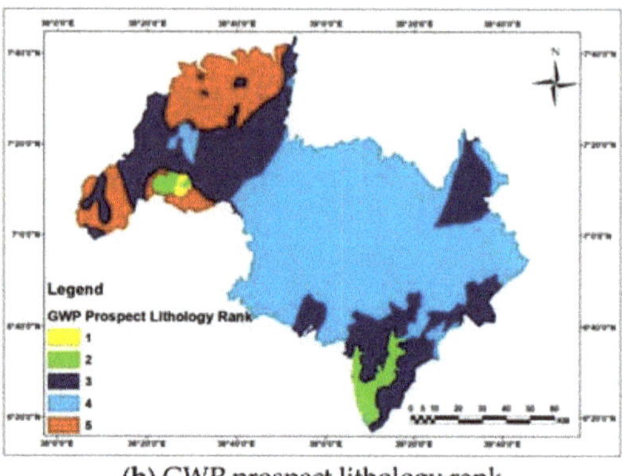

(b) GWP prospect lithology rank

Figure 12. Lithology and groundwater potential prospect rank of the lithology map.

Each influence layer was ready and reclassified in to 5 categories in an exceeding manner, which will support the general goal of groundwater potential and recharge zone mapping. These maps layers were projected onto an equivalent reference system, resampled into an equivalent formation layer of 30 m cell size, and reclassifying all the thematic layers' individual parameters as appropriate for groundwater potential zonation so as to be acceptable for the weight overlay analysis. All the desired thematic maps were developed from the collected datasets applying MS-Excel, ArcGIS 10.3.1 version, PCI geomatica Banff, and the ERDAS IMAGINE 2015 package. The spatial resolution of reclassified precipitation, slope, elevation, drain density, lineament density, LULCr, soil, and lithology map was 30 m × 30 m and with a 10,000 m^2 to hectare conversion factor. Accordingly, the area coverage of these categories will be calculated using the formula

$$\text{Area(ha)} = \frac{\text{Pixel Count} \times 30\ \text{m} \times 30\ \text{m}}{10{,}000\ \text{m}^2} \tag{5}$$

Area in percent will be calculated using the formula

$$\text{Area}(\%) = \frac{\text{Row Area}}{\text{Total Area}} \times 100 \tag{6}$$

In groundwater potential influencing parameters (rainfall, slope, elevation/altitude, drain density, and lineament density layers classification), there is no onerous and quick rule for groundwater potential and recharge or runoff generation. Hence, merely the natural break on the ArcGIS ArcMap classified, which show the kind that existed by default, was applied.

2.2.3. Analytical Hierarchy Process to Assign Weight

Among the numerous techniques, the Analytical Hierarchy Process (AHP) enables plenty to systematically discover the maximum influencing parameters [12,22,34–36]. The Analytical Hierarchy Process (AHP), proposed by [37], is the regularly used approach for groundwater potential mapping. The eight criteria/elements (rainfall, lithology, lineament density, land use/land cover, soil group, slope, elevation, and drain density) predicted to affect groundwater distribution of the West Arsi sector were separated and set for weight overlay. A pairwise comparison matrix, P(m × m), of which m is the number of parameters to be in comparison, was changed to being built primarily based totally on the quantity of the entered elements for delineation of the groundwater potential and recharge zones [38]. Ordering and assigning a scale for parameters (elements) affecting the groundwater potential calls for a review of the variety of literature, personal judgments, and professional opinion.

The main purpose of AHP in this study was to determine the appropriate Normalized Principal Eigen Vector (NPEV) or Percent Weight in ArcGIS environment in the weight overlay analysis. Generally, the procedure followed to determine and validate the normalized principal eigenvector is shown in (Figure 13).

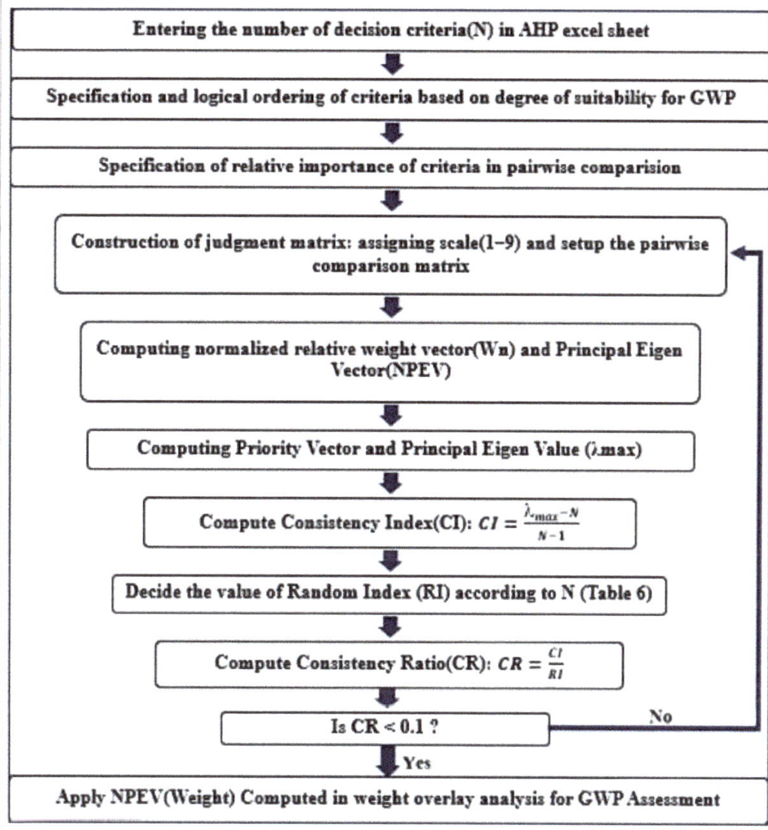

Figure 13. AHP Procedure to determine and validate the percent weight (NPEV).

Before placing the criterion into pairwise comparison and assigning the scale for groundwater potential assessment, first all of the elements need to be in a logical order in the AHP Excel sheet, primarily based on the degree of suitability for groundwater potential and recharge zone contribution (Table 4). In this study, primarily based on the features of the area under study and suitability of the criteria elements for groundwater potential and recharge zone contribution, all of the parameters are ordered. Slope and elevation decide the destiny of the water that reaches the floor of the earth. From slope conduct of the west Arsi Zone, about 72.41% of the vicinity is appropriate for groundwater potential and recharge of surface water whilst in comparison to the rest of the criteria under consideration. This suggests that about 72.41% of a place is almost flat to mildly sloped, which permits extra rainfall or surface water to percolate and infiltrate.

Table 4. Analytical Hierarchy Process pairwise comparison matrix of the thematic layers and scale assigned.

		Criteria		More Important?	Scale
i	j	A	B	A or B	(1–9)
1	2	Slope	LULC	A	1
1	3		Lithology	A	1
1	4		Elevation	A	3
1	5		Drainage Density	A	3
1	6		Lineament Density	A	5
1	7		Rainfall	A	7
1	8		Soil	A	9
2	3	LULC	Lithology	A	1
2	4		Elevation	A	3
2	5		Drainage Density	A	3
2	6		Lineament Density	A	5
2	7		Rainfall	A	7
2	8		Soil	A	9
3	4	Lithology	Elevation	A	3
3	5		Drainage Density	A	3
3	6		Lineament Density	A	5
3	7		Rainfall	A	7
3	8		Soil	A	9
4	5	Elevation	Drainage Density	A	1
4	6		Lineament Density	A	3
4	7		Rainfall	A	3
4	8		Soil	A	5
5	6	Drainage Density	Lineament Density	A	1
5	7		Rainfall	A	3
5	8		Soil	A	5
6	7	Lineament Density	Rainfall	A	1
6	8		Soil	A	3
7	8	Rainfall	Soil	A	1

Therefore, consistent with this study, slope is located at the start order. Area protected through water body, vegetation, and agricultural location are maximally appropriate for surface water percolation. Especially agricultural and vegetation protected areas trap the water, reduces runoff, and will increase infiltration. The overall sum of area protected through a water body, agriculture, and vegetation makes a contribution to about 71.67% of a place and puts LULC at the second order subsequent to slope in phrases of suitability of the criteria for groundwater potential and recharge zone. Geology/lithology performs an essential function in the occurrence and distribution of groundwater in any terrain [39] due to the fact water might recharge aquifers directly. A 66.2% lithology of a place is appropriate for groundwater potential, which places it in the third order in terms of suitability of the criteria for groundwater potential. Elevation was 49.02% appropriate for groundwater potential and thus placed in the fourth order accompanied by drainage density, which contributes 46.32% to high and very high groundwater potential. In phrases of lineament density, only 27.94% of a place is appropriate for surface water percolation, recharge, and groundwater potential formation. Therefore, groundwater potential is low regarding lineament density areas and consequently lineament density is located in the 6th order.

Rainfall performs an essential function for hydrologic cycle and controls groundwater potential [40]. Rainfall performs an important function in the occurrence of groundwater. It is clear that greater rainfall might also additionally reason greater recharge ability, even though that ability is challenged through different constraining elements, including slope, geology, land use/cover, drainage density, lineament density, and others. Therefore, excessive recharge vicinity does now no longer always mean excessive groundwater potential areas [41]. Knowing the nature and characteristics of rainfall might also additionally allow one to conceptualize and predict its outcomes on runoff, infiltration, and groundwater potential and recharge [42]. The opportunity of groundwater recharge might be excessive on the location in which the rainfall is excessive and is low in which rainfall is low [41,43–45]. Regions that obtain greater rainfall have greater possibility of infiltration than districts with low precipitation [40]. From a rainfall factor view, only 13.85% of a place gets high and very high rainfall; consequently, in terms of rainfall, only 13.85% of a place will have high and very high groundwater potential. This displays that rainfall contribution for groundwater potential formation and recharge sector could be very low and positioned in the 7th place, as compared to slope, LULC, lithology, elevation, drainage density, and lineament density.

Soil kind and texture additionally determine the infiltration ability and permeability. In phrases of the soil group, only 7.3% is anticipated to have high and very high groundwater potential and recharge, and is thus positioned in the last place in this study. This displays that about 92.7% of the soil group of a place is impermeable and will increase surface runoff and reduce infiltration. Factor effects on every different one, based on Saaty's one to nine factor scale, were used, where 1 represents both parameters being similarly essential and nine suggests one parameter is extraordinarily essential over the alternative in phrases of goal influence [40]. The summary of this hierarchy and pairwise comparison and the assigned scale using the AHP Excel sheet is given in Table 5, generated from the AHP Excel sheet and pairwise comparison matrix.

Table 5. Analytic hierarchy process pairwise comparison matrix and the assigned scale.

Matrix	SL	LULC	Lith	El	DD	LD	RF	SG
SL	1	1	1	3	3	5	7	9
LULC	1	1	1	3	3	5	7	9
Lith	1	1	1	3	3	5	7	9
El	1/3	1/3	1/3	1	1	3	3	5
DD	1/3	1/3	1/3	1	1	1	3	5
LD	1/5	1/5	1/5	1/3	1	1	1	3
RF	1/7	1/7	1/7	1/3	1/3	1	1	1
SG	1/9	1/9	1/9	1/5	1/5	1/3	1	1
Column Total of PCM	4.120635	4.120635	4.120635	11.86667	12.53333	21.33333	30	42

Where RF represents rainfall, SG represents soil group, LD represents lineament density, Lith represents lithology, LULC represents land use/land cover, Sl represents slope, El represents elevation, and DD represents drainage density.

Normalized Relative Weight (Wn), Eigenvector, and Normalized Principal Eigenvector (NPEV) are determined as step below.

AHP employs experts' opinion; the role of Eigenvectors and Eigenvalues is to lessen noise withinside the records and additionally assist in decreasing over-fitting [46]. The Eigenvector is the ordering of parameter impact on groundwater potential and recharge with the aid of using assigning the weights [47]. The Eigenvector was computed to display the comparative weights of every parameter in the direction of groundwater potential and recharge [48] (Table 6).

Table 6. Normalized relative weight and Normalized Principal Eigen Vector (NPEV).

Matrix	Normalized Relative Weight Vector (Wn)								Eigenvector	NPEV (%)
	SL	LULC	Lith	El	DD	LD	RF	SG		
SL	0.242681	0.242681	0.242681	0.252809	0.239362	0.214286	0.233333	0.214286	0.24	24
LULC	0.242681	0.242681	0.242681	0.252809	0.239362	0.214286	0.233333	0.214286	0.24	24
Lith	0.242681	0.242681	0.242681	0.252809	0.239362	0.214286	0.233333	0.214286	0.24	24
El	0.080894	0.080894	0.080894	0.08427	0.079787	0.128571	0.1	0.119048	0.09	9
DD	0.080894	0.080894	0.080894	0.08427	0.079787	0.128571	0.1	0.119048	0.09	9
LD	0.048536	0.048536	0.048536	0.02809	0.079787	0.042857	0.033333	0.071429	0.05	5
RF	0.034669	0.034669	0.034669	0.02809	0.026596	0.042857	0.033333	0.02381	0.03	3
SG	0.026965	0.026965	0.026965	0.016854	0.015957	0.014286	0.033333	0.02381	0.02	2
Total	1	1	1	1	1	1	1	1	1	100

Steps followed to calculate Wn, Eigenvector, and NPEV:

- Row 1–9 of Table 5 was generated from the AHP Excel sheet.
- Scale values (Column 2–9 of Table 5) of the pairwise comparison matrix was summarized (Row 10 of Table 5).
- Normalized Relative Weight Vector (Wn) of Table 6 from Row 3–10 of Column 2–9 was computed from division of each column criterion value of Table 5 (Column 2–9 of Row 2–9) by column total (Row 10) of Table 5.
- Each Eigenvector value (Column 10 of Row 2–9) of Table 6 is the average of each row.
- Normalized Principal Eigen Vector (NPEV) of Table 6 from Row 3–10 of Column 11 is a multiply of the Eigenvector values by 100%.
- The column sum of the normalized relative weight vector (Wn) and Eigenvector is equal to 1 and Normalized Principal Eigen Vector (NPEV) is equal to 100% (Row 11 of Table 6).

The consistency ratio (CR) is used for assessment of matrix consistency. AHP includes a powerful approach used for checkup the consistency of the evaluations made through the decision maker whilst constructing every of the pairwise comparison matrix concerned within the process. Inconsistencies in pairwise comparisons grow with the growing number of comparisons [49]. For the estimation of the consistency ratio (CR), the following stages is involved:

- Priority vector (column 4 of Table 7) for criterion is calculated by multiplying the column total of pairwise comparison matrix by Eigenvector.
- Principal Eigenvalue (λ_{max}) is summation of priority vector (Row 10 of Table 7).
- Consistency Index (CI) is the ratio of the distinction among the Principal Eigenvalue (λ_{max}) and the number of criteria (m) to number of criteria under investigation (m) less one.
- Random index (RI) was determined from Table 8 of Satty (1990), which depends on number of criteria (m) considered.
- Consistency ratio (CR) is the ratio of the consistency index (CI) to the random index (RI).

Table 7. Principal Eigenvalue (λ_{max}) and priority vector.

Thematic Criterion	Column Total of PCM	Eigenvector	Priority Vector
SL	4.120635	0.24	0.969440451
LULC	4.120635	0.24	0.969440451
Lith	4.120635	0.24	0.969440451
El	11.86667	0.09	1.118962872
DD	12.53333	0.09	1.181825955
LD	23.33333	0.05	1.169888994
RF	30.00000	0.03	0.970094234
SG	42.00000	0.02	0.971951575
	Principal Eigenvalue (λ_{max})		8.321044984

Table 8. Random Index (RI) belongs to the number of assessment criteria (m).

m	1	2	3	4	5	6	7	8	9	10
RI	0	0	0.58	0.9	1.12	1.24	1.32	1.41	1.45	1.49

The sum of the priority vector, known as the Principal Eigenvalue (λ_{max}), is a degree of matrix deviation from consistency [48].

According to [49], a pairwise contrast matrix exists only if the Principal Eigenvalue (λ_{max}) is extra than or equal to the number of the parameters investigated/criteria (m). In any other case a brand-new matrix is required. If there may be any inconsistency within the experts' opinions, a difference among m and λ_{max} is indicated. Therefore, λ_{max}—n may be classed as a measure of inconsistency. A perfectly regular decision maker has to continually obtain CI = 0; however, small values of inconsistency can be tolerated if the consistency ratio (CR) < 0.1 [37]. The consistency index (CI) for groundwater potential and recharge zone parameters investigated in this study was calculated by the equation below.

$$CI = \frac{\lambda_{max} - m}{m - 1} \quad (7)$$

where m is the number of assessment criteria (thematic layers in the case of this study) and λ is the Principal Eigenvalue of judgment matrix as set through Satty (1995). RI relies on the range of the criteria being compared, as shown in Table 8 [49].

From this table for m = 8, RI = 1.41. Analytical Hierarchy Process takes the consistency ratio (CR) figure among zero and 0.1 or 10%; a value greater than 10% invites for modification of comparisons.

$$CR = \frac{CI}{RI} \quad (8)$$

Consistency Ratio (CR) calculation is to confirm the consistency of the judgements. Saaty (1995) advised a different consistency ratio value for different consistent pairwise evaluation matrix sizes. The recommended consistency ratio value for a three × three matrix is much less than 0.05, a four × four matrix is 0.09, and for large matrices it is recommended 0.1 [38].

2.2.4. Weighted Overlay Analysis

During the weighted overlay analysis, the ranks were given for all parameters of all thematic layers set for the study and the weight is assigned in line with their relative effect of the different parameters on groundwater potential applying the Analytic Hierarchical Process (AHP) technique [48]. After assigning weights to all thematic layers, ranks/scale values from 1 to 5 were given for the sub-variable of every thematic layer, in line with their significance for groundwater potential occurrence. According to this study, 1 represents less vital and 5 represents more vital for groundwater potential and recharge. The most worth is given to the feature expected with maximum groundwater potentiality and the minimal given to the lowest groundwater potentiality feature (Tables 9–16).

Table 9. Groundwater potential prospect rainfall map classes.

Rainfall Classes (mm)	Value (Rank)	GWP Prospect Rank	Count	Area (ha)	Percent Area
700.5–940.7	1	Very low	4,092,491	368,324.2	29.54
940.7–1090.8	2	Low	4,551,345	409,621.1	32.85
1090.8–1281	3	Moderate	3,241,868	291,768.1	23.4
1281–1561.3	4	High	1,220,442	109,839.8	8.81
1561.3–1976.7	5	Very High	747,753	67,297.77	5.4

Table 10. Groundwater potential prospect reclassified slope map information.

Slope Classes (Degree)	GWP Prospect Rank	Value (Rank)	Count	Area (ha)	Area (%)
27.7–75.9	Very Low	1	40,4616	36,415.44	2.93
17.9–27.7	Low	2	1,166,148	104,953.32	8.43
10.4–17.9	Moderate	3	2,243,511	201,915.99	16.23
4.5–10.4	High	4	4,274,118	384,670.62	30.91
0–4.5	Very High	5	5,737,163	516,344.67	41.50

Table 11. Groundwater potential prospect reclassified elevation map information.

Elevation Classes (m)	GWP Prospect Rank	Value (Rank)	Count	Area (ha)	Area (%)
3295–4171	Very Low	1	928587	83,572.83	6.70
2810–3295	Low	2	1681472	151,332.48	12.14
2433–2810	Moderate	3	4452576	400,731.84	32.14
1995–2433	High	4	2589305	233,037.45	18.69
1464–1995	Very High	5	4202042	378,183.78	30.33

Table 12. Groundwater potential prospect reclassified drainage density map information.

DD Class (km/km^2)	GWP Prospect Rank	Value (Rank)	Count	Area (ha)	Area (%)
1.3–1.8	Very Low	1	1,286,764	115,808.76	9.29
1.0–1.3	Low	2	2,926,141	263,352.69	21.12
0.7–1	Moderate	3	3,223,861	290,147.49	23.27
0.4–0.7	High	4	3,645,417	328,087.53	26.31
0–0.4	Very High	5	2,771,782	249,460.38	20.01

Table 13. Groundwater potential prospect reclassified lineament density map information.

L.d Classes (km/km^2)	GWR Prospect Rank	Value (Rank)	Count	Area (ha)	Area (%)
0–0.4	Very Low	1	3663565	329,720.85	26.45
0.4–0.8	Low	2	3706704	333,603.36	26.76
0.8–1.2	Moderate	3	2611845	235,066.05	18.85
1.2–1.6	High	4	2617700	235,593	18.89
1.6–2.7	Very High	5	1254151	112,873.59	9.05

Table 14. Groundwater potential prospect reclassified LULC map information.

Class Name	GWP Prospect Rank	Value (Rank)	Count	Area (ha)	Area (%)
Built Up Area	Very Low	1	403,696	36,332.64	2.92
Barren Landscape	Low	2	456,928	41,123.52	3.30
Forest	Moderate	3	3,062,368	275,613.12	22.11
Agriculture & Vegetation	High	4	9,206,823	828,614.07	66.48
Water Body	Very High	5	718,378	64,654.02	5.19

Table 15. Groundwater potential prospect reclassified soil group map information.

Soil Group	GWR Prospect Rank	Value	Count	Area (ha)	Area (%)
Eutric Cambisols/Pellic Vertisols/Orthic Solonchaks/Chromic Vertisols/Chromic Cambisols/Mollic Gleysols	Very Low	1	5,905,315	531,478.4	42.63
Dystric Histosols	Low	2	16,214	1459.26	0.12
Vitric Andosols/Mollic Andosols/Chromic Luvisols/Eutric Nitisols/Eutric Regosols/Orthic Luvisols	Moderate	3	6,921,369	622,923.2	49.96
Leptosols/Calcic Xerosols	High	4	197,857	17,807.13	1.43
Calcic Fluvisols/Calcaric Fluvisols/Eutric Fluvisols	Very High	5	813,210	73,188.9	5.87

Table 16. Groundwater potential prospect reclassified lithology map information.

Lithology Name	GWR Rank Prospect	Value	Count	Area (ha)	Area (%)
Qwo	Very Low	1	42,032	3782.88	0.30
gt/Pra/PRr/Qwpu	Low	2	418,592	37,673.28	3.02
Pna/PRw/Qdi/Qdp/Qwa	Moderate	3	4,221,989	379,979.01	30.48
Nc/Ncb/Nn/Qb/Qwbp	High	4	7,358,588	662,272.92	53.12
QI	Very High	5	1,811,535	163,038.15	13.08

2.2.5. Groundwater Potential Map Development

The groundwater potential layer was developed via way of means of overlapping the determinant groundwater contributing thematic layers. A weighted overlay analysis device was used to develop the groundwater potential map and to compute the groundwater potential index values. The reclassified layers of rainfall, lithology, slope, elevation, lineament density, drain density, soil group, land use/land cover, and their corresponding percentage, have an impact on groundwater potential, and have been included to produce a map of the spatial distribution of the groundwater potential districts inside the West Arsi space with the help of the weighted overlay tool in ArcGIS software. Weighted Overlay analysis device reclassifies values within the enter raster layers right into a common assessment scale of 1, 2, 3, 4, and 5—very low, low, moderate, high, and very high, respectively—via way of means of multiplying the cell values (rank) of every factor class via way of means of the factor weight and sums the resulting cell values collectively to produce a map of groundwater potential zones, as given by the following equation (Raviraj 2017; ESRI 2015).

$$\mathrm{GWPI} = \mathrm{RFwRFr} + \mathrm{LDwLDr} + \mathrm{LULCwLULCr} + \mathrm{SwSr} + \mathrm{EwEr} + \mathrm{SGwSGr} + \mathrm{LiwLir} + \mathrm{DDwDDr} \tag{9}$$

where GWPI represents groundwater potential, RF represents rainfall, LD represents lineament density, LULC represents land use/land cover, S represents slope, E represents elevation, SG represents soil group, Lith represents lithology, DD represents the drain density index and the subscript w and r represent weight and rank, respectively [50]. The GWPI values were used to categorize whether or not a place may be very high, high, moderate, low, or very low with respect to groundwater potential [1,51].

2.2.6. Validation of Groundwater Potential Occurrence Zone Map

For validating the anticipated groundwater potential zone map, an attempt was made to acquire current data from different sources for validation. Overall, 113 current groundwater inventory borehole data points were accrued from West Arsi water, mineral and energy office. Additional sources from Dodola-Goba, Hosaina, Dila, and Asela hydrogeology annexes and notes were also used. For the cause of evaluation or assessment of the qualitative consequences of the groundwater potential zones, well yield was decided on as a higher candidate than different current data. Although there is no standard category scheme, well yields may be grouped into a few category schemes, considering the particular site conditions.

According to the hydrogeology notes of the above cited maps and others, the aquifer yields that exist on this hydrogeology are in a different way categorized. Some classify as zero–three L/s: low, three—6 L/s: moderate, 6–20 L/s: high and greater than 20 L/s as very high groundwater potential zones. Others classify as 0.5–1 L/s low, 1–five L/s moderate to low and five—25 L/s high groundwater potential areas. For this study with a few changes the water point inventory categorized as less than 2 L/s was classified as low (33 boreholes), 2 to 10 L/s moderate (69 boreholes), and more than 10 high yield (11 boreholes). Borehole inventory facts were mapped on groundwater potential map; percentage of agreement was calculated and validation of the groundwater thematic map with groundwater inventory facts was done (Figure 14d). Groundwater potential prediction accuracy was summarized as poor for 0.5 to 0.6; average for 0.6 to 0.7; good for 0.7 to 0.8; very good for 0.8 to 0.9; and excellent for 0.9 to 1 [3,52].

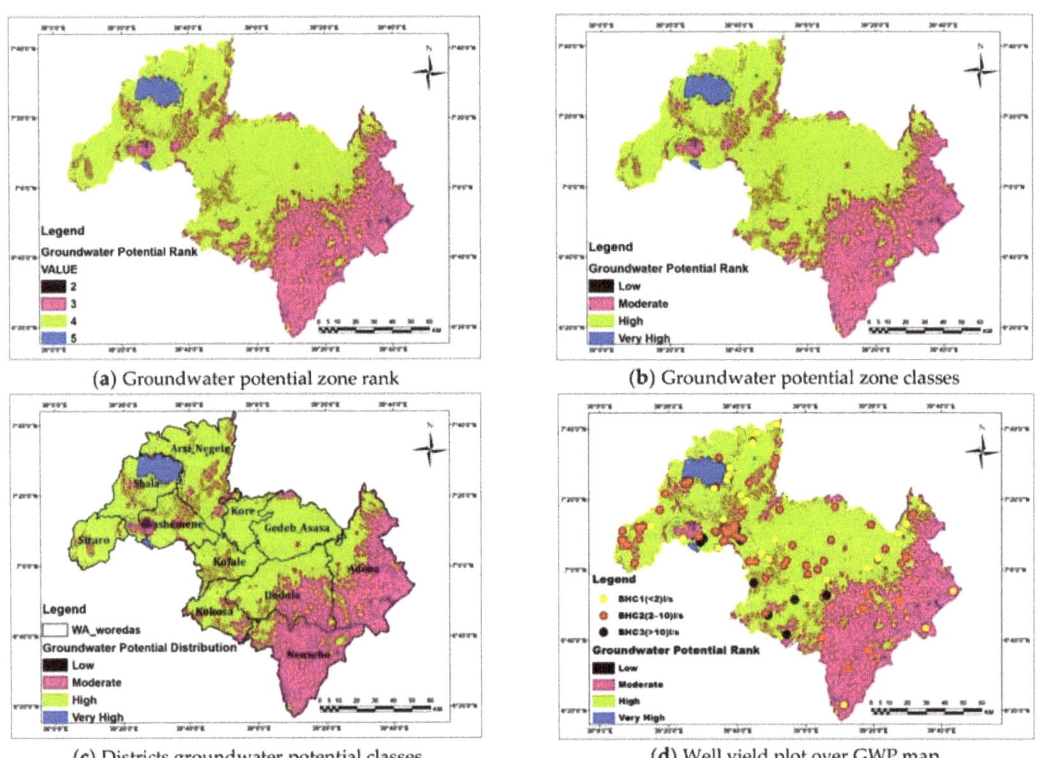

(a) Groundwater potential zone rank

(b) Groundwater potential zone classes

(c) Districts groundwater potential classes

(d) Well yield plot over GWP map

Figure 14. Groundwater potential map and well yield plot over it for validation.

3. Result and Discussions

3.1. Rainfall and Reclassified Rainfall Layer

Groundwater potential phenomena are the end result of the long time-period effect. The annual rainfall of the study area ranges from 700.5–1976.7 mm (Figure 6a).

The opportunity of groundwater potential and recharge could be excessive in the region where the rainfall is excessive and is low where rainfall is low [41,43–45]. The rainfall distribution in the course of the vicinity varies and consequently groundwater potential may also be varying. Only 14.21% of a place is anticipated to have excessive to very excessive groundwater potential with respect to rainfall sample prospect (Table 9).

3.2. Slope and Reclassified Slope Layer

Slope is a crucial terrain parameter that has an effect on groundwater potential and recharge. Slope governs the amount of infiltration and runoff [53]. About 72.41% of a place is predicted to have excessive to very excessive groundwater ability with respect to slope.

In the nearly level slope (gentle slope) vicinity, the surface runoff is sluggish, which trap precipitation and allow rainwater to percolate/infiltrate via the soil and is considered a good groundwater potential zone, while a steep slope vicinity enables excessive runoff, permitting much less lag time for rainfall and therefore relatively much less infiltration and poor groundwater potential. However, slope classes have been identified based on their degree of significance to groundwater potential and recharge in GIS [34,54] and reclassified according to groundwater potential prospect (Table 10).

3.3. Elevation and Reclassified Elevation Layer

Elevation or altitude could have an indirect and inverse impact at the groundwater potential of a given area. Therefore, excessive altitudes favor extra recharge and make certain the provision of groundwater in lowland regions in a watershed. Mountainous regions are regularly favorable for recharge in deep-seated confined aquifers located at lowland regions [55,56]. Water has a tendency to store at lower topography than on the higher topography [42]. Therefore, the higher the elevation, the smaller the groundwater potential and vice versa.

The area covered by very low elevation (1464–1995 m) is expected to have a very high groundwater potential and recharge zone and the one covered by very high elevation (3295 m–4171 m) is expected to have a very low groundwater potential and recharge zone with respect to elevation, as shown in Table 11. About 49.02% of an area is expected to have high to very high ground water potential with respect to elevation.

3.4. Drainage and Reclassified Drainage Density Layer

A drainage network is to a degree a panorama dissection with the aid of using streams and may be expressed as drainage density, indicating the entire length of streams associated with an area (km/km^2) [57]. Drainage density has an inverse relation with the permeability of aquifers and performs an important position within the runoff distribution and degree of infiltration. Drainage density is one of the parameters affecting the groundwater potential, recharge, and play an essential position in groundwater potential zoning. Groundwater potential is poor in regions with a very excessive drainage density because it misplaces the majority in the form of runoff while regions with low drainage density permit extra infiltration to recharge the groundwater and, therefore, have extra for groundwater potential occurrence. According to [58], additionally cited, the low drainage density area has better infiltration and it yields higher groundwater potential zones, as compared to an excessive drainage density area. A dense drain is the consequence of feeble or impervious subsurface formations, light plant life, and mountainous relief. The drainage density of the looked at region starts from 0 to 1.8 km/km^2 (Figure 9a). About 46.32% of the place is predicted to have excessive to very excessive groundwater potential with respect to drainage density as shown in Table 12.

3.5. Lineament and Reclassified Lineament Density Layer

Geological formations that deliver rise to lineaments encompass faults, shear zones, fractures, dykes, and veins—in addition to bedding planes and stratigraphic contacts. The lineament density map shows the quantitative length of linear formations expressed in (km/km^2). An excessive lineament length density suggests excessive secondary porosity, therefore representing a sector with excessive groundwater potential [59]. Lineament density is a crucial geological formation that have an effect on groundwater potential and recharge. Areas having a better lineament density facilitate infiltration and recharge of groundwater and, therefore, is proper for groundwater potential development. In turn, the ones having a low lineament density have low groundwater potential. Accordingly, only 27.94% of the place is predicted to have excessive to very excessive groundwater potential with respect to lineament density (Table 13).

3.6. LULC and Reclassified LULC Layer

The land-use and/or land-cover map is the principal issue for controlling the groundwater potential and recharge method. Its situations have an effect on the hydrologic cycle and hydrologic manner by changing the evapotranspiration, transpiration, infiltration, interception, and surface runoff, and thereby the groundwater potential distribution conduct in lots of ways. Due to population increase and different anthropogenic impacts in lots of watersheds, there is change in land use and/or land cover from one shape to the other [60–63]. Some land uses service the groundwater potential while others bring negative result to groundwater potential and recharge. Accuracy evaluation or validation is an essential step within the processing of remote sensing data, which determines the information value of the resulting data to a user [64]. The study had an average category accuracy of 87% and kappa coefficient (K) of 0.84. The kappa coefficient is rated as almost perfect and subsequently the classified image discovered to be a match for further research [27].

Built-up and rocky surfaces have much less opportunity of groundwater potential prevalence through growing runoff during rainfall while the surfaces protected through vegetation such as agricultural plants and forests have a better chance of groundwater opportunity due to higher infiltration through trapping and protecting the rainwater in roots of plants and cracks [65–68]. About 71.67% of a place is predicted to have excessive to very excessive groundwater potential with respect to land use/land cover (Table 14).

3.7. Soil and Reclassified Soil Group Layer

The water-conserving capability of a place relies upon the soil sorts and their permeability [69]. Soil mainly influences the rainfall infiltration and percolation strategies that, in the long run, impact the groundwater recharge after which the groundwater potential of a given area [70,71]. Soil properties influence the connection among runoff and infiltration rates, which, in turn, controls the degree of permeability that determines the groundwater potential [72]. The permeability of the soil sorts relies upon their texture.

Therefore, to identify the soil group permeability of the study area, a different soil group was categorized into six soil texture families with its permeability rate and classes. In the reclassification of the soil map, soil group according to its texture and permeability classes was grouped in to five classes according to is contribution for groundwater potential and recharge (Table 15). About 7.3% of the area is expected to have high to very high groundwater potential with respect to soil group.

3.8. Lithology and Reclassified Lithology Layer

Geology or lithology is one of the groundwater potential and recharge controlling parameters taken into consideration in groundwater research, which plays a vital role in the distribution and prevalence of groundwater. Lithology is the bodily makeup of rocks and sediments and consists of mineral configuration, grain quartz, and grain packing [39]. Geology affects both the porosity and permeability of the aquifer material [1,73]. The Lithology of West Arsi area is grouped into distinctive kinds of formations/geological units [33,74–79]. However, every one of these lithological units do not have the same significance in determining and controlling groundwater potential and recharge. According to [74,75], the West Arsi lithological unit is categorized into 5 classes, depending on groundwater potential and recharge ability (Table 16). About 66.2% of the area is predicted to have excessive to very excessive groundwater potential with respect to the lithological units.

3.9. Weight Overlay Analysis

A main task within the proposed GIS-based multicriteria selection evaluation is the choice of criteria for groundwater potential area mapping wherein the criteria choice calls for suitable information of the site, the correct weighting of criteria through hydrogeologist experts, and correct cooperation among the decided-on factors. Pairwise comparison and Normalized Principal Eigen Vector (NPEV) found out that slope, LULC, and lithology are the most influential parameters, accounting for 24%. The percent weight of elevation, drainage density, lineament density, rainfall, and soil group is 9, 9, 5, 3, and 2, respectively (Table 6).

In computing the consistency ratio, the Principal Eigenvalue of 8.321 became done for an eight*eight matrix. Hence, the Principal Eigenvalue (λ_{max}) needs to constantly be greater than or identical to the number of the criteria (m) as in the above end result of λ_{max}, paving the way for the calculation of the consistency index (CI). Hence, pairwise comparison is reliable due to the fact the Principal Eigen value (λ_{max}), 8.321044984 calculated in Table 7, is greater than the number of criteria investigated (m), which was 8. Consistency index (CI) is 0.046.

The Random Index (RI) for the eight criteria is 1.41. The Consistency Ratio (CR) computed is 0.033, which is much less than 0.1 for large matrices more than 4 × 4 [38] and, therefore, the CR gained is suitable and the weight or percentage influence assigned for every thematic layer is acceptable.

3.10. Groundwater Potential Occurrence District Map

The groundwater potential sector was delineated with the aid of preparing maps, reclassifying, weighting, and ranking eight groundwater potential-influencing parameters (rainfall, lithology, slope, elevation, lineament density, drainage density, soil group, and LULC) in an ArcGIS environment applying weighted overlay analysis guided with AHP MCDM pairwise comparison techniques. The groundwater potential district layer was generated from superimposed thematic layers applying the weighted overlay approach with the help of spatial analysis tools in ArcGIS. In the weight overlay analysis and groundwater potential layer development, as found in Table 17 soil is assigned the lowest percentage of influence (weight), while slope, LULC, and lithology were assigned the higher weight/percentage influence. The spatial dispersal of groundwater potential and recharge throughout the study sector is the sum of the products of factors percentage influence (weight) and the corresponding reclassified parameters rank.

Table 17. Thematic maps rank and weight in terms of groundwater potential prospect.

Thematic Maps	Classes	Gw Prospect	Weight (%)	Rank
Rainfall	700.5–955.4	Very Low	3	1
	955.4–1210.8	Low		2
	1210.8–1466.2	Moderate		3
	1466.2–1721.6	High		4
	1721.6–1976.7	Very High		5

Table 17. *Cont.*

Thematic Maps	Classes	Gw Prospect	Weight (%)	Rank
Soil Group	Eutric Cambisols/Pellic Vertisols/Orthic Solonchaks/Chromic Vertisols/Chromic Cambisols/Mollic Gleysols	Very Low	2	1
	Dystric Histosols	Low		2
	Vitric Andosols/Mollic Andosols/Chromic Luvisols/Eutric Nitisols/Eutric Regosols/Orthic Luvisols	Moderate		3
	Leptosols/Calcic Xerosols	High		4
	Calcic Fluvisols/Calcaric Fluvisols/Eutric Fluvisols	Very High		5
Lineament Density	0–0.4	Very Low	5	1
	0.4–0.8	Low		2
	0.8–1.2	Moderate		3
	1.2–1.6	High		4
	1.6–2.7	Very High		5
Lithology	Qwo	Very Low	24	1
	gt/PRa/PRr/Qwpu	Low		2
	PNa/PRw/Qdi/Qdp/Qwa	Moderate		3
	Nc/Ncb/Nn/Qb/Qwbp	High		4
	QI	Very High		5
LULC	Built Up Area	Very Low	24	1
	Barren Landscape	Low		2
	Forest	Moderate		3
	Agriculture & Vegetation	High		4
	Water Body	Very High		5
Slope	27.7–75.9	Very Low	24	1
	17.9–27.7	Low		2
	10.4–17.9	Moderate		3
	4.5–10.4	High		4
	0–4.5	Very High		5
Elevation/Altitude	3295–4171	Very Low	9	1
	2810–3295	Low		2
	2433–2810	Moderate		3
	1995–2433	High		4
	1464–1995	Very High		5
Drainage Density	1.3–1.8	Very Low	9	1
	1.0–1.3	Low		2
	0.7–1.0	Moderate		3
	0.4–0.7	High		4
	0–0.4	Very High		5

From Table 18 above, about 61.27% of the area is anticipated to have high groundwater potential and 2.47% of the area is anticipated to have very high groundwater potential.

Table 18. West Arsi Zone groundwater potential distribution and its area coverage.

Value	GWP Prospect Rank	Count	Area (ha)	Area (%)
2	Low	109,176	9825.84	0.79
3	Moderate	4,896,961	440,726.49	35.46
4	High	8,460,429	761,438.61	61.27
5	Very High	341,652	30,748.68	2.47

The reclassified thematic layers have a rank of 1 to 5, where 1 is very low, 2 low, 3 moderate, 4 high, and 5 very high groundwater prospects (Figures 5–12). However, the weight overlay analysis produced ranks 2, 3, 4, and 5 only, which display the West Arsi region groundwater potential distribution being categorized into four classes: very high, high, moderate, and low (Figure 14a,b) groundwater potential and recharge zones. Figure 14c indicates regions of high groundwater potential are concentrated in the central and western a part of the study region, which covers an area of 61.27% (761,438.61 ha). Areas having very high groundwater potential is located at Shala, Arsi Negele, and the boundary of Shashemene and Siraro, which accounts about 2.47% (30,748.68 ha). The remaining part of the study region, 35.46% (440,726.49 ha), is classified as moderate groundwater potential in which the distribution is throughout the complete district; however, it is dominant on the eastern a part of the study location, and 0.79% (9825.84 ha) of the land is classified as low groundwater potential region, positioned in the Nensebo, Adeba, Shala, Siraro, Gedeb Asasa, and Shashemene Western and Southern sector of the study area in small amounts (Figure 14c).

In validating the groundwater potential distribution region layer, while the inventory data were plotted over the groundwater potential map (Figure 14d), from the 33 total number of wells categorized to low yield, 23 (69.70%) wells failed under the low groundwater potential region; from 69 total number of wells categorized to moderate yield, 67 (97.10%) wells failed under moderate groundwater potential; and from 11 total number of wells categorized to high yield, 9 (81.82%) wells failed under high groundwater potential region (Table 19). In cross validation evaluation from 113 borehole yields, 99 (87.61%) conform to the corresponding groundwater potential region classifications from the qualitative evaluation.

Table 19. Agreement (%) between the groundwater recharge zone map and borehole yield.

Groundwater Potential Zone Classes	Yield (L/s)	Total Number of Wells in Groundwater Potential Zone	Number of Wells Fall in Groundwater Potential Zone	Percent Agreement
Low	<2	33	23	69.70
Moderate	2–10	69	67	97.10
High and Very high	>10	11	9	81.82
Sum		113	99	87.61

The validation results, 87.61% or 0.876, confirm that there may be a very good agreement among the groundwater inventory data and groundwater potential zones delineated applying GIS and RS techniques. Therefore, the results of the groundwater potential maps obtained with the support of the AHP technique and weight overlay analysis were taken into consideration as a very good prediction. This indicates that the findings developed from the study are proper compared with the well yield of the point inventory acquired from the field.

4. Conclusions

Increased availability of remotely sensed information has helped in analyzing the natural surroundings without direct measuring in field. Remote Sensing (RS) and Geographic Information System (GIS) are possible in terms of cost, time, and resources, which make it fast, correct, and economical strategies to be utilized in groundwater potential and recharge sector evaluation than the traditional ground survey and resistivity methods. It incorporates comparing groundwater potential zones of the study area by applying the different groundwater spatial parameters.

The occurrence and distribution of groundwater potential specifically relies upon on the groundwater potential-affecting parameters. The most vital parameters affecting groundwater potential and the recharge sector decided on for this study were rainfall, lithology/geology, lineament density, LULC, soil group and texture, slope, elevation/altitude/geomorphology, and drainage density. Therefore, assessment of the groundwater potential sector of a place was carried out via way of means of analysis of those parameters. Those parameter factors were reclassified and ranked according to its significance in influencing groundwater potential.

Weighted value or percent influence determination was done with the reclassified thematic prepared and the organized maps. Weighted value or percentage influence dedication were completed with reclassified thematic organized and prepared maps. Weight overlay analysis considers all parameters that have an impact on groundwater potential; it offers the correct weight for parameters; it offers ranking of the characteristics of the parameters and exercising through the cell. Therefore, it better estimated the groundwater potential distribution throughout the region. Groundwater potential and recharge district affecting parameters taken into consideration in this study do not have a similar influence on groundwater potential distribution and the recharge zone. The percent influence (weight) of those parameters was decided applying Analytic Hierarchy Process (AHP) Multi-Criteria Decision Making (MCDM) pairwise comparison matrix techniques.

Weight overlay analysis produced 24% for slope, lithology, and LULC; 9% for elevation; 9% for drainage density; 5% for lineament density; 3% for rainfall; and 2% for soil group. Accordingly, slope, LULC, and lithology were found the most significant parameters. Whereas, lineament density, rainfall, and soil group were the least groundwater potential-influencing parameters.

Groundwater distribution throughout the West Arsi Zone is not uniform according to this study and therefore classified into very high (2.47%, which is 30,748.68 ha), high (61.27%, which is 761,438.61 ha), moderate (35.46%, which is 440,726.49 ha), and low (0.79%, which is 9825.84 ha) groundwater potential distribution.

The suitable groundwater potential and recharge areas are found with lithological formation of QI (lacustrine sediments deposits, silt clays, diatomite, and minor ignimbrites), Qwbp (Pleistocene basalt), Qb (alkaline basalt and trachyte), Nn (ignimbrite, un-welded tuffs, ash flows, rhyolites, domes, and trachyte), Ncb (alkaline basalts and trachyte), and Nc (basalt and peralkaline rhyolite with minor alkaline basalt). LULC areas covered with a water body, agricultural practices, and area covered with vegetation are suitable for groundwater potential formation. Areas acquired excessive annual common rainfall were given excessive groundwater potential. A location with a low slope, elevation, and drainage density are good for groundwater potential and recharge. In turn, locations with excessive lineament density were discovered to be the most essential regarding groundwater potential and recharge. Soil groups of leptosols, calcic xerosols, calcic fluvisols, calcaric fluvisols, and eutric fluvisols, and those with a sandy loam and loamy sand texture, permit infiltration and percolation, which will increase groundwater potential and recharge.

The groundwater potential distribution assessed and the map generated were validated using borehole inventory data. Accordingly, the percent agreement between the groundwater potential recharge zone map generated and inventory borehole yield found in the rank analysis was 87.61%, which is in very good agreement.

Author Contributions: Conceptualization, J.K.; methodology, J.K.; software, J.K.; validation, J.K., D.A. and M.S.R.; formal analysis, J.K.; investigation, J.K.; data curation, J.K.; writing—original draft preparation, J.K.; writing—review and editing, J.K., D.A., M.S.R. and M.K.L.; visualization, J.K.; supervision, J.K., D.A., M.S.R. and M.K.L. All authors have read and agreed to the published version of the manuscript.

Funding: This research received no external funding.

Institutional Review Board Statement: Not applicable.

Informed Consent Statement: Not applicable.

Data Availability Statement: The data used in this study can be available from the authors on reasonable request.

Acknowledgments: The authors would like to acknowledge the University of Rostock for their willingness to finance the publication under the funding program Open Access Publication.

Conflicts of Interest: The authors declare that they have no conflict of interest or financial conflict to disclose.

References

1. Chowdhury, J.; Jhan, M.K.; Chowdary, V.M.; Mal, B.C. Integrated Remote Sensing and GIS Based Approach for Assessing Groundwater Potential in West Mendinipur district, West Bengal, India. *Int. J. Remote Sens.* **2010**, *30*, 231–250. [CrossRef]
2. Jhan, M.; Chowdary, V.; Chowdhury, A. Groundwater Assessment in Salboni Block, West Bengal, India Using Remote Sensing, Geographic Information System and Multi-criteria Decision Analysis Techniques. *Hydrogeol. J.* **2012**, *18*, 1713–1728. [CrossRef]
3. Naghibi, S.A.; Moradi, D.M. Evaluation of four supervised learning methods for groundwater spring potential mapping in Khalkhal region, Iran, using GiS- based features. *Hydrogeol. J.* **2017**, *25*, 169–189. [CrossRef]
4. Naghibi, S.A.; Moghaddam, D.D.; Kalantar, B.; Pradhan, B.; Kisi, O. A comparative assessment of GIS-based data mining models and a novel ensemble model in groundwater well potential mapping. *J. Hydrol.* **2017**, *548*, 471–483. [CrossRef]
5. Maniar, H.H.; Bhatt, N.J.; Prakash, I.; Mahmood, K. Application of Analytical Hierarchy Process (AHP) and GIS in the Evaluation of Groundwater Recharge Potential of Rajkot District, Gujarat, India. *Int. J. Tech. Innov. Mod. Eng. Sci.* **2017**, *5*, 1078–1087.
6. Zeinolabedini, M.; Esmaeili, A. Groundwater Potential Assessment Using Geographic Information Systems and AHP Method (Case Study: Baft City, Kerman, Iran). *Int. Arch. Photogramm. Remote Sens. Spat. Inf. Sci.* **2015**, *1*, W5. [CrossRef]
7. Krishnamurthy, J.; Mani, A.; Jayaraman, V.; Manivel, M. Groundwater resources development in hard rock terrain: An approach using remote sensing and GIS techniques. *Int. J. Appl. Earth Obs. Geoinf.* **2000**, *3*, 204–215. [CrossRef]
8. Saraf, A.K.; Choudhury, P.R. Integrated remote sensing and GIS for groundwater exploration and identification of artificial recharge sites. *Int. J. Remote Sens.* **1998**, *19*, 1825–1841. [CrossRef]
9. Carver, S. Integrating multi-criteria evaluation with geographic information systems. *Int. J. Geogr. Inf. Syst.* **1991**, *5*, 321–339. [CrossRef]
10. Bahunguna, I.M.; Nayak, S.; Tamilarsan, V.; Moses, J. Groundwater prospective zones in basaltic terrain using remote sensing. *J. Indian Soc. Remote Sens.* **2003**, *31*, 101–105. [CrossRef]
11. Das, S.; Behera, S.C.; Kar, A.; Narendra, P.; Guha, S. Hydro geomorphological mapping in groundwater exploration using remotely sensed data: Case study in Keonjhar district, Orissa. *J. Indian Soc. Remote Sens.* **1997**, *25*, 245–259. [CrossRef]
12. Das, S.; Pardeshi, S.D. Integration of different influencing factors in GIS to delineate groundwater potential areas using IF and FR techniques: A study of Pravara basin, Maharashtra, India. *Appl. Water Sci.* **2018**, *8*, 197. [CrossRef]
13. Dar, I.A.; Sankar, K.; Dar, M.A. Remote sensing technology and geographic information system modeling: An integrated approach towards the mapping of potential groundwater recharge zones in hard rock terrain, Mamundiyar basin. *J. Hydrol.* **2010**, *394*, 285–295. [CrossRef]
14. Singh, P.; Thakur, J.K.; Kumar, S. Delineating groundwater potential zones in a hard-rock terrain using geospatial tool. *Hydrol. Sci. J.* **2013**, *58*, 213–223. [CrossRef]
15. Saaty, T. Decision making with the analytic hierarchy ptocess. *Int. J. Serv. Sci.* **2008**, *1*, 83–89.
16. Doke, A.B.; Zolekar, R.B.; Patel, H.; Das, S. Geospatial mapping of groundwater potential zones using multi-criteria decision-making AHP approach in a hardrock basaltic terrain in India. *Ecol. Indic.* **2021**, *127*, 107685. [CrossRef]
17. Sresto, M.A.; Siddika, S.; Haque, M.N.; Saroar, M. Application of fuzzy analytic hierarchy process and geospatial technology to identify groundwater potential zones in north-west region of Bangladesh. *Environ. Chall.* **2021**, *5*, 100214. [CrossRef]
18. Karimi-Rizvandi, S.A.; Goodarzi, H.V.; Afkoueieh, J.H.; Chung, I.-M.; Kisi, O.; Kim, S.; Linh, N.T.T. Groundwater-Potential Mapping Using a Self-Learning Bayesian Network Model: A Comparison among Metaheuristic Algorithms. *Water* **2021**, *13*, 658. [CrossRef]
19. Maity, B.; Mallick, S.K.; Das, P.; Rudra1, S. Comparative analysis of groundwater potentiality zone using fuzzy AHP, frequency ratio and Bayesian weights of evidence methods. *Appl. Water Sci.* **2022**, *12*, 63. [CrossRef]
20. Tamiru, H.; Wagari, M. Comparison of ANN model and GIS tools for delineation of groundwater potential zones, Fincha Catchment, Abay Basin, Ethiopia. *Geocarto Int.* **2021**, 1–19. [CrossRef]
21. Nguyen, P.T.; Ha, D.H.; Jaafari, A.; Nguyen, H.D.; van Phong, T.; Al-Ansari, N.; Prakash, I.; van Le, H.; Pham, B.T. Groundwater Potential Mapping Combining Artificial Neural Network and Real AdaBoost Ensemble Technique: The DakNong Province Case study, Vietnam. *Int. J. Environ. Res. Public Health* **2020**, *17*, 2473. [CrossRef]

22. Razavi-Termeh, S.V.; Sadeghi-Niaraki, A.; Choi, S.-M. Groundwater Potential Mapping Using an Integrated Ensemble of Three Bivariate Statistical Models with Random Forest and Logistic Model Tree Models. *Water* **2019**, *11*, 1596. [CrossRef]
23. Mallick, J.; Khan, R.A.; Ahmed, M.; Alqadhi, S.D.; Alsubih, M.; Falqi, I.; Hasan, M.A. Modeling Groundwater Potential Zone in a Semi-Arid Region of Aseer Using Fuzzy-AHP and Geoinformation Techniques. *Water* **2019**, *11*, 2656. [CrossRef]
24. Farzin, M.; Avand, M.; Ahmadzadeh, H.; Zelenakova, M.; Tiefenbacher, J.P. Assessment of Ensemble Models for Groundwater Potential Modeling and Prediction in a Karst Watershed. *Water* **2021**, *13*, 2540. [CrossRef]
25. Congalton, R. A review of assessing the Accuracy of classifications of remotely sensed data. *Remote Sens. Environ.* **1991**, *37*, 35–46. [CrossRef]
26. Jensen, J. *Introductory Digital Image Processing: A Remote Sensing Perspective*, 2nd ed.; Prentice Hall, Inc.: Upper Suddle River, NJ, USA, 1996.
27. Rwanga, S.S.; Ndambuki, J.M. Accuracy Assessment of land use land cover classification using remote sensing and GIS. *Int. J. Geosci.* **2017**, *8*, 611–622. [CrossRef]
28. ESRI. *ArcGIS Desktop: Release 10.3.1*; Environmental Systems Research in USA: Redlands, CA, USA, 2015.
29. Singh, P.; Gupta, A.; Singh, M. Hydrological Inferences from watershed analysis for water resource management using remote sensing and GIS techniques, Egypt. *J. Remote Sens. Space Sci.* **2014**, *17*, 111–121.
30. Rusli, N.; Majid, M.R. *Digital Elevation Model (DEM) Extraction from Google Earth: A Study in Sungai Muar Watershed*; Applied Geoinformatics for Society and Environment: Garching bei München, Germany, 2012; pp. 32–36; ISBN 978-3-943321-06-7.
31. Hobbs, W. Lineaments of the Atlantic border region. *Geol. Soc. Am. Bull.* **1904**, *15*, 483–506. [CrossRef]
32. Clark, C.D.; Wilson, C. Spatial Analysis of Lineaments. *Comput. Geosci.* **1994**, *20*, 1237–1258. [CrossRef]
33. GSE. *Geological Map of Oromia*; Ethiopian Mapping Authority: Addis Ababa, Ethiopia, 1999.
34. Yeh, H.F.; Cheng, Y.S.; Lin, H.I.; Lee, C.H. Mapping groundwater recharge potential zone using a GIS approach in Hualian River, Taiwan. *Sust. Environ. Res.* **2016**, *26*, 33–43. [CrossRef]
35. Hussein, A.A.; Govindu, V.; Nigusse, A.G.M. Evaluation of Groundwater potential using geospatial techniques. *Appl. Water Sci.* **2017**, *7*, 244–246. [CrossRef]
36. Singh, L.K.; Jha, M.K.; Chowdary, V.M. Multi-criteria analysis and GIS modeling for identifying prospective water harvesting and artificial recharge sites for sustainable water supply. *J. Clean. Prod.* **2017**, *142*, 1436–1456. [CrossRef]
37. Saaty, T.L. *The Analytical Hierarchy Process*; WS Publication: Manila, PH, USA, 1990; ISBN1 0962031720. ISBN2 9780962031724.
38. Saaty, T.L. *Decision Making for Leaders*; RWS Publications: Pittsburgh, PA, USA, 1995.
39. Freeze, R.A.; Cherry, J.A. *Groundwater*; Prentice-Hall, Inc.: Hoboken, NJ, USA, 1979.
40. Abdel Rahman, R. A Modified Analytical Hierarchy Process Method to Select Sites for Groundwater Recharge in Jordan. Ph.D. Dissertation, University of Leicester, Leicester, UK, 2015.
41. Kotchoni, D.O.V.; Vouillamoz, J.; Lawson, F.M.A. Relationship between rainfall and groundwater recharge in seasonal humid Benin: A comparative analysis of long-term hydrographs in sedimentary and crystalline aquifers. *Hydrogeol. J.* **2019**, *27*, 447–457. [CrossRef]
42. Ramu, M.; Vinay, M. Identification of groundwater potential zones using GIS and remote sensing techniques: A case study of Mysore taluk Karnataka. *Int. J. Geomat. Geosci.* **2015**, *5*, 393–403.
43. Zomlot, Z.; Verbeiren, B.; Huysmans, M.; Batelaan, O. Spatial distribution of groundwater recharge and base flow: Assessment of controlling factors. *J. Hydrol. Reg. Stud.* **2015**, *4*, 349–368.
44. Shakya, B.M.; Nakamura, T.; Shrestha, S.D.; Nishida, K. Identifying the deep groundwater recharge Processes in an intermountain basin using the hydrogeochemical and water isotope characteristics. *Nord. Hydrol.* **2019**, *50*, 1216–1229. [CrossRef]
45. Wang, J.; Huo, A.; Zhang, X. Prediction of the response of groundwater recharge to climate changes in Heihe river basin, China. *Environ. Earth Sci.* **2020**, *79*, 1–16. [CrossRef]
46. Dabrala, S.; Bhatt, B.; Joshi, J.P.; Sharma, N. Groundwater Suitability recharge zones modelling: A GIS application. *ISPRS* **2014**, *8*, 347–352. [CrossRef]
47. Saaty, T. Decision making with the AHP: Why is the principal eigenvector necessary. *Eur. J. Oper. Res* **2003**, *145*, 85–91. [CrossRef]
48. Brunelli, M. *Introduction to the Analytic Hierarchy Process*; Springer: New York, NY, USA, 2015; p. 83. ISBN 978-3-319-12501-5. [CrossRef]
49. Saaty, T. The Analytic Hierarchy Process (AHP) for Decision Making. In Kobe, Japan. 1980, pp. 1–69. Available online: http://www.cashflow88.com/decisiones/saaty1.pdf (accessed on 5 May 2022).
50. Senanayake, I.P.; Dissanayake, D.M.; Mayadunna, B.B.; Weerasekera, W.L. An Approach to delineate groundwater recharge potential sites in Ambalantota, Sri Lanka using GIS techniques. *Geosci. Front.* **2016**, *7*, 115–124. [CrossRef]
51. Sahoo, S.; Jha, M.K.; Kumar, N.; Chowdary, V.M. Evaluation of GIS based Multicriteria decision analysis and probabilistic modeling for exploring groundwater prospects. *Environ. Earth Sci.* **2015**, *74*, 2223–2246.
52. Yesilnacar, E.; Topal, T. Landslide susceptibility mapping: A comparison of logistic regression and natural networks methods in a medium scale study, Hendek region (Turkey). *Eng. Geol.* **2005**, *79*, 251–266. [CrossRef]
53. Lerner, D.N.; Issar, A.S.; Simmers, I. *Groundwater Recharge: A Guide to Understanding and Estimating Natural Recharge*; IAH International Contributions to Hydrogeology, 8; Taylor and Francis: Rotterdam, The Netherlands, 1990.
54. Andualem, T.G.; Demeke, G.G. Groundwater potential assessment using GIS and Remote sensing: Study of guna Tana landscape, Upper Blue Nile basin, Ethiopia. *J. Hydrol. Reg. Stud.* **2019**, *24*, 3. [CrossRef]

55. Dufy, C.J.; Al-Hassan, S. Groundwater circulation in a closed desert basin: Topographic scaling and climatic forcing. *Water Resour. Res.* **1988**, *24*, 1675–1688. [CrossRef]
56. Todd, D.K.; Mays, L.W. *Groundwater Hydrology*, 3rd ed.; John Wiley & Sons, Inc.: New York, NY, USA, 2005.
57. Schillaci, C.; Braun, A.; Kropacek, J. Terrain Analysis and Landform Recognition. In *Geomorphological Techniques*; British Society for Geomorphology: London, UK, 2015; Chapter 2.
58. Murasingh, S. *Analysis of Groundwater Potential Zones Using Electrical Resistivity: Remote Sensing and GIS Techniques in a Typical Mine Area of Odisha*; National Institute of Technology: Rourkela, India, 2014.
59. Al-Abadi, A.M.; Al-Shamma, A. Groundwater potential mapping of the major aquifer in Northeastern Missan Governorate, South of Iraq by using AHP and GIS. *J. Environ. Earth Sci.* **2014**, *10*, 125–149.
60. Kenji, J.; Atsushi, T.; Othoman, A.; Susumu, S.; Ronny, B. Effects of land use change on groundwater recharge model parameters. *Hydrol. Sci. J. Des. Sci. Hydrol.* **2009**, *54*, 300–315. [CrossRef]
61. Pan, Y.; Gong, H.; Zhou, D.; Li, X.; Nakagoshi, N. Impact of land use change on groundwater recharge in Guishui river basin, China. *Chin. Geogr. Sci.* **2011**, *21*, 734–743. [CrossRef]
62. Owuor, S.O.; Butterbach-Bahl, K.; Guzha, A.C.; Rufino, M.C.; Pelster, D.E.; Díaz-Pinés, E.; Breuer, L. Groundwater recharge rates and surface runoff response to land use and land cover changes in semi-arid environments. *Ecol. Processes* **2016**, *5*, 16. [CrossRef]
63. Riley, D.; Mieno, T.; Schoengold, K.; Brozović, N. The impact of landcover on groundwater recharge in the high plains: An application to the conservation reserve program. *Sci. Total Environ.* **2019**, *696*, 133871. [CrossRef]
64. Abubaker, H.M.; Elhag, A.M.H.; Salih, A.M. Accuracy assessment of land use land cover classification: Case study of Shomadi area-Renk county, upper Nile State, South Sudan. *Int. J. Sci. Res. Publ.* **2013**, *3*, 2250–3153. Available online: http://citeseerx.ist.psu.edu/viewdoc/summary?doi=10.1.1.414.8771 (accessed on 5 May 2022).
65. Shaban, A.; Khawlie, M.; Abdallah, C. Use of remote sensing and GIS to determine recharge potential zone: The case of Occidental, Lebanon. *Hydrogeol. J.* **2006**, *14*, 433–443. [CrossRef]
66. Singh Sk Singh, C.K.; Mukherjee, S. Impact of land use and land cover change on groundwater quality in the lower Shiwalik hills: A remote sensing and GIS based approach. *Cent. Eur. J. Geosci.* **2010**, *2*, 124–131. [CrossRef]
67. Fenta, A.A.; Kife, A.; Gebreyohannes, T.; Hailu, G. Spatial analysis of groundwater potential using remote sensing and GIS based multi-criteria evaluation in Raya Valley, northern Ethiopia. *Hydrogeol. J.* **2015**, *23*, 195–206. [CrossRef]
68. Li, S.; Yang, H.; Lacayo, M.; Liu, J.; Lei, G. Impacts of land use and land cover changes on water yield: A case study in Jing-jin-ji., China. *Sustainability* **2018**, *10*, 960. [CrossRef]
69. Kumar, P.; Herath, S.; Avtar, R.; Takeuchi, K. Mapping of groundwater potential zones in Killinochi area, Sri Lanka, using GIS and remote sensing techniques. Sustain. *Water Resour. Manag.* **2016**, *2*, 419–430. [CrossRef]
70. Anuraga, T.S.; Ruiz, L.; Mohan Kumar, M.S.; Sekhar, M.; Leijnse, A. Estimating groundwater recharge using land use and soil data: A case study in south india. *Agric. Water Manege.* **2006**, *84*, 65–76. [CrossRef]
71. Rukundo, E.; Dogan, A. Dominant influencing factors of groundwater recharge spatial patterns in Ergene river catchment, Turkey. *Water* **2019**, *11*, 653. [CrossRef]
72. Tesfaye, T. Ground Water Potential Evaluation Based on Integrated GIS and Remote Sensing Techniques, in Bilate River Catchment: South Rift Valley of Ethiopia. *Am. Acad. Sci. Res. J. Eng. Technol. Sci.* **2014**, *10*, 85–120.
73. Yazi, A.; Pirasteh, S.; Arvin, A.; Pradhan, B.; Nikouravan, B.; Mansor, S. Dsasters and Risk reduction in Groundwater: Zagros Mountain southwest Iran using geoinformatics techniques. *Interdiscip. Neurosurg.* **2010**, *3*, 51–57.
74. GSE. Integrated Hydrological and Hadrochemical Mapping of Yiag Map Sheet. 2003. Available online: https://gis.gse.gov.et/hg-maps/ (accessed on 15 April 2020).
75. Halcrow, G. *Rift Valley Lakes Basin Integrated Resources Development Master Plan Study Project*; Draft Phase 2 Report Part II Prefeasibility Studies; Unpublished Report; Halcrow Group Limited: London, UK; GIRD Consultants: Mumbai, India, 2008.
76. Kefale, T.; Jiri, S. *Hydrogeological and Hydro Chemical Maps of Hosaina Explanatory Notes (NB 37-2)*; Antonin Orgon: Prague, Czech Republic, 2013.
77. Astatike, K.; Sima, J. *Hydrogeological and Hydro Chemical Maps of Asela Explanatory Notes (NB37-3)*; Antonin Orgon: Prague, Czech Republic, 2012.
78. Astatike, K.; Sima, J. *Hydrogeological and Hydro Chemical Maps of Dodola Explanatory Notes (NB 37-7)*; Antonin Orgon: Prague, Czech Republic, 2011.
79. Thomas, A.; Tegist, R.; Sima, J. *Hydrogeological and Hydro Chemical Maps of Dila Explanatory Notes (NB 37-6)*; Antonin Orgon: Prague, Czech Republic, 2014.

MDPI

Article

Minimizing Errors in the Prediction of Water Levels Using Kriging Technique in Residuals of the Groundwater Model

Alireza Asadi [1] and Kushal Adhikari [2,*]

[1] Department of Civil, Environmental, & Construction Engineering, Texas Tech University, 911 Boston Ave, Lubbock, TX 79409, USA; alireza.asadi@ttu.edu
[2] Department of Environmental Resources Engineering, Cal Poly Humboldt, 1 Harpst Street, Arcata, CA 95521, USA
* Correspondence: kushal.adhikari@humboldt.edu or adhikari.kushal1992@gmail.com; Tel.: +1-806-283-0788

Abstract: Groundwater monitoring and water level predictions have been a challenging issue due to the complexity of groundwater movement. Simplified numerical simulation models have been used to represent the groundwater system; these models however only provide the conservative approximation of the system and may not always capture the local variations. Several other efforts such as coupling groundwater models with hydrological models and using geostatistical methods are being practiced to accurately predict the groundwater levels. In this study, we present a novel application of a geostatistical tool on residuals of the groundwater model. The kriging method was applied on the residuals of the numerical model (MODFLOW) generated by the TWDB (Texas Water Development Board) for the Edwards–Trinity (Plateau) aquifer. The study was done for the years 1995 through 2000 where 90% of the observation data was used for model simulation followed by cross-validation with the remaining 10% of the observations. The kriging method reduced the average absolute error of approximately 31 m (for MODFLOW simulation) to less than 5 m. Furthermore, the residuals' average standard error was reduced from 9.7 to 4.7. This implies that the mean value of residuals over the entire period can be a good estimation for each year separately. The use of the kriging technique thus can provide improved monitoring of groundwater levels resulting in more accurate potentiometric surface maps.

Keywords: groundwater monitoring; modeling; MODFLOW; kriging; residuals; water levels; Edwards–Trinity aquifer

Citation: Asadi, A.; Adhikari, K. Minimizing Errors in the Prediction of Water Levels Using Kriging Technique in Residuals of the Groundwater Model. *Water* **2022**, *14*, 426. https://doi.org/10.3390/w14030426

Academic Editor: Cristina Di Salvo

Received: 22 December 2021
Accepted: 27 January 2022
Published: 29 January 2022

1. Introduction

The exponential growth of population, rapid socio-economic development, increasing food demand, and changing climatic factors have led to a decline in both the quality and quantity of freshwater resources. The decreasing available resources have posed serious challenges in the agricultural sector with limited water available for irrigation. As an alternative resource, the reuse of wastewater in irrigation has been increasingly recognized as an essential, and economical strategy [1–3]. However, only a small fraction of wastewater with less than 6% in the US [4] and less than 3% globally [5] is reclaimed; and the irrigation largely relies on groundwater sources. Groundwater is one of the primary sustainable water resources, especially during the high demand seasons, due to its lower susceptibility to sudden changes. About 70% of groundwater withdrawal worldwide is used for agriculture while irrigating nearly 38% of irrigated lands [6]. Likewise, nearly 50% of irrigated lands in the United States are based on groundwater sources [6].

Over the past centuries, extreme drought events have significantly affected both surface and groundwater resources [7,8]. While low surface water levels might be an immediate indicator of drought, changes to groundwater levels indicate long-term water scarcity. Further, it is straightforward to monitor and assess surface water changes while measuring variations in groundwater resources is very challenging and time-consuming. Water movement in porous media is sufficiently complex to simulate and even the most elaborate models cannot estimate all details of this complexity, even if attempted, it ends with a parsimonious estimation of the reality.

The most common approach which has been used for decades is to estimate the groundwater head variable, mostly by using numerical models [9–11]. The numerical models use mathematical equations to describe the physics of the groundwater flow; the model accuracy thus depends on how precisely the conceptual models describe the real-world system. These models largely rely on the available water-level data [12,13]. However, groundwater-level data are often irregularly sampled, leading to temporal gaps in the record, and are not adequately distributed spatially across an aquifer [14–16]. The spatial sparseness of data presents challenges when spatially interpolating potentiometric surfaces and creating groundwater maps [14,17,18].

In addition to numerical models, statistical-based or regression-based approaches like kriging, spline interpolation, and neural networks have also been used in predicting water levels [19,20]. The most vivid method applied in the spatially auto-correlated variable like groundwater level is the kriging method namely, the ordinary kriging technique. Aboufirassi et al. [21] employed the universal kriging to estimate the water table for the Souss aquifer in central Morocco. Pucci and Murashige [22] used kriging for optimizing data collection and utility in a regional groundwater investigation in central New Jersey and confirmed that kriging is a useful tool especially in areas lacking enough data for developing a water table management network. Hoeksema et al. [23] applied the co-kriging method to estimate the groundwater level at unknown points. A similar study was done by other researchers where the kriging method was used to estimate the water levels at wells [24,25]. Theodossiou and Latinopoulos [13] used the kriging method on 31 wells in evaluating and optimizing the groundwater level observation networks. Ahmadi and Sedghamiz [26] evaluated the spatial and temporal variations of groundwater level of 39 observation wells using kriging. In their later article [27], the kriging and co-kriging methods were applied for groundwater depth mapping in southern Iran. Tapoglou et al. [12] used Artificial Neural Networks (ANNs) to estimate the temporal prediction of the water level and applied the kriging method to spatial parts. Ruybal et al. [17] used spatiotemporal kriging in evaluating groundwater levels in the Arapahoe aquifer. These approaches come with a major limitation in that they fail to explain the physics of the groundwater flow [28].

This study presents an integrated approach where the groundwater model is coupled with a geo-spatial kriging tool. This way, we expect to integrate the strong aspects of both approaches while reducing their demerits. The numerical model will explain the physics of groundwater flow. The statistical interpolation method, kriging will then be used on the numerical model's errors with an expectation of improving our estimation by considering the complexity not detected by a numerical model. In this study, the kriging method was applied on the residuals of the numerical model (MODFLOW) generated by the TWDB (Texas Water Development Board) for the Edwards–Trinity (Plateau) aquifer to improve the estimation of the water table spatially. The study was done for the years 1995 through 2000 where 90% of the observation data was used for model simulation followed by cross-validation with the remaining 10% of the observation data. To the authors' knowledge, no prior efforts have been done using the technique adopted in this paper. Most importantly, the significant improvement in groundwater level predictions makes this study a promising approach for the sustainable management of water resources.

2. Methodology

2.1. Study Area

The Edwards–Trinity (Plateau) Aquifer, as shown in Figure 1, expands over west-central Texas between 97° and 105° west longitudes and between 29° and 33° north latitudes. The topology of the aquifer is known as a plateau, lightly leveling from about 610 m (2000 ft) above sea level in the southeast to about 915 m (3000 ft) in the northwest. The precipitation in the region ranges between 86 cm (34 in) in the east to 30 cm (12 in) in the west. The maximum average annual temperature for the study area ranges between 23 °C (73° F) in the Trans-Pecos uplands to 26 °C (79° F) in southern Val Verde County. The study area had numerous drought events during the past hundred years [29]. Yet, the drought events are expected to have minimal impacts for the study period (1995 through 2000) as the last drought was during the 1950s.

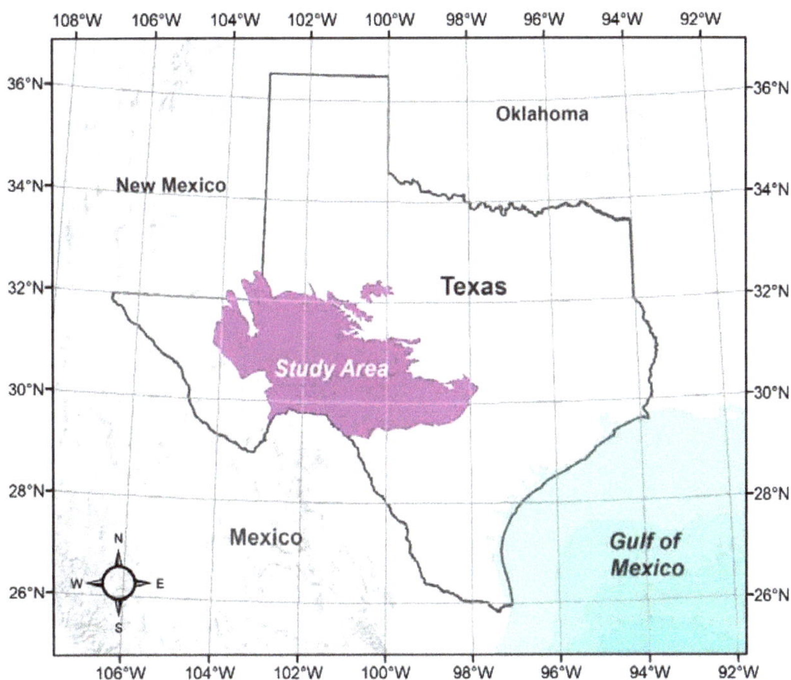

Figure 1. Location of the study area—Edwards–Trinity (Plateau) Aquifer.

2.2. Data

Data on groundwater levels, climate data, boundaries shapefiles, data on hydraulic properties of the aquifer, elevation data, and MODFLOW simulated heads were required. All these data were obtained from the TWDB website [30]. The water levels were obtained from the Groundwater Database (GWDB:1, accessed on 1 April 2020) consisting of shapefiles with geospatial information on water level and quality for management, monitoring, and characterization of the water in the Edwards–Trinity aquifers. The precipitation data was gathered using raster data for the years 1995 through 2001. The boundaries and hydraulic properties were gathered from separate shapefiles consisting of aquifer boundaries, model boundaries, Texas county boundaries, hydraulic conductivity for each one sq. km. The elevation data was obtained from the raster files consisting of DEM (Digital Elevation Model), top and bottom elevation of the Edwards and Trinity aquifers. Lastly, MODFLOW generated heads were obtained as publicly accessible binary files produced from the MODFLOW model developed by TWDB. Figure 2 shows the location of observation wells for all the years (1995–2000). Not all years have observation data available for the same locations, thus the locations for each year were selected randomly as can be seen in Figure 2.

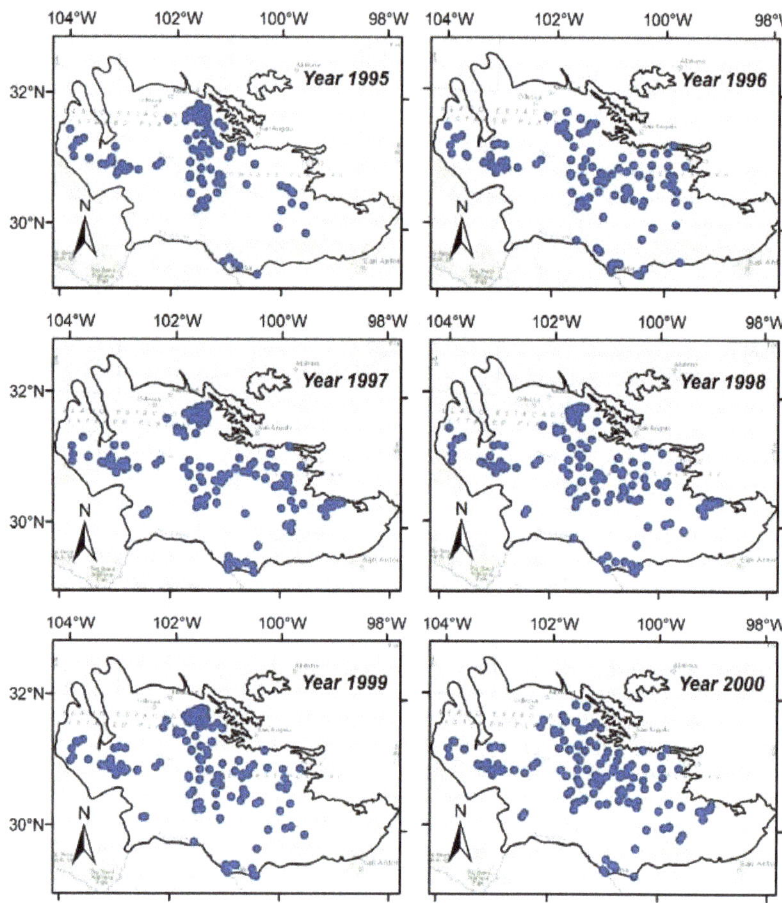

Figure 2. Location of observation wells in the Edwards–Trinity aquifer.

2.3. Modeling

Figure 3 represents a schematic diagram of the research methodology adopted. As shown in the figure, the general scheme of this study consists of three major components—(1) Data Preparation, (2) Calibration, and (3) Cross-Validation and includes a series of steps used interactively as listed below:

- Mapping MODFLOW simulated groundwater heads (model imported from TWDB) into their corresponding coordinates and overlap with observation data to find the MODFLOW estimated values in the observation point.
- Subtracting the observed groundwater head with MODFLOW simulated head and consider as the model residuals.
- Dividing the residuals into two separate datasets, 90 percent of data for fitting kriging methods (calibrating residuals) and 10 percent for validating part (validating residuals), in a random selection.
- Pre-evaluating the calibrating residuals and fit kriging method to generate the estimated residual map for the study domain.
- Comparing the validating residuals with an estimated one to evaluate the accuracy of the kriging method.

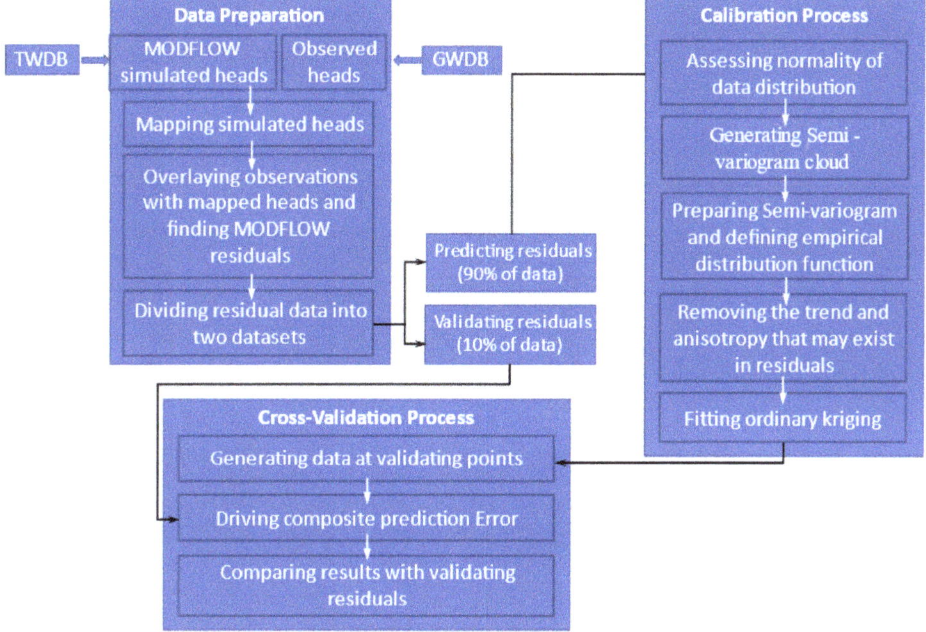

Figure 3. Schematic diagram of the MODFLOW + kriging model.

2.3.1. Data Preparation

To minimize the distortion of the area and appropriate evaluation of the data, the NAD 1983 Texas Centric Mapping System Albers was used. First, the data were categorized into two groups—observed data and MODFLOW generated output. The observed data consists of the mean values for the water tables during the winter season (winter months were selected as the water tables are expected to be more stable during winter and thus minimize the anthropogenic effects due to uncertainties in demand seasons). The observed data were then provided with coordinates and transformed into a shapefile. For the MODFLOW generated data, the model grid shapefile produced by the MODFLOW program was employed in the study area and saved based on the needed attributes for future analysis. The binary files consisting of groundwater head data were converted to a text file using python programming. The extracted head data were matched with the corresponding observed data by locating the observation data in the grid shapefile. Finally, the shapefile of observation data was overlaid on the MODFLOW grid shapefile to determine the value of the MODFLOW generated head for each observation point. The MODFLOW generated heads were compared with the observed data for the corresponding locations thus obtaining the residuals.

2.3.2. Kriging Method

Kriging is one of the well known methods of predicting spatial characteristics and it has been used in a variety of fields (e.g., soil science, ecology, mining, and water resources) to provide a robust unbiased estimation of geo-distributed variables from small scales like X-ray scattering experiments [31] to large ones like traffic behavior pattern [32], soil properties' profile [33], and anticipating the metrological variables [34]. The main advantage of the kriging method over the other spatial interpolation techniques is that the method is driven based on the statistical theory, comparing others that are mostly deterministic and they have the lack of ability to use for prediction. Furthermore, studies over geo-dependent variables showed that kriging has outstanding performance compared with its similar existing interpolation methods [35].

Equation (1) shows the stochastic function that is considered in the simple kriging method:

$$Z(s) = \mu + \varepsilon(s) \tag{1}$$

where $Z(s)$ is the estimating function—which is simulated groundwater head elevation in our study, μ is a constant value representing the mean value of the groundwater head and $\varepsilon(s)$ is a random error function regarding the model error from the observed records.

Estimating the value of $Z(s)$ relies on two primary assumptions of considering that variables are randomly distributed, and they preserve second-order stationary condition respect to the location [36], which means:

$$E[Z(s+h)] = E[Z(s)] \tag{2}$$

$$cov[Z(s+h),\ Z(s)] = C[h] \tag{3}$$

where h is a vector that connects point s to point s + h. Equation (2) implies that the expected value $E[Z(s)]$ is constant in all domains as represented in Equation (1) by the constant μ. However, in reality, this value fluctuates from place to place due to inherent trends and variability in data. To deal with this problem, it is usually assumed that the estimating variable (groundwater head) comprises two components

$$Z(s) = m(s) + e(s) \tag{4}$$

where $m(s)$ presents the deterministic part and $e(s)$ is the statistical component of the estimating variable which includes the spatially correlated random variable. Since $m(s)$ is treated as a deterministic part, it can be determined in a separate process and summed up with residuals random function component $e(s)$. In this study, the deterministic part of the head variable is estimated by existing numerical models. We then applied the ordinary kriging (OK) to the residuals of the numerical model for the statistical part assuming the mean value of error is not known and needs to be obtained over optimization process by minimizing the variance of errors.

3. Results and Discussion

3.1. Data Investigation

In standard statistical problems, the first step before starting to model the phenomena is to examine the data. In this study, a commonly used statistical tool, histogram, was used to see if the MODFLOW residuals are normally distributed. As shown in Figure 4, the residuals in general, follow the normality pattern. However, some level of negative skewness and distortion of normality was observed in the data which could be attributed to the complexity of nature.

Next, the residuals were checked for spatial correlation as the data tends to share some information with their neighbors. A variogram method was used to estimate how strongly data are related to each other. The variogram in space is usually implemented to check two major assumptions in the application of the kriging method—stationary in space (variance is independent of location) and isotropy (variance is independent of direction). Thus, a semi-variogram cloud or plot was used to provide a better visual understanding of the data distribution and to detect any possible trends or geometric anisotropic behavior.

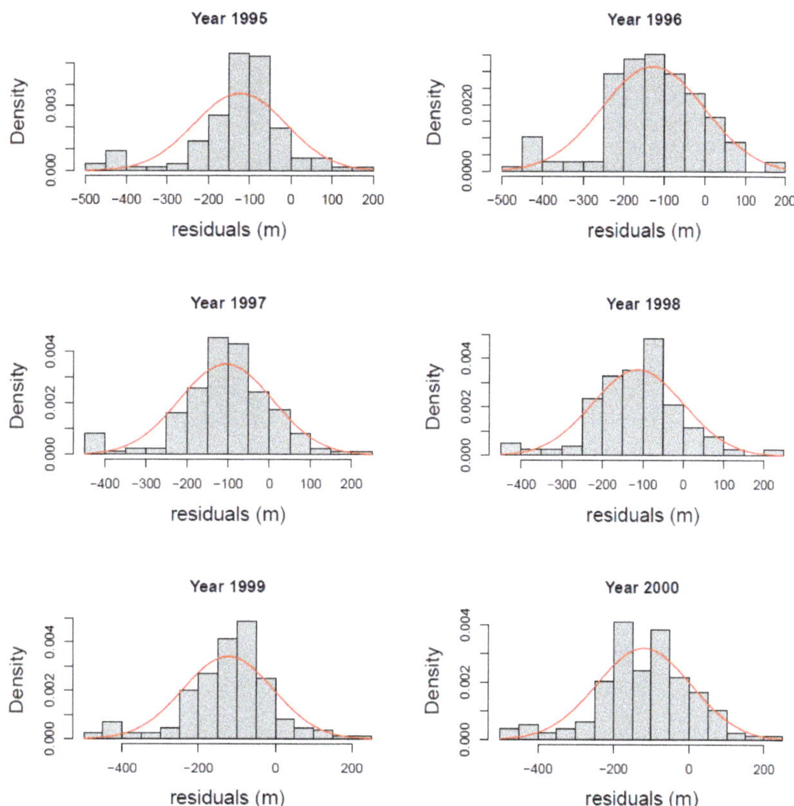

Figure 4. Histogram presenting the water level residuals for Edwards–Trinity aquifer (years 1995–2000).

Figure 5 shows the semi-variogram clouds of water level residuals calculated from subtraction between observation points and the MODFLOW outputs for the years 1995 to 2000. The difference squared (γ) in Figure 5 represents the MODFLOW residuals dissimilarity which is defined as:

$$\gamma = \frac{1}{2}\left(Z(s_i) - Z(s_j)\right)^2 \quad (5)$$

where s is sample location and Z is the residual value.

The reddish circles in the figure indicate that there is a strong gradient observed at short distances, as the value of the semi-variogram for each pair in these areas is significantly high. This significant change can be an indication of non-stationarity in higher ranks caused by neighboring drainage areas or rivers. However, to get a proper conclusion over the sources that resulted in non-stationary, more investigation needs to be considered which is beyond the scope of this study.

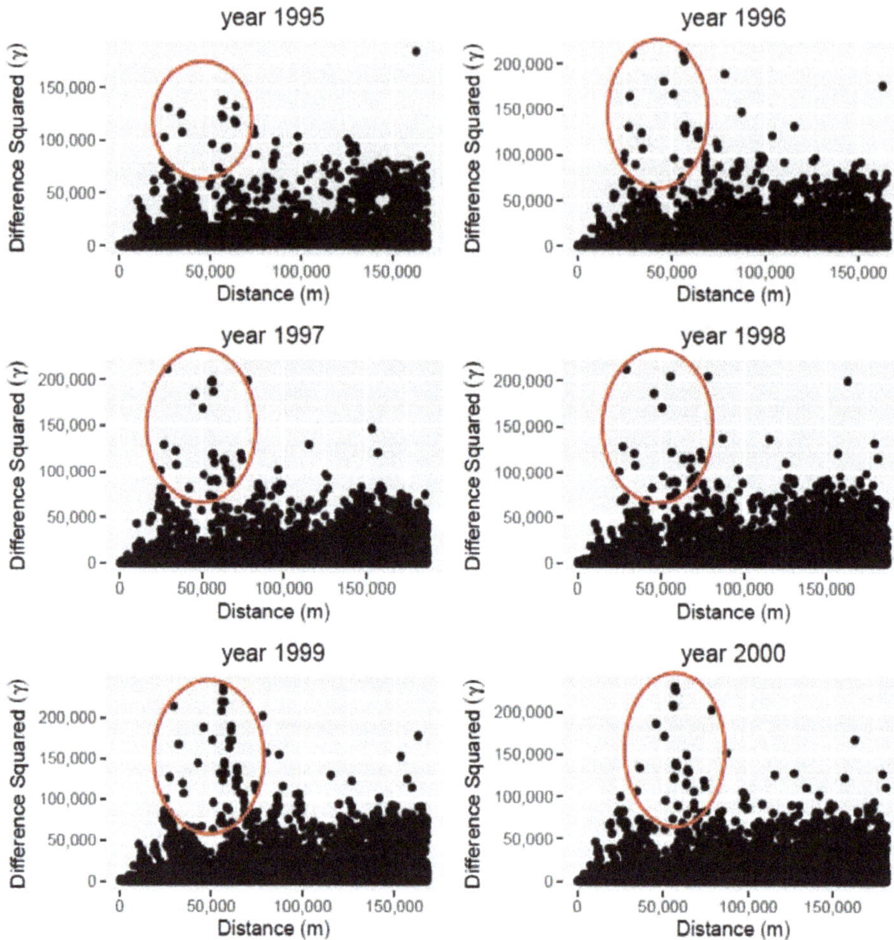

Figure 5. Semi-variogram cloud for water level residuals during the years 1995 through 2000.

The semi-variogram was further used for interpolation where semi-variance was used to represent the expected value of the residuals' dissimilarity. A theoretical model was fit into the sample data. In this study, a commonly used Spherical function was used. The spherical function shows a progressive decrease of spatial autocorrelation until some distance (radius of influence), beyond which autocorrelation is zero [37]. As observed from the sample variogram in Figure 6, the Nugget variance of 50 and range of 50,000 was defined for fitted theoretical semi-variogram.

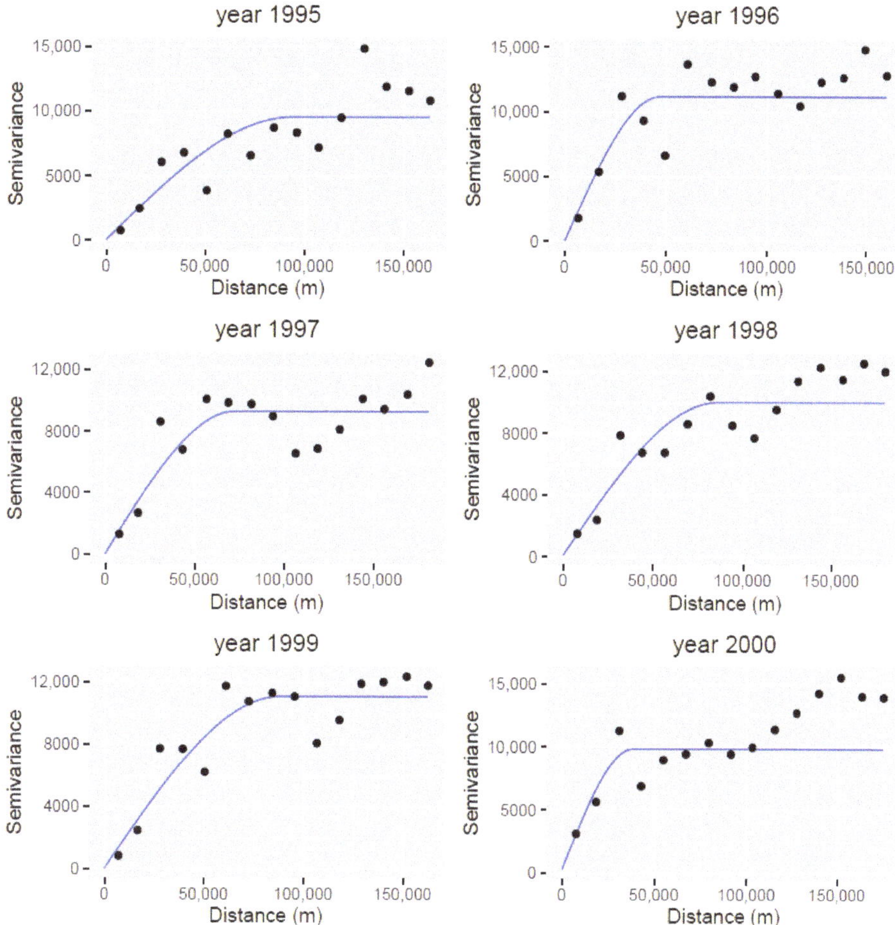

Figure 6. Sample variogram and fitted model applied on Edwards aquifer dataset for years 1995 through 2000.

3.2. Model Simulation

The ordinary Kriging method was applied to the MODFLOW residuals. In the process, the existing trend arising from uncertainties of the simulated model was removed using first-order trend and the exponential Kernel function was applied to weight the values of the neighbors closed to sampled values [38].

The spherical model was chosen to fit on a semi-variogram, and the maximum number of neighbors affecting the predicted data was limited to ten points. The kriging then applied as a single model resulted in a higher mean root square error (not presented in this paper), thus the area was divided into four quadrants with 45 offsets to minimize the distortion due to anisotropy as observed in Figure 5.

Figure 7 shows the prediction for OK applied to Edwards–Trinity aquifer between the years 1995 to 2000. High residuals (as shown by dark red and blue color) were observed for the areas outside the boundary while the residual values are minimal inside the study area. Furthermore, similar observations were observed in deviations as shown in Figure 8. The high residuals or larger deviation (as shown by purple color) outside the boundary is attributed to the missing observation data. Furthermore, some areas inside the study area showed relatively higher residuals with larger deviations which could also be due to the missing observation data.

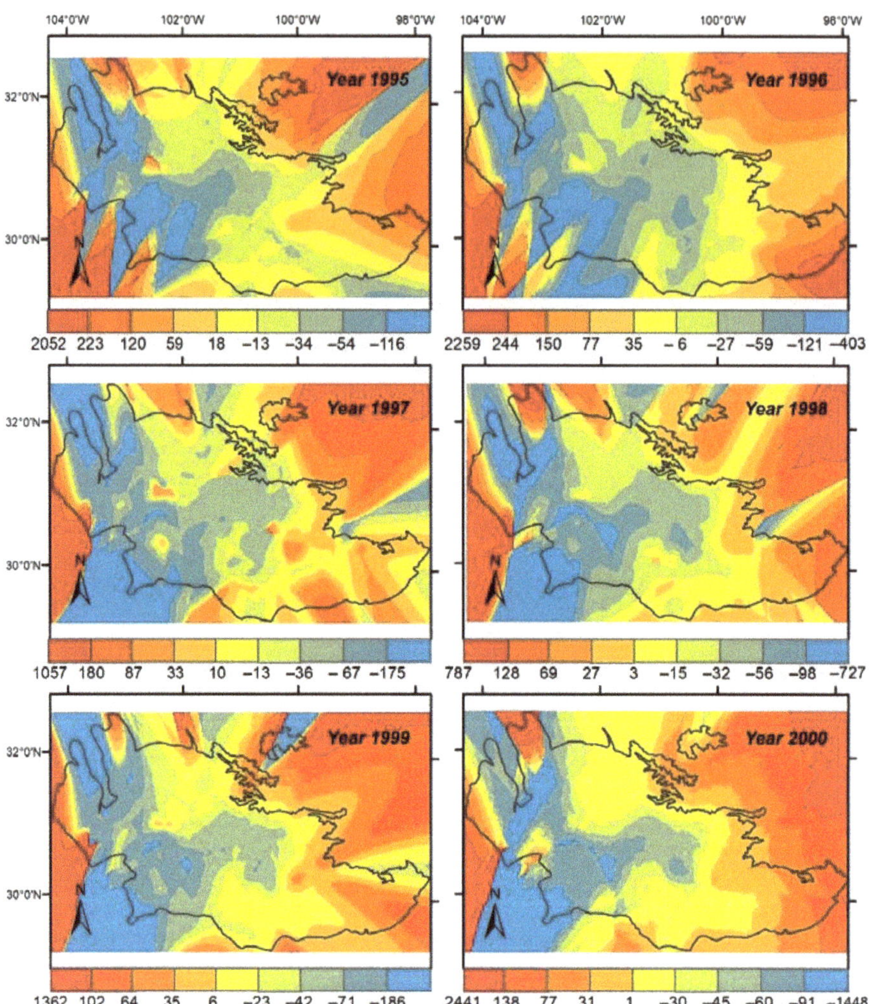

Figure 7. Predicted residuals (in meters) after application of kriging on MODFLOW residuals for years 1995 through 2000.

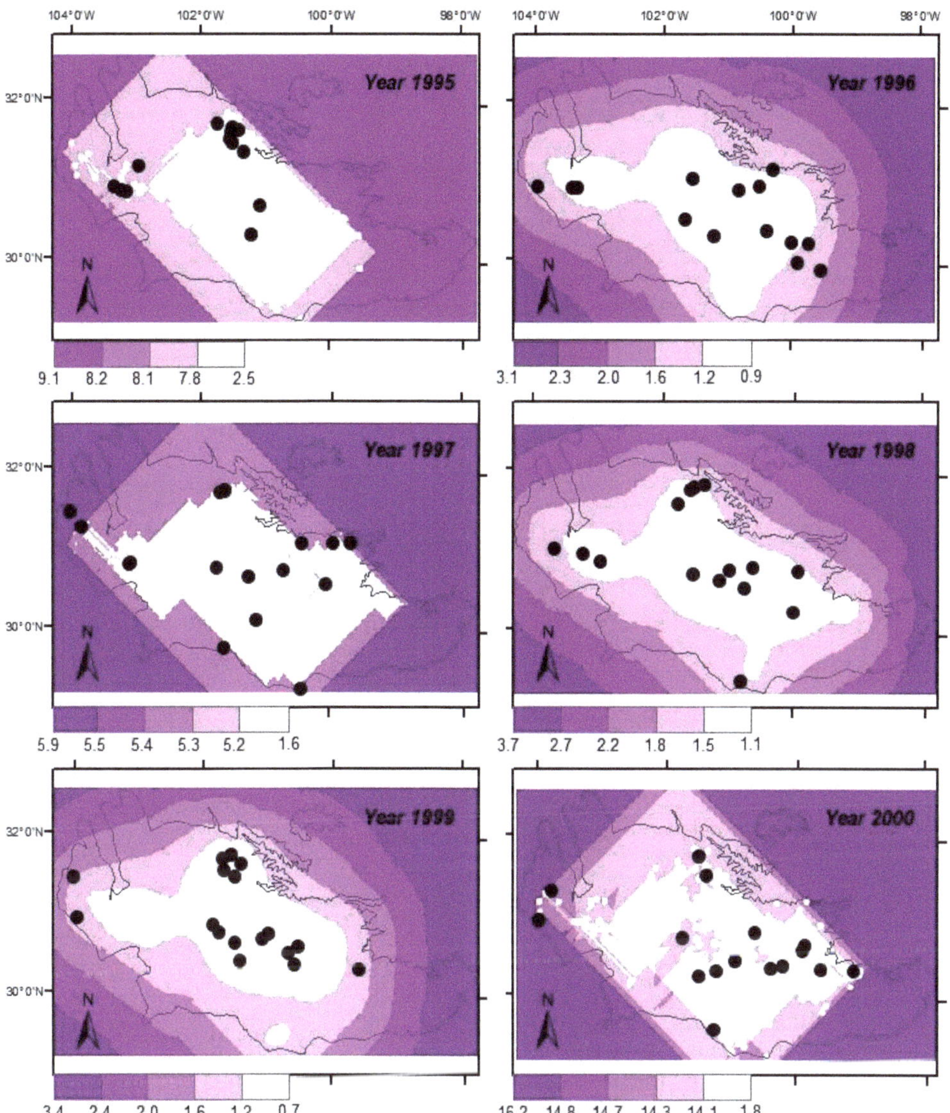

Figure 8. Standard Deviation (in meters) for ordinary kriging applied to Edwards–Trinity aquifer for the years 1995 through 2000 (the black dots in the figure represent the observation points used during cross-validation).

3.3. Model Validation

As mentioned before, the entire dataset was divided into two sets—a simulation dataset (90% data) and a validation dataset (remaining 10% data). First, kriging was applied on 90% of the data where the validation was performed automatically using the ArcGIS Geostatistical Analyst toolbar [38]. The Geostatistical Analyst toolbar provides the measured and predicted values and the standard error for each point. During the process, the software keeps any individual points separate from other data (referred to as dataset) for estimating the spherical model parameters. The parameters are then estimated using every such dataset and the process continues until optimal parameters that best fit the entire datasets are obtained. Table 1 presents the average water head for observation points along with predicted water heads using MODFLOW and kriging, while Table 2 provides a detailed summary of kriging application on the MODFLOW residuals. As observed in Table 1, the residuals obtained from the MODFLOW simulation are much higher as compared to the residuals obtained after the application of kriging. The residuals significantly reduced from an average value of approximately 37 m to less than 1 m, with a standard error of approximately 0.5 m.

Table 1. Average water heads using MODFLOW and kriging method (simulation dataset).

Year	Observations [1]	MODFLOW (m)			MODFLOW + Kriging (m)			
		Mean a [2]	Observed Residuals [3]	Standard Error	Mean b [4]	Predicted Residuals [5]	Error [6]	Standard Error
1995	727.5	764.7	−37.2	3.1	728.7	−36	1.2	0.7
1996	665.5	706	−40.5	3.3	667.1	−38.9	1.6	0.4
1997	692.4	725.5	−33.1	2.9	693.5	−32	1.1	0.4
1998	682.4	716.6	−34.2	2.9	682.8	−33.8	0.4	0.5
1999	708.3	747.1	−38.8	2.9	708.6	−38.5	0.3	0.3
2000	686	724.7	−38.7	3.2	685.3	−39.4	−0.7	1.1

[1] Observed average value for water level for the given year (winter season). [2] MODFLOW simulated average value for water level for the given year (winter season). [3] Difference between observed values and the MODFLOW simulated values (Observations—Mean a). [4] Simulated average value for water level for the given year after application of kriging (Mean a + Predicted Residuals). [5] Residuals obtained after application of kriging on the observed residuals. [6] Difference between predicted and observed residuals for each point.

Table 2. Detailed summary of kriging application on MODFLOW residuals (simulation dataset).

Year		Min.	1st Qu.	Median	Mean	3rd Qu.	Max.
1995	Observed Residuals	−141.6	−53.1	−32.2	−37.2	−20.8	55.2
	Predicted Residuals	−147.1	−52.1	−31.9	−36	−22.5	52.3
	Error	−97.3	−6.7	−0.1	1.2	5.2	84.5
	Standard Error	0.4	0.7	0.8	0.7	0.8	0.8
1996	Observed Residuals	−137.1	−60.8	−40.9	−40.5	−16.1	30.4
	Predicted Residuals	−132.7	−56.8	−43.3	−38.9	−14.3	33.6
	Error	−49.6	−6.4	0.8	1.6	11.6	72.8
	Standard Error	0.4	0.4	0.4	0.4	0.4	0.4
1997	Observed Residuals	−135.4	−51.7	−32.6	−33.1	−11.3	62.5
	Predicted Residuals	−148.1	−47	−31.7	−32	−9.1	45.7
	Error	−59	−6.4	0.5	1	7.7	78
	Standard Error	0.2	0.4	0.4	0.4	0.4	0.5
1998	Observed Residuals	−131	−54.3	−32.3	−34.2	−16.9	64
	Predicted Residuals	−145.8	−53.2	−33.4	−33.8	−19.3	43.8
	Error	−50	−8.7	0.4	0.4	7.1	61.3
	Standard Error	0.4	0.5	0.5	0.5	0.5	0.5
1999	Observed Residuals	−144.6	−54	−31.7	−38.8	−18.4	52.5
	Predicted Residuals	−150	−53.1	−31.1	−38.5	−20.9	52.2
	Error	−55.2	−8.2	0.4	0.3	7.7	75.9
	Standard Error	0.3	0.3	0.3	0.3	0.3	0.3
2000	Observed Residuals	−141.1	−59.3	−35.7	−38.7	−15.8	48
	Predicted Residuals	−139.3	−56.4	−39.5	−39.4	−19.1	36.1
	Error	−79.7	−8.7	0	−0.7	8.3	66.2
	Standard Error	0.4	1.2	1.2	1.1	1.2	1.3

As a next step, a manual cross-validation was adopted where the calibrated model was applied to the remaining 10% of the data. Table 3 presents the average water heads for observation points along with predicted water heads using MODFLOW and kriging. Similar to observations in Table 1, higher residuals were obtained for the MODFLOW simulation for all the years. However, after the application of kriging on MODFLOW residuals, the average absolute error of approximately 31 m (from MODFLOW simulation) was reduced to less than 5 m. A similar reduction was observed in the residuals' average standard error where the values reduced from 9.7 to 4.7. The observed reduction in standard error thus indicates that the average value of residuals over the entire period can be a good estimation for each year separately. Table 4 provides a detailed summary of the kriging application on the residuals.

Table 3. Average water heads using MODFLOW and kriging method (validation dataset).

Year	Observations	MODFLOW (m)			MODFLOW + Kriging (m)			
		Mean	Observed Residuals	Standard Error	Mean	Predicted Residuals	Error	Standard Error
1995	758.7	802.1	−43.5	8.7	753.6	−48.5	−5.1	3
1996	697.6	729	−31.4	14.7	689.1	−39.9	−8.5	5.3
1997	683	707.6	−24.6	5.9	677.4	−30.1	−5.5	3.9
1998	694.8	733	−38.1	10.7	694.6	−38.3	−0.2	8.1
1999	716.5	744.8	−28.3	8.4	714	−30.8	−2.5	3.8
2000	644.7	665.5	−20.8	10	646.5	−19	1.8	4.6

Table 4. Detailed summary of kriging application on MODFLOW residuals (validation dataset).

Year		Min.	1st Qu.	Median	Mean	3rd Qu.	Max.
1995	Observed Residuals	−131.8	−50	−33.9	−43.5	−24	−15
	Predicted Residuals	−132.5	−55	−31.8	−48.5	−24.8	−18.9
	Error	−7.1	−1.3	0.8	5.1	8.9	29.3
	Standard Error	-	-	-	3	-	-
1996	Observed Residuals	−138.1	−55.2	−30.1	−31.4	−1.8	61
	Predicted Residuals	−143.1	−46.5	−37.2	−40	−5.3	40.3
	Error	−20.5	−5.2	5	8.5	20.3	52.1
	Standard Error	-	-	-	5.3	-	-
1997	Observed Residuals	−82.3	−36.6	−28.1	−24.6	−6.9	16.2
	Predicted Residuals	−80.9	−44.9	−31.1	−30.2	−7.3	16.7
	Error	−18.5	−3.1	0	5.5	20.6	38.7
	Standard Error	-	-	-	3.9	-	-
1998	Observed Residuals	−134.5	−56.6	−33.7	−38.1	−25.2	62
	Predicted Residuals	−76.3	−51.7	−45	−38.3	−22.5	−10.1
	Error	−87.3	−6	0.8	0.2	10.4	72.9
	Standard Error	-	-	-	8.1	-	-
1999	Observed Residuals	−68.1	−54.8	−33.6	−28.3	−19.3	65.5
	Predicted Residuals	−71	−52.1	−39.1	−30.8	−22.8	32
	Error	−38.6	−2.7	0.3	2.5	9.7	35.8
	Standard Error	-	-	-	3.8	-	-
2000	Observed Residuals	−120.5	−46.4	−16.2	−20.8	−3.2	65.6
	Predicted Residuals	−71.7	−45.9	−20.4	−19.1	4.2	41
	Error	−48.9	−4	0.5	−1.7	10.2	24.7
	Standard Error	-	-	-	4.6	-	-

For all the years, the water head for observation points obtained after kriging is much closer to the observed values. However, for the year 2000, three observation points (11, 12, and 15) showed higher residuals after the application of kriging as shown in Table 5. This could possibly be due to the higher prediction accuracy of the MODFLOW simulation in those points. The table with water heads for all the years (1995 through 2000) has been provided as a supplementary file in Table S1.

Table 5. Predicted water heads for observation points for the year 2000.

Observation Well	Observed Head (m)	MODFLOW (m)		MODFLOW + Kriging (m)	
		Simulated Head	Residuals	Predicted Head	Residuals
1	769	786.9	−17.9	764.4	4.6
2	745.7	760.1	−14.4	741.8	3.8
3	748.3	868.8	−120.5	797.2	−48.9
4	1059.2	993.6	65.6	1034.6	24.7
5	662.2	713.2	−50.9	652.3	10
6	548.9	595.9	−47	552.5	−3.6
7	586.6	608.2	−21.6	574.7	12
8	648.8	695	−46.2	647.2	1.6
9	585	617.7	−32.6	570.1	14.9
10	616.1	663.8	−47.7	618.4	−2.3
11	613.4	619.5	−6.2	602.2	11.1
12	618.5	623.3	−4.8	629.4	−10.9
13	623.6	620.6	3	624.1	−0.5
14	588.9	568.3	20.5	594	−5.1
15	555.1	568.4	−13.3	593	−37.8
16	346.3	344.7	1.6	347.4	−1.2

3.4. Comparison to Other Studies

The integrated approach used in this study satisfactorily described the general pattern of residuals generated by numerical model MODFLOW, thereby improving the prediction of the groundwater level at ungauged areas. Results showed that the integrated kriging and MODFLOW method can be used as an alternative approach in improving the existing numerical models whilst reducing the underlying model uncertainties. Most of the previous studies focused on decreasing the uncertainties by improving the resolution, or/and including local hydraulic variables like pumping and recharge/discharge to the large-scaled models [39].

However, the uncertainties in residuals can have sources of randomness that the improvements in the modeling process may not address, and rather could be represented by statistical methods such as kriging. The applied ordinary kriging in this study showed that the MODFLOW model's residuals are locally correlated noises and can be estimated from their neighbors to some extent (with the range of 50 km) by using the spherical method as a correlation function. The numerical model used in this study was developed and improved by Anaya & Jones [29] and is the MODFLOW model approved by TWDB for the Edwards–Trinity Aquifer. In this study, MODFLOW was used as a deterministic part and the model uncertainties were presented as trends during the kriging model. The use of kriging reduced the average absolute error from approximately 31 m (from MODFLOW simulation) to less than 5 m after the application of kriging, which aligns with the findings from a previous study by Liu et al., [40]. They calibrated the MODFLOW model and were able to reduce the average absolute error from 7.7 m to 3.44 m through updates in the numerical modeling process. During the calibration process, they revised the recharge and used the actual value for pumping. However, the improvement was only performed for a smaller region (only the San Antonio segment of the aquifer) in contrast to improvement over the entire Edwards–Trinity as in the study presented by the authors.

4. Conclusions

The kriging method was applied to improve spatial confidence in groundwater-level predictions at unsampled locations. Kriging was applied on the MODFLOW residuals for the groundwater levels in the Edwards–Trinity aquifer in Texas. The study was done for the years 1995 through 2000 where 90% of the observation data was used for model simulation followed by validation with the remaining 10% of the observations. The kriging method significantly improved water level predictions. The average absolute error of approximately 31 m (from MODFLOW simulation) was reduced to less than 5 m after the application of kriging on MODFLOW residuals. Furthermore, the average residuals' standard error decreased from 9.7 to 4.7, which indicates that the average value of residuals over the entire period can be a good estimation for each year separately. With improved water level predictions, geostatistical tools such as kriging can be used to produce more accurate potentiometric surface maps. Such improved results and accurate monitoring of groundwater resources will lead to the sustainable use of groundwater resources while also aiding in efficient and effective conjunctive management of surface and groundwater resources.

Supplementary Materials: The following are available online at https://www.mdpi.com/article/10.3390/w14030426/s1, Table S1: Predicted water head for observation points for the the years 1995 through 2000.

Author Contributions: Conceptualization, A.A. and K.A.; methodology, A.A.; validation, K.A.; formal analysis, A.A. and K.A.; investigation, A.A. and K.A.; data curation, A.A.; writing—original draft preparation, A.A. and K.A.; writing—review and editing, A.A. and K.A.; visualization, A.A. and K.A.; funding acquisition, K.A. All authors have read and agreed to the published version of the manuscript.

Funding: The manuscript hasn't receive external founding. Publishing fee for this article has been covered by College of Natural Sciences at Cal Poly Humboldt.

Data Availability Statement: Data available on request.

Acknowledgments: The authors would like to thank the Department of Civil, Environmental and Construction Engineering at Texas Tech University for their support in conducting this research with assistantship support. And, sincere thanks to Guofeng Cao fromthe Department of Geography at University of Colorado, Boulder, and Adjunct Faculty of the Department of Geography at Texas Tech University, for his guidance with the research.

Conflicts of Interest: The authors declare no conflict of interest.

References

1. Adhikari, K.; Fedler, C.B. Water sustainability using pond-in-pond wastewater treatment system: Case studies. *J. Water Process Eng.* **2020**, *36*, 101281. [CrossRef]
2. Adhikari, K.; Fedler, C.B. Pond-In-Pond: An alternative system for wastewater treatment for reuse. *J. Environ. Chem. Eng.* **2020**, *8*, 103523. [CrossRef]
3. Mehan, G.T., III; Scientific American. The EPA Says We Need to Reuse Wastewater. 2019. Available online: https://blogs.scientificamerican.com/observations/the-epa-says-we-need-to-reuse-wastewater/ (accessed on 11 December 2019).
4. FAO. AQUASTAT Main Database, Food and Agriculture Organization of the United Nations (FAO). 2018. Available online: http://www.fao.org/nr/water/aquastat/didyouknow/index.stm (accessed on 26 January 2022).
5. FAO. AQUASTAT Main Database, Food and Agriculture Organization of the United Nations (FAO). 2016. Available online: http://www.fao.org/nr/water/aquastat/data/query/index.html?lang=en (accessed on 26 January 2022).
6. NGWA. The Groundwater Association. 2020. Available online: https://www.ngwa.org/what-is-groundwater/About-groundwater/information-on-earths-water (accessed on 26 January 2022).
7. Fendekova, M.; Demeterova, B.; Slivova, V.; Macura, V.; Fendek, M.; Machlica, A.; Gregor, M.; Jalcovikova, M. Surface and groundwater drought evaluation with respect to aquatic habitat quality applied in Torysa river catchment, Slovakia. *Ecohydrol. Hydrobiol.* **2011**, *11*, 49–61. [CrossRef]
8. Lange, B.; Holman, I.; Bloomfield, J.P. A framework for a joint hydro-meteorological-social analysis of drought. *Sci. Total Environ.* **2017**, *578*, 297–306. [CrossRef]
9. Chakraborty, S.; Maity, P.K.; Das, S. Investigation, simulation, identification and prediction of groundwater levels in coastal areas of Purba Midnapur, India, using MODFLOW. *Environ. Dev. Sustain.* **2020**, *22*, 3805–3837. [CrossRef]
10. Hutchison, W.R.; Jones, I.; Anaya, R. *Update of the Groundwater Availability Model for the Edwards-Trinity (Plateau) and Pecos Valley Aquifers of Texas*; Unpublished Report; Texas Water Development Board: Austin, TX, USA, 2011.
11. Xue, S.; Yu Liu, S.L.; Li, W.; Wu, Y.; Pei, Y. Numerical simulation for groundwater distribution after mining in Zhuanlongwan mining area based on visual MODFLOW. *Environ. Earth Sci.* **2018**, *77*, 400. [CrossRef]

12. Tapoglou, E.; Karatzas, G.P.; Trichakis, I.C.; Varouchakis, E.A. A spatio-temporal hybrid neural network-Kriging model for groundwater level simulation. *J. Hydrol.* **2014**, *519*, 3193–3203. [CrossRef]
13. Theodossiou, N.; Latinopoulos, P. Evaluation and optimisation of groundwater observation networks using the Kriging methodology. *Environ. Model. Softw.* **2006**, *21*, 991–1000. [CrossRef]
14. Marchant, B.P.; Bloomfield, J.P. Spatio-temporal modelling of the status of groundwater droughts. *J. Hydrol.* **2018**, *564*, 397–413. [CrossRef]
15. Oikonomou, P.D.; Alzraiee, A.H.; Karavitis, C.A.; Waskom, R.M. A novel framework for filling data gaps in groundwater level observations. *Adv. Water Resour.* **2018**, *119*, 111–124. [CrossRef]
16. Varouchakis, E.; Hristopulos, D. Comparison of stochastic and deterministic methods for mapping groundwater level spatial variability in sparsely monitored basins. *Environ. Monit. Assess.* **2013**, *185*, 1–19. [CrossRef] [PubMed]
17. Ruybal, C.J.; Hogue, T.S.; McCray, J.E. Evaluation of groundwater levels in the Arapahoe aquifer using spatiotemporal regression kriging. *Water Resour. Res.* **2019**, *55*, 2820–2837. [CrossRef]
18. Ziolkowska, J.R.; Reyes, R. Groundwater level changes due to extreme weather—An evaluation tool for sustainable water management. *Water* **2017**, *9*, 117. [CrossRef]
19. Kuria, D.N.; Gachari, M.K.; Macharia, M.W.; Mungai, E. Mapping groundwater potential in Kitui District, Kenya using geospatial technologies. *Int. J. Water Resour. Environ. Eng.* **2011**, *4*, 15–22.
20. Minor, T.B.; Russell, C.E.; Mizell, S.A. Development of a GIS-based model for extrapolating mesoscale groundwater recharge estimates using integrated geospatial data sets. *Hydrogeol. J.* **2007**, *15*, 183–195. [CrossRef]
21. Aboufirassi, M.; Mariño, M.A. Kriging of water levels in the Souss aquifer, Morocco. *J. Int. Assoc. Math. Geol.* **1983**, *15*, 537–551. [CrossRef]
22. Pucci, A.A., Jr.; Murashige, J.A.E. Applications of universal kriging to an aquifer study in New Jersey. *Groundwater* **1987**, *25*, 672–678. [CrossRef]
23. Hoeksema, R.J.; Clapp, R.B.; Thomas, A.L.; Hunley, A.E.; Farrow, N.D.; Dearstone, K.C. Cokriging model for estimation of water table elevation. *Water Resour. Res.* **1989**, *25*, 429–438. [CrossRef]
24. Desbarats, A.J.; Logan, C.E.; Hinton, M.J.; Sharpe, D.R. On the kriging of water table elevations using collateral information from a digital elevation model. *J. Hydrol.* **2002**, *255*, 25–38. [CrossRef]
25. Olea, R.; Davis, J. *Optimaizing the High Plains Aquifer Water-Level Observation Network*; Kansas Geological Surveying: Lawrence, KS, USA, 1999.
26. Ahmadi, S.H.; Sedghamiz, A. Geostatistical analysis of spatial and temporal variations of groundwater level. *Environ. Monit. Assess.* **2007**, *129*, 277–294. [CrossRef]
27. Ahmadi, S.H.; Sedghamiz, A. Application and evaluation of kriging and cokriging methods on groundwater depth mapping. *Environ. Monit. Assess.* **2008**, *138*, 357–368. [CrossRef] [PubMed]
28. Rivest, M.; Marcotte, D.; Pasquier, P. Hydraulic head field estimation using kriging with an external drift: A way to consider conceptual model information. *J. Hydrol.* **2008**, *361*, 349–361. [CrossRef]
29. Anaya, R.; Jones, I. *Groundwater Availability Model for the Edwards-Trinity (Plateau) and Pecos Valley Aquifers of Texas*; Texas Water Development Board: Austin, TX, USA, 2009.
30. Texas Water Development Board, Groundwater Models. Available online: https://www.twdb.texas.gov/groundwater/models/alt/eddt_p_2011/alt1_eddt_p.asp (accessed on 11 May 2020).
31. Noack, M.M.; Yager, K.G.; Fukuto, M.; Doerk, G.S.; Li, R.; Sethian, J.A. A kriging-based approach to autonomous experimentation with applications to X-ray scattering. *Sci. Rep.* **2019**, *9*, 11809. [CrossRef] [PubMed]
32. Pinto, J.A.; Kumar, P.; Alonso, M.F.; Andreao, W.L.; Pedruzzi, R.; Espinosa, S.I.; Albuquerque, T.T. Kriging method application and traffic behavior profiles from local radar network database: A proposal to support traffic solutions and air pollution control strategies. *Sustain. Cities Soc.* **2020**, *56*, 102062. [CrossRef]
33. Gotway, C.A.; Ferguson, R.B.; Hergert, G.W.; Peterson, T.A. Comparison of kriging and inverse-distance methods for mapping soil parameters. *Soil Sci. Soc. Am. J.* **1996**, *60*, 1237–1247. [CrossRef]
34. Venäläinen, A.; Heikinheimo, M. Meteorological data for agricultural applications. *Phys. Chem. Earth Parts A/B/C* **2002**, *27*, 1045–1050. [CrossRef]
35. Laslett, G.M. Kriging and splines: An empirical comparison of their predictive performance in some applications. *J. Am. Stat. Assoc.* **1994**, *89*, 391–400. [CrossRef]
36. Wackernagel, H. *Multivariate Geostatistics: An Introduction with Applications*; Springer Science & Business Media: Berlin, Germany, 2013.
37. esri. How Kriging Works. 2021. Available online: https://pro.arcgis.com/en/pro-app/latest/tool-reference/3d-analyst/how-kriging-works.htm (accessed on 26 January 2022).
38. ArcGIS Desktop. 2021. Available online: https://desktop.arcgis.com/en/arcmap/latest/extensions/geostatistical-analyst/what-is-arcgis-geostatistical-analyst-.htm (accessed on 26 January 2022).
39. Sharp, M.J., Jr.; Green, R.T.; Schindel, G.M. *The Edwards Aquifer: The Past, Present, and Future of a Vital Water Resource*; Geological Society of America: Boulder, CO, USA, 2019; Volume 215.
40. Liu, A.; Troshanov, N.; Winterle, J.; Zhang, A.; Eason, S. *Updates to the MODFLOW Groundwater Model of the San Antonio Segment of the Edwards Aquifer*; Edwards Aquifer Authority: San Antonio, TX, USA, 2017.

 water

MDPI

Review

Improving Results of Existing Groundwater Numerical Models Using Machine Learning Techniques: A Review

Cristina Di Salvo

CNR-Institute of Environmental Geology and Geoengineering, Area della Ricerca di Roma 1—Montelibretti Via Salaria km 29,300, 00015 Monterotondo, Italy; cristina.disalvo@igag.cnr.it

Abstract: This paper presents a review of papers specifically focused on the use of both numerical and machine learning methods for groundwater level modelling. In the reviewed papers, machine learning models (also called data-driven models) are used to improve the prediction or speed process of existing numerical modelling. When long runtimes inhibit the use of numerical models, machine learning models can be a valid alternative, capable of reducing the time for model development and calibration without sacrificing accuracy of detail in groundwater level forecasting. The results of this review highlight that machine learning models do not offer a complete representation of the physical system, such as flux estimates or total water balance and, thus, cannot be used to substitute numerical models in large study areas; however, they are affordable tools to improve predictions at specific observation wells. Numerical and machine learning models can be successfully used as complementary to each other as a powerful groundwater management tool. The machine learning techniques can be used to improve calibration of numerical models, whereas results of numerical models allow us to understand the physical system and select proper input variables for machine learning models. Machine learning models can be integrated in decision-making processes when rapid and effective solutions for groundwater management need to be considered. Finally, machine learning models are computationally efficient tools to correct head error prediction of numerical models.

Keywords: groundwater; physically-based models; machine learning models; artificial neural network; random forest; support vector machine

Citation: Di Salvo, C. Improving Results of Existing Groundwater Numerical Models Using Machine Learning Techniques: A Review. *Water* **2022**, *14*, 2307. https://doi.org/10.3390/w14152307

Academic Editor: Adriana Bruggeman

Received: 21 May 2022
Accepted: 22 July 2022
Published: 25 July 2022

1. Introduction

1.1. Physically Based Models in Groundwater Management

Physically-based models are the most commonly used tools in quantitative groundwater flow and solute transport analysis and management. Traditionally, the conceptual or numerical models are applied to hydrological modelling in order to understand the physical processes characterising a particular system, or to develop predictive tools for detecting proper solutions to water distribution, landscape management, surface water–groundwater interaction, or impact of new groundwater withdrawals. Along with the ever rising accessibility of computational power, field measurements, and improved understanding of the dynamics of hydrogeological systems, the accuracy required for these models is increasing. This brings some practical limitations of physically-based based models, including the need for large amount of data and input parameters [1,2]. In order to solve the equations describing the dynamics of flow, the physical properties as well as the boundary conditions of the system must be suitably defined within the time and space domains of the model in order to achieve acceptable accuracy. Quantifying these properties and conditions can be expensive and time-consuming; thus, very few field measurements are often available, and the accurate estimate of model parameters across the study area can be challenging [3].

1.2. Uncertainty and Error Types of Physically-Based Models

Many modellers recognised the inherent uncertainty of physically-based models (e.g., refs. [4–6]. They are subject to three types of errors: (1) model structural error introduced by misrepresentation of the real system, as well as from the numerical implementation, for example, spatial and temporal discretisation [4,7,8]; (2) parameter error due to indirect estimation (e.g., prior knowledge or calibration) [9,10]; (3) errors in input data [11] and measurements used to evaluate the model. Alternatively, when the target is to obtain accurate predictions rather than understanding the underlying groundwater system, conventional statistical techniques, such as autoregressive (AR), AR moving average (ARMA), and AR integrated moving average (ARIMA) have been applied invariably to modelling groundwater resources [12,13]. However, the abovementioned methods do not take into account the nonstationary and non-linear characteristics of the data structure [14,15].

1.3. Machine Learning Models

The need to address groundwater problems through alternative, relatively simpler modelling techniques pushed authors in different parts of the world to explore machine learning models. Machine learning methods have been widely used in recent years in many fields (i.e., bioinformatics, biomedicine [16,17], biochemical engineering [18], civil engineering problems, see refs. [19–21] and references therein), transportation networks [22–24], geosciences and environmental applications [25–27], and environmental risk prediction [28,29]. Their largely diffused uses are due to the fact that they are simple and provide acceptable results. Recently, the modelling of non-linear and non-stationary problems has been provided with great ability by machine learning techniques compared with traditional statistical approaches [30–34]. Dealing with machine learning techniques, modellers do not need to introduce the mathematical relationships among variables because machine they are capable of learning the relationships from the input data. Of course, these methods have some limitations, such as overtraining leading to low generalisability [35], risk of using unrelated data, incorrect modelling with inappropriate methods, their dependency on data for training [36], and so on. However, their simplicity of use, high-speed run and reasonable accuracy without the need to know the physics of the problem have led many researchers to apply them.

1.4. Machine Learning for Groundwater Level Forecasting: Current State of the Research

Many recently published review papers have explored the use of machine learning models in hydrology (e.g., refs. [37–41] and references therein, refs. [42,43], or in many water resources fields (e.g., refs. [44–47] specifically for groundwater level (GWL) modelling and forecasting, refs. [48,49] and references therein)). However, there is not yet a complete review paper examining the application of machine learning methods in GWL modelling in comparison to numerical models. The development of better approaches for GWL modelling makes it necessary to look at what has been done in the field of the comparison of numerical and machine learning models and current research.

1.5. Aim of This Work

This paper presents a review of those papers specifically focused on the use of both numerical and machine learning methods for groundwater modelling to estimate the groundwater levels. The aim of the paper was to furnish information to orient modellers which want to explore machine learning approaches starting from an already developed numerical model, highlighting the advantages and disadvantages of both modelling techniques. Moreover, it attempts to clarify some common questions such as: which machine learning techniques are appropriated to solve a specific problem; which is the optimal input data range for machine learning modelling; and which software is suitable for a specific machine learning model. In the following chapters, the types of physically based models used in the reviewed papers are briefly described. Then, some commonly used machine learning methods for modelling GWL are addressed. The methods include Artificial Neural Networks, Radial Basis Function, Adaptive-Neuro Fuzzy Inference System, Time Lagged Recurrent neural networks, Extreme Learning Machine, Bayesian Network, Instance-Based Weighting, Inverse-Distance Weighting, Support Vector Machine, Decision Tree, Random Forest, Gradient-Boosted Regression Tree, and some hybrid models such as wavelet-machine learning models. Radial Basis Function, Adaptive-Neuro Fuzzy Inference System, Time Lagged Recurrent neural networks and Extreme Learning Machine, Bayesian Network, Instance-Based Weighting, Inverse-Distance Weighting, Support types of Artificial Neural Networks, Random Forest, and Gradient-Boosted Regression Tree are types of Decision Trees; however, in this review, each technique was treated individually. The most frequently used machine learning techniques used are Artificial Neural Networks, Bayesian Network, Decision Tree, and Support Vector Machine. At first, each method is briefly described and thereafter the related studies are reviewed. This is followed by general and specific results, discussions, and conclusions, including recommendations for future research.

2. Modelling Techniques Explored in This Review

2.1. Physically Based Numerical Groundwater Flow Models

Numerical groundwater flow models simulate the distribution of head by solving the equations of conservation of mass and momentum. Because these equations represent the physical flow system, in order to obtain accurate results accuracy, the physical properties of the aquifer (e.g., hydraulic conductivity, specific storage) as well as the initial and boundary conditions of the system must be properly assigned within the time and space domains of the model [3]. The physically based models used in the reviewed papers are briefly described as follows.

MODFLOW [50,51] is the modular finite difference flow model distributed by the U.S. Geological Survey. It is one of the most popular groundwater modelling programs. Thanks to its modular structure, MODFLOW integrates many modelling capabilities to simulate most types of groundwater modelling problems. The corresponding packages (e.g., solute transport, coupled groundwater/surface-water systems, variable-density flow, aquifer-system compaction and land subsidence, parameter estimation) are well structured and documented and can be activated and used to solve required modelling problems. The source code is free and open source, and can be fixed and modified by anyone with the necessary mathematical and programming skills to improve its capabilities [52].

SUTRA (Saturated-Unsaturated Transport) [53] is a 3D groundwater model that simulates solute transport (i.e., salt water) or temperature. The model employs a grid that is based on a finite element and integrated finite difference hybrid method framework. The program then computes groundwater flow using Darcy's law equation, and solute or transport modelling use similar equations. It is very frequently used for calculation of salinity of infinite homogeneous, isotropic unconfined aquifer.

The Princeton Transport Code (PTC, [54,55] is a 3D groundwater flow and contaminant transport simulator. It uses a hybrid coupling of the finite-element and finite-difference methods. The domain is discretised by the algorithm into parallel horizontal layers; the elements within each layer are discretised by finite-element method. The vertical connection between layers is allowed by a finite-difference discretisation. During any iteration, all the horizontal finite-element discretisations are firstly solved independently of each other; then, the algorithm solves the vertical equations connecting the layers using the solution of the horizontal equation.

SHETRAN is a physically-based distributed modelling system for simulating water flow, sediment, and contaminant transport in river basins [56]. It is often used to model integrated groundwater–surface water systems. SHETRAN simulates surface flows using a diffusive wave approximation to the Saint–Venant equations for 2D overland flow and 1D flow through channel networks. Subsurface flows are modelled using a 3D extended Richards equation formulation, where the saturated and unsaturated zones are represented as a continuum. Surface and subsurface flows exchange is allowed in either direction. The partial differential equations for flow and transport are solved on a rectangular grid by the finite difference methods; the soil zone and aquifer are represented by cells which extend downwards from each of the surface grid elements. Precise river–aquifer exchange flows can be represented by using the local mesh refinement option near river channels.

2.2. Machine Learning Models

2.2.1. Artificial Neural Networks (ANNs)

An artificial neural network (ANN) model is a data-driven model that simulates the actions of biological neural networks in the human brain. Typically, an ANN comprises a variable number of elements, called neurons, which are linked by connections. Generally, an ANN is composed of three separate layers: input, hidden, and output layers. Each single layer contains neurons with similar properties. The input layer takes input variables (e.g., past GWL, temperature, precipitation time series); a relative weight (i.e., an adaptive coefficient) is given to each input, which modifies the impact of that input. In the hidden and output layers, each neuron sums its input, and then applies a specific transfer (activation) function to calculate its output. By processing historical time series, the ANN learns the behaviour of the system. An ANN learns by relating a given number of input data with a resulting set of outputs [57], which is the training process. Training means modifying the network architecture to optimise the network performance, which involves tuning the adjustable parameters: tuning the weights of the connections among nodes, pruning or creating new connections, and/or modifying the firing rules of the single neurons [58]. The training process can be conducted with various training (learning) algorithms. ANN learning is iterative, comparable to the human learning from experience [59]. ANNs are very popular for hydrologic modelling and is used to solve many scientific and engineering problems. These models may be ascribed to two categories: feed-forward, which is the most common, and feed-back networks [60,61]. The most frequently used family of feed-forward networks is the multilayer perceptron [62,63]; it contains a network of layers with unidirectional connections between the layers.

2.2.2. Radial Basis Function Network (RBF)

RBF network is commonly a three-layer ANN which uses RBF as activation functions in the hidden layer; the network architecture is the same as multilayer perceptron. The number of neurons in the input layer is the same as the input vectors. The radial basis functions in the hidden layer map the input vectors into a high-dimension space [64]. A linear combination of the hidden layer outputs is used to calculate the neurons in the output layer of the network. The distinctive characteristic of RBF is that the responses increase (or decrease) monotonically with Euclidean distance between the centre and the input vectors [65].

2.2.3. Adaptive Neuro-Fuzzy Inference System (ANFIS)

ANFIS, first described by Jang [36], combines the neural networks with the fuzzy rule-based system. In the fuzzy systems, relationships are represented explicitly in the form of if-then rules [66,67]. Different from a typical ANN, which uses sigmoid function to convert the values of variables into normalises values, an ANFIS network converts numeric values into fuzzy values. Firstly, a fuzzy model is developed, where input variables are derived from the fuzzy rules. Then, the neural network tweaks these rules and generates the final ANFIS model [68]. Usually, an ANFIS model is structured by five layers named according to their operative function, such as 'input nodes', 'rule nodes', 'average nodes', consequent nodes', and 'output nodes', respectively [69].

2.2.4. Time Lagged Recurrent Neural Networks (TLRNs)

TLRN are multilayer perceptrons extended with "short-term" memory structures that have local recurrent connections. The approach in TLRNs differs from a regular ANN approach in that the temporal nature of the data is taken into account [69], allowing accurate processing of temporal (time-varying) information. The most common structure of a TLRN comprises an added feedback loop which introduces the short-term memory in the network [70] so that it can learn temporal variations from the dataset [71]. TLRN uses a more advanced training algorithm (back propagation through time) than standard multilayer perceptron [72]. The main advantage is that the network size of TLRNs is lower than multilayer perceptrons that use extra inputs to represent the past state of the system. Furthermore, TLRNs have a low sensitivity to noise.

2.2.5. Extreme Learning Machine (ELM)

ELM is a training algorithm for the single-layer feed-forward-neural network (SLFFNN). Input weights and biases values of the nodes in the hidden layer are randomly determined according to continuous probability distribution with probability of 1, so as to be able to train N separate samples. Compared with conventional neural networks, in ELM, only the number of hidden layer neurons needs to be tuned, and no adjustments are required for parameters such as learning rate and learning epochs. Training of ELM is conducted quickly and is considered a universal approximator [73–75].

2.2.6. Bayesian Network (BN)

The Bayesian networks (Figure 1) are statistical-based models which compute the conditional probability associated with the occurrence of an event by using the Bayes' rule. A typical Bayesian network is composed of a set of variables where their conditional dependencies are represented by a directed acyclic graph.

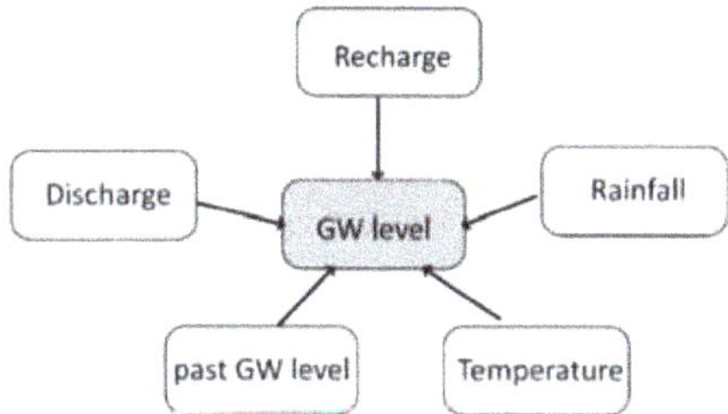

Figure 1. Example of the structure of a Bayesian model applied to groundwater-level study.

Connections define the conditional dependencies among variables (i.e., nodes) [76]. The dependencies are quantified by conditional probabilities for each node through a conditional table of probabilities. Usually, BNs are built by software that generates many network structures with the input parameters.

2.2.7. Instance-Based Weighting (IBW)

Instance-based algorithms derive from the nearest-neighbour pattern classifier [77], which is modified and extended by introducing a weighting function. IBW models are also inspired by exemplar-based models of categorisation [78]. Different from other machine learning algorithms, which return an explicit target function after learning from the training dataset, instance-based algorithms simply save the training dataset in memory [79]. For any new data, the algorithm first finds its n nearest neighbour in the training set and delays the processing effort until a new instance needs to be classified. IBW has many advantages such as the low training cost, the efficiency gained through solution reuse [80], ability to model complex target functions, and the capability to describe probabilistic concepts [81]. However, when irrelevant features are present, their performance decreases; an accurate distinction of relevant features can be achieved through feature weighting to ensure acceptable performance. IBW does not need to be trained and the results are less influenced by the training data size. Inverse-distance weighting is a special case of instance-based weighting with the weighting factor p = 2 [82].

2.2.8. Support Vector Machine (SVM)

SVM are kernel-based neural networks developed by Vapnik [83] to overcome the several weaknesses which affect the ANNs' overall generalisation capability [84], including possibilities of getting trapped in local minima during training, overfitting the training data, and subjectivity in the choice of model architecture [85]. The SVM is based on statistical learning theory [86]; in particular, it is based on structural risk minimisation (SRM) instead of empirical risk minimisation (ERM) of ANNs. The SVM minimises the empirical error and model complexity simultaneously, which can improve the generalisation ability of the SVM for classification or regression problems in many disciplines. This is achieved by minimising an upper bound of the testing error rather than minimising the training error [79]; the solution of SVM with a well-defined kernel is always globally optimal, while many other machine learning tools (e.g., ANNs) are subjected to local optima; finally, the solution is represented sparsely by Supporting Vectors, which are typically a small subset of all training examples [87]. For further details, see refs. [63,86,88,89].

2.2.9. Decision Trees (DT)

Decision tree models [90] are based on the recursive division of the response data into many parts along any of the predictor variables in order to minimise the residual sum of squares (RSS) of the data within the resulting subgroups (i.e., "nodes" in the terminology of tree models) [91]. The number of nodes increases during the process of splitting along predictors. The tree-growing process stops when the within-node RSS is below a specified threshold or when a minimum specified number of observations within a node is reached [92]. However, the modeller places minimal limitations upon tree-fitting process, and fitted trees may be more complex than is actually warranted by the data available. The problem of overfitting results is then managed by the 'pruning' algorithms, which aid the modeller in the selection of a parsimonious description of interactions between response and predictors, fitting trees for the optimum structure for any level of complexity [91]. Because no prior assumptions are made about the nature of the relationships among predictors, and between predictors and response, decision trees are extremely flexible.

2.2.10. Random Forest (RF)

Random forests work by constructing groups of decision trees during the training process, representing a distinct instance of the classification of data input. Each tree is developed by independently sampling the values of a random vector with the same distribution for all trees in the forest [93].

The random forest technique considers the instances individually so that the trees are run in parallel; there is no interaction between these trees while building the trees. The prediction with the majority of votes or an average of the prediction is taken as the selected prediction (Figure 2). The RF algorithm was created to overcome the limitations of DT, reducing the overfitting of datasets and increasing prediction accuracy. The decision tree grows to the largest possible size without being pruned in accordance with the number of trees and the number of predictor variables [94].

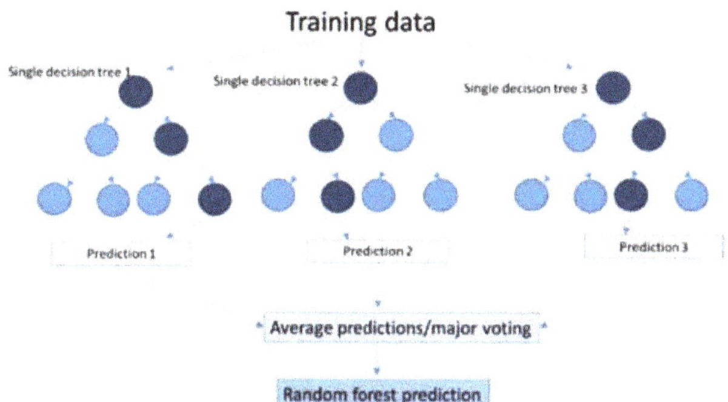

Figure 2. Scheme of Random Forest.

2.2.11. Gradient-Boosted Regression Trees (GBRT)

Gradient-Boosted Regression trees are ensemble techniques in which weak predictors are grouped together to enhance their performance [95]. Learning algorithms are combined in series to achieve a strong learner ("boosting") from different weak learners (i.e., the decision trees) connected sequentially. Each tree attempts to minimise the errors of the previous tree. After the initial tree is generated from the data, subsequent trees are generated using the residuals from the previous tree. At each step, trees are weighted, with the lower-performing trees weighted the highest; this allows the improvement of performance at each iteration. A variety of loss functions can be used to detect the residuals.

3. Bibliographic Review

The following section describes the reviewed papers. Throughout our research, few papers were found in the literature that examine the use of both numerical models and machine learning models in GWL forecasting. Here, 16 papers dealing with the use of both models for the prediction of GW levels, which were published in 10 international journals and 1 book from 2003 to 2020, were reviewed. Each paper was analysed in detail; for each one, the author provided a description of the study area and the geological context, the area of model use (e.g., groundwater planning and supply, management in farming systems, coastal water management), the machine learning technique, and any details of its application in the specific case study. Finally, the statistical indicators used to compare the performance of numerical and machine learning models were reported. In the reviewed papers, machine learning models are always used to improve the results of physically-based models in GWL forecasting and to overcome the problem of long computational time of regional models. This is accomplished by comparing results of a physically-based model and a surrogate machine learning model (i), comparing results of a physically-based model and different machine learning models (ii), testing hybrid or ensemble models (iii), and reducing and correcting physically-based model errors by means of machine learning approaches (iv). In the cases (i) and (ii), each model is run independently. In the case (iii), machine learning techniques are applied at different stages of the modelling procedure, such as data pre-processing; in some papers, numerical model output is used to train machine learning models, obtaining statistical models capable of speeding up the numerical model runs. In case (iv), numerical model errors are used as training datasets for machine learning models. Details of the selected papers are given in Table 1, which includes information such as the region of study, the key area of model use, the used machine learning model, the hydrologic input variables of machine learning models, the time step, the range of total data, the total simulation time, time step, and the grid size of the physically-based model (Journal Citation Reports, Clarivate Analytics). In some cases, lacking information was integrated with literature complementary to the reviewed papers (i.e., same study areas).

Table 1. List of the reviewed studies. P: precipitation. T: temperature. E: evapotranspiration. SWL: surface water level. EP: effective precipitation.

n	Reference	Region of Study (Country)	Key Area of Model Use	Used ML Models	Input Variables to ML Model	ML Model Time Step	Range of Total Data (Number of Data or Observation Wells)	PB Simulation Time (Time Step)	Size of the PB Model Domain	Field of Application of the ML Technique	Journal (202+C:L0 IF)	Aquifer or Basin Hydrostratigraphy
1	Mohammadi, 2009 [59]	Chamchamal plain (Iran)	Farming and agricoltural systems	ANN	MODFLOW output	monthly	1986–1998 (144 sets)	1 year (monthly)	145.7 Km^2	using the results of PB models to train a single ML model	Pratical hydroinformatics (book)	Alluvial (karst bedrock)
2	Coppola et al., 2003 [3]	Northwest Hillsborough Wellfield (USA)	Planning and supply	ANN	GWL, pumping rates, P, T, dew point, wind speed conditions, stress period lenghts	weekly	January 1995–August 2000, (212 sets)	20 years (monthly)	10,359.9 Km^2	using the results of PB models to train a single ML model	Journal of hydrologic engineering (2.064)	Highly permeable limestone overlain by low permeability clay and, above, sand with interbedded clay
3	Banajeree et al., 2011 [96]	Kavaratti, island of the Lakshadweep archipelago (India)	Coastal water management	ANN	not mentioned in the paper	monthly	2005–2007 (23 sets)	5 years (monthly)	2D model, with section lenght = 2650 m and depth 1000 m	using the results of PB models to train a single ML model	Journal of hydrology (5.722)	Coastal
4	Mohanty et al., 2013 [97]	Kathajodi-Surua Inter-basin of Odisha (India)	Coastal water management	ANN, TLRNs	GWL, P, E, river stage, SWL, pumping rates	weekly	February 2004–May 2007 (174 sets)	3 years, (weekly)	114.5 m^2	using the results of PB models to train a single ML model	Journal of hydrology (5.722)	Alluvial
5	Parkin et al., 2007 [98]	Winterbourne stream, Thames Basin, Berkshire (UK)	aquifer-river interaction	ANN	GWL, river flow depletion	daily	Not specified (1 well)	25 years (daily)	Regional aquifer: 200 Km^2. Valley aquifer: 2 Km^2	using the results of PB models to train a single ML model	Journal of hydrology (5.722)	Alluvial
6	Moghaddam et al., 2019 [76]	BirjandAquifer, South Khorasan (Iran)	Drought-prone regions	ANN, BN	GWL, E, T, EP, Discharge	monthly	2002–2014 (1872 sets)	12 years	277.8 Km^2 ([99])	using the results of PB models to train and compare different ML models	Groundwater for sustainable development (no IF)	Alluvial

Table 1. *Cont.*

n	Reference	Region of Study (Country)	Key Area of Model Use	Used ML Models	Input Variables to ML Model	ML Model Time Step	Range of Total Data (Number of Data or Observation Wells)	PB Simulation Time (Time Step)	Size of the PB Model Domain	Field of Application of the ML Technique	Journal (202+C:L0 IF)	Aquifer or Basin Hydrostratigraphy
7	Almuhaylan at al., 2020 [68]	Saq Aquifer in Quassim (Saudi Arabia)	Drought-prone regions	ANN, ANFIS	GWL, pumping rates	not specified	1980–2018 (55 wells)	not specified	600 Km^2	using the results of PB models to train and compare different ML models	Water (3.103)	Sandstone
8	Chen et al., 2020 [63]	Heihe River Basin (China)	Drought-prone regions	ANN, RBF, SVM	pumping rates, recharge, stream-flow rates	monthly	1986–2008 (11,088 sets)	22 years (monthly)	21,120 Km^2	using the results of PB models to train and compare different ML models	Scientific reports (4.380	Alluvial
9	Fienen et al., 2016 [95]	Lake Michigan Basin (USA)	Planning and supply	ANN, GBRT, BN	parameters expected to have predictive power to the source of water to wells	not specified	1864–2005, (4911 sets)	141 years (variable, [100])	204,764.4 Km^2	using the results of PB models to train and compare different ML models	Environmental modelling and software (5.288)	Glacial deposits
10	Miro et al., 2021 [101]	San Bernardino and Rialto-Colton basins, San Bernardino Valley Municipal Water District - Valley District (USA)	Drought-prone regions	RF, SVM, ANN	Recharge, pumping rates	not specified	2015–2050 (not specified)	35 years (monthly)	3000 Km^2	using the results of PB models to train and compare different ML models	Climate risk management (4.090)	Basin comprising ancient metamorphic bedrock, eolic sands, ancient fans, recent alluvium
11	Malekzadeh et al., 2019 [102]	Kabodarahang Plain, Hamadan (Iran)	Farming and agricoltural systems	ELM, WA-ELM	decomposed sub-series of observed GWL	monthly	August 1990–September 2015 (301 sets)	10 years (monthly)	not specified	using the results of PB models to train a single ML model	Groundwater for sustainable development (no IF)	Alluvial (limestone bedrock)

Table 1. *Cont.*

n	Reference	Region of Study (Country)	Key Area of Model Use	Used ML Models	Input Variables to ML Model	ML Model Time Step	Range of Total Data (Number of Data or Observation Wells)	PB Simulation Time (Time Step)	Size of the PB Model Domain	Field of Application of the ML Technique	Journal (202+C:L0 IF)	Aquifer or Basin Hydrostratigraphy
12	Nikolos et al., 2008 [103]	Northern Rhodes Island (Greece)	Coastal water management	ANN combined with DE algorithm	GWL, pumping rates	daily	1997–1998 (3125 sets)	1 year (2 seasonal stress periods)	217 Km^2	using the results of PB models to test hybrid or ensamble modelling approaches	Hydrological processes (3.565)	Coastal
13	Sahoo et al., 2017 [104]	High Plains aquifer and Mississippi River Valley aquifer (USA)	Farming and agricoltural systems	Automated hybrid artificial neural network (HANN)	GWL, P, T, streamflow, climate indexes, irrigation demand, NAO index	monthly	1980–2012, (HPA: 263,808 sets. MRVA: 115,368 sets)	33 years (monthly)	MRVA: 405,720 Km^2 ([105]). HPA: 3.34 Km^2 ([106])	using the results of PB models to test hybrid or ensamble modelling approaches	Water resource research (5.240)	High Plain Aquifer: ancient alluvial fans and quaternary deposits. Mississippi River Valley Alluvial Aquifer: Tertiary and Quaternary clay, silt, sand and gravel deposits.
14	Michael et al., 2005 [82]	Argonne National Laboratory, Illinois (USA)	Contaminant/phytoremediation	DT, IDW, ANN	GWL, P	quartely	November 1999–March 2001 (22 wells with quarterly data); May 2001 (7 wells with hourly data)	6 years, (monthly)	4.8 Km^2 ([107])	using the results of PB models to test hybrid or ensamble modelling approaches	Water resource research (5.240)	Glacial deposits
15	Xu et al., 2014 [79]	Republican River Basin and Spokane Valley-Rathdrum Prairie aquifer (USA)	Planning and supply	Cluster analysis, IBW, SVM	GWL, well location, observation time	monthly	RRCA: 1918–2007 (300,000 sets). SVRP: 1990–2005 (2191 sets)	RRCA: 89 years. SVRP: 15 years, (monthly)	RRCA: 79,396 Km^2. SVRP: 844.3 Km^2	using ML techniques for PB models errors reduction/correction	Groundwater (2.671)	Alluvial
16	Demissie et al., 2009 [85]	Argonne National Laboratory, Illinois (USA)	Contaminant/phytoremediation	ANN, DT, SVM, IBW	GWL, EVP, stress periods	monthly	2000–2005, (3600 sets) [107]	6 year (monthly)	0.75 Km^2	using ML techniques for PB models errors reduction/correction	Journal of hydrology (5.722)	Glacial deposits

3.1. Comparing Results of a Physically-Based Model and a Surrogate Machine Learning Model

Many authors compared results of physically-based models and machine learning models run independently. The two approaches are then compared in terms of GWL prediction performance.

Mohammadi et al. [59] investigated the applicability of ANN models in simulating GWL for aquifers with limited data. The study area was the Chamchamal plain (Iran), an alluvial plain surrounded by a karstic formation. Groundwater flow was simulated by MODFLOW and hundreds of data sets were generated from the calibrated model to train the ANN model. Another purpose was to detect ANN models capable of simulating the complex dynamics of GWL, even with relatively short lengths of training data of the ANN model. To achieve this objective, different ANN models were implemented, with different combinations of input data. Furthermore, different network architectures, with different number of hidden layers and activation functions, were evaluated. The models' performances were evaluated by means of MODFLOW outputs and measured groundwater levels through the coefficient of determination (R^2), mean squared error (MSE), and normalised mean squared error (NMSE). The water table was estimated with reasonable accuracy by all the models, but the ANN required lesser input data and took less time to run. However, the authors remarked two disadvantages of these networks: (i) the water table cannot be predicted in all observation wells by a single model with similar input parameters; and (ii) models are static and inputs and outputs from previous time steps are not considered (unless these are introduced explicitly). This results in a high difference between the observed and calculated GWL at some points. In order to overcome these difficulties, the authors tested TLRN to simulate the entire groundwater system with one model. The aim of TLRN is to predict a multivariate time series using past values and available covariates. Instead of using static feed-forward ANNs to model nonlinear relationships in water table level forecasting, the TLRNs approach takes into account the temporal nature of the data (i.e., the lagged inputs, see Section 2.2.4), and in this respect compares favourably with ANN multilayer perceptron networks. The model used in the TLRNs is the gamma model [71], which is characterised by a memory structure that is a cascade of leaky integrators. The neural network can control the depth of the memory by changing the value of the feedback parameter, instead of changing the number of inputs. Since the feedback parameter is recursive, a backpropagation through time algorithm was used to apply a more powerful learning rule. Considering the reduced computational costs and the lower data requirements, the authors concluded that a TRLN model can be effectively used in the field of GWL simulation.

In the work of Coppola et al. [3] ANNs are used to accurately forecast transient water levels in a complex groundwater system under variable aquifer stresses. The model was tested in the Northwest Hillsborough Wellfield near Tampa Bay, Florida, USA, the model area being represented by the Upper Floridian aquifer (consisting of high permeability karst limestone overlain by a low permeability semiconfining unit, with a surficial sandy unconfined aquifer above). Results of numerical and machine learning models were compared for representative monitoring wells by using root mean square error and absolute mean error. The oscillation of the water levels was modelled with much more accuracy by ANN than the numerical flow model. The Absolute Mean Error of numerical model exceeded the maximum ANN prediction error at any single observation during each stress period. The authors concluded that for certain problems, ANN represents a better option to numerical modelling approaches because it does not require difficult-to-quantify aquifer parameters and time- and space-variable conditions. Then, three types of sensitivity of ANN were evaluated: (1) the sensitivity of ANN prediction performance to training set size; (2) sensitivity analysis of selected ANN inputs on water level responses; (3) sensitivity of ANN performance to data noise and measurement error.

(1) The sensitivity of ANN performance to data availability was assessed by using different sizes of training sets. The results showed that, during validation, acceptable prediction accuracy was achieved with a relatively small number of training sets.

(2) Input parameters groundwater withdrawals and rainfall were included in the sensitivity analysis. Results showed that, in the unconfined aquifer, short-term oscillations were correlated most strongly to rainfall, while in the underlying semiconfined aquifer the water level was mostly influenced by withdrawals. Since these results are in accordance with the hydrological conditions, the authors concluded that the physical dynamics of the system must be sufficiently understood by the modeller in order to identify the important predictor input variables.
(3) The effect of measurement error and data noise (inherently present in most hydrologic data set) on ANN performance was assessed by introducing normally distributed random noise into the input variables of the training set. The results demonstrated that the ANN can filter out noise in the training data and effectively learn groundwater system behaviour.

Banerjee et al. [96] evaluated the use of ANN simulation over mathematical modelling as a management tool for coastal aquifers. The aim of the models was to forecast the increase in the salinity of groundwater due to pumping at different rates in the island of Kvaratti, Lakshadweep archipelago (India) and to detect management strategies to avoid the increase of salinity of groundwater. A physically-based 2D finite element model was developed with SUTRA [53]. The study demonstrated the superiority of ANN with respect to the physically-based model, evaluated by mean of root mean squared error (RMSE) and mean absolute error (MAE). Its non-linear nature makes it a formidable tool for analysing real-world data, allowing modelling of complex dependencies. With respect to traditional models such as SUTRA, ANN requires a lesser number of input parameters and avoids the model building and parameter estimation phases. While only a few seconds are needed for the training in the ANN models, modelling in SUTRA is very time-consuming.

Mohanty et al. [97] compared the results of the finite difference-based numerical MODFLOW model and the ANN model in simulating GWL in an alluvial aquifer system (Kathajodi-Surua Inter-basin of Odisha, India) for improving the efficiency of planning and management of groundwater resource at the basin scale. To evaluate the results, 6 statistical criteria were used: bias, coefficient of determination (R^2, MAE, RMSE), Nash–Sutcliffe efficiency (NSE), and mean percent deviation (Dv). Results revealed that the ANN model performed better for short-time predictions that require high accuracy, while numerical models were more appropriate for long-term predictions. Furthermore, the authors highlighted that physically based models provide the total water balance of the system, whereas the ANN models do not involve a description of the entire physics of the system. In the case of ANNs, a new model must be developed from the beginning to include any changes in the input or output parameters, differently from numerical models. Thus, the type of model should be selected in accordance with the type of problem.

Parkin et al. [104] developed and tested an approach in which numerical and ANN models were used to evaluate the impacts of groundwater withdrawals on river flows in areas representing the hydrogeologic settings of most of England and Wales. Several ANN hidden node structures were tested. The ANN model was trained using the input and output data from about 2000 simulations of the SHETRAN numerical modelling system. The outputs of ANN model were compared against analytical models, and tested using a field data from a case study site: the Winterbourne stream within the Thames Basin near Reading, Berkshire, flowing across a chalk fractured aquifer. The parameters used for the ANN model come from many sources and comprise the distance of borehole from river, the aquifer transmissivity and storage coefficient, the valley-fill transmissivity and specific yield, the river width, the hydraulic conductivity and thickness of riverbed, and the mean annual recharge and the date of peak recharge. The performance was evaluated by comparing root mean square errors of normalised outputs. The results showed the successful application of the approach for modelling river–aquifer interactions and its potential for modelling complex hydrological systems. The good correspondence between the simulated and observed flow depletion using independently-derived parameter values demonstrates that this approach can be applied for modelling realistic field conditions.

3.2. Comparing Results of a Physically-Based Model and Different Machine Learning Models

Even if ANNs are widely used among machine learning technique by groundwater modellers, their limitations encouraged authors to explore alternative models to achieve better performance in GWL prediction.

Moghaddam et al. [76] compared a MODFLOW, an ANN, and a BN model to determine the most accurate method for simulating GWL in the alluvial Birjand aquifer, located in an arid region of the eastern Iran, and solve the problem of GWL overestimation of the MODFLOW model. Both BN and ANN models provided a reliable prediction for GWL. The BN model showed the best match between the measured and the predicted groundwater level values, evaluated by comparing R^2, RMSE, and NASH, and the best performance evaluated by a 2-year period groundwater hydrograph. BN models showed many advantages, such as the easier implementation, the higher forecasting accuracy, and the ability to deal with missing or incomplete data. Moreover, in the BN models, the variables were modelled by means of probability distributions; this allowed the authors to estimate uncertainty more accurately compared with other models other models [108–110].

Almuhaylan et al. [68] compared a MODFLOW model, three ANNs, and one adaptive neuro-fuzzy inference system (ANFIS) model developed in the Saq-Aquifer, Al-Qassim region (Saudi Arabia), an aquifer mainly characterised by medium-to-coarse sandstone. The modelling framework was implemented for assessing the impact of different groundwater pumping scenarios on aquifer depletion. The performance of ANN/ANFIS models for long-term future predictions of GWL and for finding a simple solution to the problem of undefined boundary conditions was examined. Deep learning models, e.g., recurrent neural network or convolutional neural network, are usually required for long-term predictions. The authors instead adopted a simple approach by changing the targets and predictions into GWL changes instead of GWL to develop a standard ANN/ANFIS simulation problem. Additionally, the training of the ANN/ANFIS model was handled with the prediction of changes in GWL instead of the direct simulation of GWL. The authors optimised the use of ANN model by choosing different combinations of architecture (number of hidden neurons and number of layers). The authors obtained a lower mean-square-error and a higher NSE in the training stage of ANN and ANFIS models compared with the calibration of the MODFLOW. Despite the hydraulic model being comparatively more reliable, ANN and ANFIS showed excellent performance, better than the MODFLOW model in terms of NSE. The authors did not simply remark any performance improvement of ANFIS with respect to ANN; they showed better performance in both with respect to the numerical model.

Chen et al. [63] applied a physically based model developed with MODFLOW and three ANN machine learning methods (ANN, RBF, SVM) to simulate the groundwater dynamics of the middle reaches of Heihe River, northwest China. The objectives were to assess the efficacy of machine learning models on reproducing groundwater dynamics in arid basins and to compare results of machine learning and numerical models to verify their applicability. The performance was evaluated by Root mean square Error (RMSE) and Coefficient of determination (R^2). As for the multilayer perceptron, the hyperbolic tangent sigmoid transfer function was applied in the neurons of the hidden layer and the linear transfer function was applied in the output layer; the number of hidden neurons was identified by trial-and-error procedure. Trial-and-error was used also to identify the number of hidden neurons for the RBF network. In RBF, the Gaussian radial basis function was applied in the neurons of the hidden layer and linear transfer function was applied in the output layer, respectively. As for the SVM, Gaussian function (i.e., radial basis function) was used as a kernel function to compute the Gram matrix. Furthermore, for each of the machine learning models, the ratio between RMSE in the prediction stage times RMSE the in training stage was calculated as a measure of the models' generalisation ability (GA). Machine learning models simulated historical data with higher performance with respect to numerical model, with the RBF model performing the best. In particular, SVM performed the best in the training stage, while RBF in the verification stage. Machine learning models showed much less computation cost in training and prediction stages than those of numerical model in calibration and verification stages. However, because of the physical based mechanism, the numerical model showed a better generalisation ability. Therefore, authors concluded that machine learning models are applicable to problems that require a high number of model runs without considering the physical mechanisms (e.g., optimisations, real-time models, sensitivity/uncertainty analysis).

3.3. Testing Hybrid or Ensemble Models

Hybrid modelling approaches including data-preprocessing and/or combination of different machine learning techniques in different stages of the modelling have also been developed in the recent years to improve the efficiency of the machine learning methods [49].

Malekzadeh et al. [100] modelled the GWL in a well located in the arid agricultural area of Kabodarahang Plain (Hamadan, Iran) using MODFLOW and a hybrid artificial intelligence model. They compared an extreme learning machine model (ELM) and a combination of ELM with the wavelet transform (WA-ELM), intending to improve MODFLOW model calibration and optimise the prediction of GWL. Wavelet analysis is commonly executed for de-noising, compressing, and decomposing input data time series in the stage of data pre-processing. Similar to the Fourier transform, the Wavelet transform considers time series as a linear combination of multiple base functions, and has the ability to obtain time, frequency, and situation data simultaneously [111]. Malekzadeh et al. [100] divided time series into several sub-series using the discrete wavelet transform (DWT), and then used the decomposed components as input for the ELM model, instead of the main time series. Different families of the wavelet model were evaluated by comparing the values of R, RMSE, and BIAS, finding the mother wavelet used for the further steps. For each of the ELM and WA-ELM models, 10 different models were defined; the best-performing activation function and topology were chosen. As a result, the best models among the ELM and the WT-ELM models were selected. Then, the results of the hybrid method were compared to ELM and MODFLOW based on the MAE and RMSRE. They found that the WA-ELM model simulates GWL with higher accuracy with respect to both ELM and MODFLOW models.

Nikolos et al. [101] utilised ANNs to approximate a finite element model and combined it with a Differential Evolution algorithm (DE) to determine the best operational strategy for the productive pumping wells located in the northern part of Rhodes Island in Greece. A 3D finite-element simulation model of the study area was initially implemented using the Princeton transport code (PTC). The DE optimisation algorithm was successfully used for solving the optimisation problem, since it provides a solution close to the global optimum in a fully automated way. In the work of Nikolos et al. [101], the calls of the PTC model were replaced with an ANN in order to overcome the time-consuming integration of the PTC model within an evolution-based optimisation procedure. The training/evaluation data for an ANN model were produced by the PTC model. Several numbers of hidden nodes and training epochs were tested to adopt an optimum ANN topology. Then, the ANN was combined with the DE algorithm to solve two different water table elevation scenarios at the observation wells. The classic DE algorithm evolves a fixed size population npop that is randomly initialised [112]. After initialising the population, an iterative process was started which produces a new population until a given condition is satisfied. At each iterative step, a newly generated element can replace each element of the population. At the end of each run, the optimum solution was used as an input to both the PTC and the ANN models to test the accuracy of the ANN predictions and the effectiveness of the constraints.

The results of this procedure demonstrated that the ANN can be used as a quick surrogate model, providing very close to optimal solutions and allowing us to run an optimisation procedure with the DE algorithm in less than a minute instead of the several hours required to run the same process with the PTC model.

Sahoo et al. [102] proposed an ensemble modelling framework (Automated hybrid artificial Neural network, HANN) comprising spectral analysis (SSA), machine learning, and uncertainty analysis to facilitate improved GWL prediction with respect to computationally expensive physically-based MODFLOW models. The method was applied in two aquifer systems exploited for agricultural production (the Mississippi River Valley alluvial aquifer and the High Plains aquifer, USA), with the aim to clarify the influence of each climate variable on the irrigation demand and streamflow and predict groundwater level change. The best input for the ANN was selected by a hybrid data pre-processing method which includes: (i) decomposing the time series using SSA to extract significant reconstructed components (RCs); (ii) selecting the best RC of inputs by mutual information and genetic algorithm; (iii) and determining time lag components using cross-correlation analysis. Then, the simulations from the HANN model during the model testing period were summed to estimate the cumulative GWL change. The HANN results were compared to regional GWL simulations coming from MODFLOW models previously developed by many authors. HANN showed better performance in terms of MSE. The authors highlighted that the HANN shows a high model structure strength since it integrated a robust data pre-processing and input variable selection technique within the ANN model for capturing the impacts of the potential predictor variables on GWL change at observation wells.

Because the model is implemented and optimised for each well, they benefit from training values at each well. On the other hand, while showing a lower prediction error than the physical models, HANN cannot furnish the outputs typical of a physically-based model, such as water balance, residence time calculations, and flux estimates. Moreover, while a numerical model can be modified to include additional input or processes (e.g., supplied water), introducing new parameters would require the building of a new ANN model. Therefore, the authors concluded that each model type excels for certain applications.

Michael et al. [82] compared three machine learning techniques (DT, IDW, and ANN), which were used in a hierarchical approach, to improve GWL forecasting by combining data from different sources, including the results of a MODFLOW numerical model. They used a collection of prewritten modules (set up for each machine learning model) composed in a "data flow" program. The MODFLOW model is incorporated into the itinerary by creating a module that returns the head prediction by MODFLOW. A hierarchy of models was then arranged, with one model used to reduce the dimensionality of the largest data set (called "specialty model") and a second model ("expert model") trained with a combination of the remaining data and the specialty model results to obtain the optimum predictions. After linking together the modules into a machine learning itinerary, a model was automatically built by the itinerary from appropriate data sets to make predictions. At first, the hierarchical approach used machine learning models as both specialty and expert models; the results demonstrated that, based on mean predicted head errors, DT provided the best prediction among the machine learning models, while neural networks provided the least accurate prediction. The best machine learning model performed better than the MODFLOW model in terms of hydraulic head predictions computed across all observations used for calibrating the MODFLOW model. Furthermore, a very short time is required to train DT, and their simplicity allows quick planning of on-site adaptive field sampling. Interestingly, IDW showed a performance nearly as good as DT and IBW when using all of the data across time. The authors concluded that the accuracy of physics-based models can be improved by using a machine learning hierarchical approach in areas with substantial data. Using this method allows identifying (i) advantages and disadvantages of different machine learning approaches and (ii) which data are most significant for long-term monitoring objectives. Secondarily, the MODFLOW model was used as a specialty model to test the potential for machine learning methods to automatically update existing numerical models. In many cases, such as groundwater remediation fields, it is not cost-effective to recalibrate numerical models whenever new data become available. Instead of updating existing models by tuning the parameters based on new data, physically based models are considered as one source of knowledge about the site and are integrated to historical data using data-driven models. Results showed that, compared with the MODFLOW model predicted errors, the mean predicted head error and the standard deviation of predicted head error were consistently reduced with the best-combined model. This means that the model learns a compromise between numerical and machine learning models. Such combined hierarchies could allow an automated update of physically-based models, expanding and adapting the prediction as new data (but also analysis and modelling techniques) become available.

Fienen et al. [95] evaluated three machine learning techniques (BN, ANNs, and GBRT) to train models simulating the source of groundwater to several wells. The aim of the work was to predict local surface water impact due to new pumping wells. The regional 215,000 km^2 groundwater model of Lake Michigan Basin [113] impedes the evaluation of local-scale impacts due to the long runtime and the too-coarse grid. The solution was to emulate the groundwater flow model using a dataset of collocated numerical model input and output to build a statistical learning model ("metamodel", [114]), providing fast decision support to water managers which need to evaluate the permission to water abstraction. In practice, the numerical model was used to generate outputs reproducing several condition of the groundwater system; then, those outputs were used to train a statistical model, which could be subsequently used to make predictions without the need to run the regional model. The ability of the three techniques to extend MODFLOW predictions to areas with few samples was evaluated. K-fold cross validation (CV) was used to assess the models performance, as well as by hold-out data. The performance of the BN model (evaluated by means of R^2 and RMSE) was lower than the other two, and this could be due to the fact that the continuous input and output variables were both discretised into a small number of bins. All the three techniques can be implemented with commonly used commercial (in the case of BN) or open source (in the case of ANN or GBRT) software. The computational time is nearly instantaneous for all the three techniques while it takes longer to perform cross-validation. ANN or GBRT may be the best options for managers who need to achieve better predictive performance when a single response is considered. BN includes estimate of the uncertainty of predictions because all variables are treated as probability distributions. The authors concluded that the metamodelling approach is valid over a wide range of conditions and, as a screening approach, is helpful. A limitation of their approach is that it assumes that the response of the system to pumping rates is linear; thus, this assumption is violated at high pumping rates.

Miro et al. [99] presented a hybrid empirical–dynamical approach application of machine learning models to a Robust Decision Making study to evaluate the effect of groundwater managed recharge. They developed an empirical model representing a high-resolution MODFLOW model previously set-up in two basins located in a drought-prone region of the American West: the San Bernardino and the Rialto-Colton basins, San Bernardino Valley Municipal Water District (Valley District, U.S.). Inputs (recharge, pumping) and outputs (resulting head) of the MODFLOW model were used to train three machine learning methods (Random Forest, Support Vector Machine regression, and Artificial Neural Network) to predict the annual change in GWL. Then, the ability of machine learning methods to simulate the output of the MODFLOW model was assessed to investigate which model is capable of reproducing the best average basin conditions. Based on R^2, the most accurate results were obtained with RF. The authors concluded that RF is able to reproduce time series trends in GWL as well as capture the variability in MODFLOW model predictions. In that way, the authors obtained a significant reduction of computational time: each MODFLOW run without the RF model would have taken approximately 36 years in a standard computing environment, instead of 24 h while simulating MODFLOW with a RF representation of the groundwater system. The procedure is integrated in a Robust Decision Making (RDM) process: the novel application of machine learning represents an improvement to the field of decision-making under deep uncertainty that allows reducing computational times and permits a greater exploration of the uncertainty space, such as future climate changes and drought conditions.

3.4. Reducing and Correcting Model Errors by Means of Machine Learning Approaches

Despite a correctly constructed and calibrated groundwater model being able to furnish valuable information about the system behaviour, the unaccounted uncertainty, which is typically associated with the phases of model development and parametrisation, can result in large localised simulation errors.

Xu et al. [79] tested two machine learning techniques (Instance-based weighting and support vector machine regression) to correct the prediction of two physically-based models, successfully improving the head prediction accuracy. The authors applied the error-correcting data-driven models to temporal, spatial, and spatiotemporal prediction. The core of the study relies on the selection of historical residuals of the physically-based models, which were used to train the data-driven models. Then, the physically-based model was used to make predictions, and the trained data-driven models were used to predict the error of the predictions. Finally, the updated head was obtained by adding the predicted error to the head simulated by the physically-based model. The procedure was applied to two real-world groundwater flow models having different data densities and extents of temporal and spatial structures in the error. The first is the regional Republic River Basin (RRCA), covering portions of eastern Colorado (USA), a 79,396 km^2 model [115] developed to resolve water conflicts as growing water demand led to dramatically increased groundwater pumping. The second is the Spokane Valley–Rathdrum Prairie aquifer (SVRP) (USA), an 844 km^2 aquifer subjected to groundwater pumping stresses. The two models differ in various aspects, including parametrisation, calibration, grid resolution, data density, and calibration strategy, leading to different spatial patterns in model residuals.

In the case of RRCA, data were pre-processed by cluster analysis: for temporal prediction, observation wells were clustered using the agglomerative hierarchical clustering algorithm according to their spatial location. In spatial and spatiotemporal prediction scenarios, input data were clustered by the k-means algorithm. Each cluster was subdivided into a training and a validation dataset, and data-driven models were applied to each subset.

In the case of the SVRP model, cluster analysis was not implemented because residuals did not show local patterns; thus, the data-driven models were applied only to the temporal prediction scenario. In the same way, to the RRCA case study, IBW and SVR models were built to forecast the error of the simulated head taking as input features the well location and MODFLOW computed head; then, the updated head was computed. For both case studies, five-fold CV was used to adjust the parameters of IBW and SVR.

The magnitude and biasedness of the prediction error (evaluated by means of ME and RMSE) were sufficiently reduced. The authors found that this complementary modelling framework was computationally efficient. New data can be easily incorporated into the training dataset. Therefore, data-driven models can be used to improve the prediction of the physically-based model for long-term prediction and under conditions different from the one used during calibration. A limitation of this methodology is that it applies only to physically based groundwater models with epistemic errors in the simulation results, while it is not suitable for models with calibration error following Gaussian distribution with zero mean and variance comparable with the observation error.

Demissie et al. [85] developed a complementary approach that integrates the calibrated groundwater MODFLOW model with data-driven models to detect and predict systematic errors in groundwater model simulation in a hypothetical test case based on the Argonne National Laboratory, Illinois (USA), a site affected by groundwater contamination by radioactive substances and volatile organic and with phytoremediation installed to clean up the soil. Using the groundwater model residual analysis results, the authors implemented four data-driven models (ANN, DT, SVM, and IBW) for simulating and correcting the groundwater head predictions both in time and in space. The data-driven models were then used to update the head predictions. The updated models showed improved performance compared to the MODFLOW head predictions at all observation wells in terms of RMSE reduction. ANN performed better in updating the future predictions but required a longer time to train the model and the definition of many parameters. IBW updates showed a better performance in the case of spatial prediction, probably because the number of spatial data was too small for the other three models to learn the spatial patterns of the residuals.

4. General Results

The general outcomes derived from the 16 reviewed studies are discussed, such as the results related to the key area of model use, input variables, simulation period of physically-based models, time step, dataset size and division, and software used.

4.1. Key Area of Groundwater Model Use

In general, machine learning models are developed to achieve a better performance in GWL forecasting in areas where groundwater management strategies are strictly required to ensure proper resource availability while protecting the environment and groundwater related ecosystems. This is especially needed in areas where the aquifers have been overexploited; where the groundwater recharge is scarce (drought-prone regions); and in coastal areas, where groundwater is threatened by saltwater wedge intrusion. Most of the reviewed papers (four, Figure 3) concern water planning and supply, usually at the catchment scale. A minor number of papers (three) focus on the groundwater management in farming systems; in coastal waters; in drought-prone regions. In two cases, machine learning models are developed in areas with contaminant pollution and phytoremediation plants. Finally, one paper attempted to use machine learning models to represent the impact of groundwater abstractions on river discharge across a wide range of conditions. From the reviewed papers, it is not possible to recommend a machine learning technique for a specific key area of model use. ANN is the most-used technique in the case of water planning and supply (also in drought-prone areas), followed by BN and SVM (Table 1).

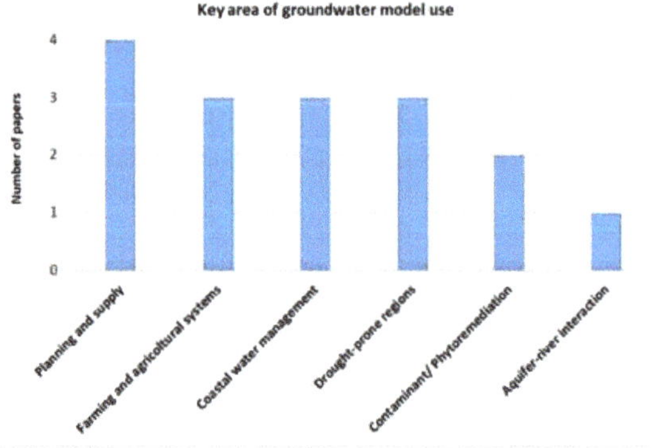

Figure 3. Key areas of groundwater models use in the reviewed papers.

4.2. Input Variables Employed for Machine Learning Modelling

Figure 4 shows the input variables that have been utilised in machine learning modelling. The past GWL time series are the most frequently used input variables to predict GWL; among 16 papers, 13 employed the GWL as an input variable. The precipitation or the net precipitation (i.e., the recharge) has been frequently used (four times the rainfall and five times the recharge, for a total of nine times) as an input variable. Moreover, other hydrological time series (e.g., pumping rates, temperature, evapotranspiration) have been also employed as the input variables in the reviewed papers. Since machine learning models can work with any data, there are many other input variables which have been used once in the reviewed papers, even if with less frequency (i.e., aquifer discharge, dew point, river stage, river flow depletion, irrigation demand). It is worth noting that the input data are commonly selected based on the availability of data rather than a physical analysis of the system. In particular, the degree of accuracy of the prediction will depend upon the spatial and temporal resolution of the monitoring network from which the model is developed for making predictions. Thus, the choice of input variables is often driven by the availability of proper time series.

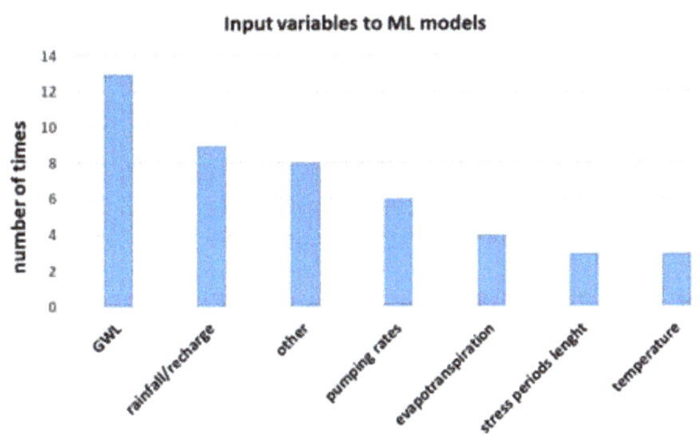

Figure 4. Input variables employed for the machine learning models.

4.3. Simulation Period of Physically-Based Models

The simulation period of the physically-based models varies from 141 to 1 year; nine physically-based models adopted a simulation period between 1 and 15 years; five physically-based models used a run simulation period between 16 and 35 years; two physically-based models adopted a run period higher than 35 years. In the reviewed papers, no mention was paid to a direct relation between the simulation period length of physically-based models and prediction accuracy of machine learning models. The size and layering of original numerical models ranges from 204,764 km^2 and 20 layers [95] to 0.75 km^2 and 1 layer [55]. Results suggest that there are no machine learning techniques nor groundwater management problems specifically suitable for a given range of physically-based size.

4.4. Time Step

The majority of the reviewed papers (8 among 16) used monthly time step for the machine learning simulations, followed by daily and weekly (both used in two papers) and quarterly (one paper). The time step selection was not declared in three of the reviewed papers. The frequent choice of monthly time steps is probably justified by the large availability of monthly recorded GWL data compared with other time steps. However, daily time steps are needed when modelling local-scale problems, such as river–aquifer interaction, or in some coastal water problems, where GWL are influenced by the tidal effects which induce daily variation to GWL.

4.5. Data Set Size

The number of total data used for groundwater modelling is highly variable. In three papers, only the number of wells was specified, without reporting the number of measure for each well. Among the papers which declared the size of data, the data set ranges from 300,000 sets [79] to 23 sets [96]. There is not a range of data set size which was more commonly used: five models used a number of dataset from 23 to 301; five models used from 1872 and 4911 datasets; four models used from 11,088 to 300,000 datasets. There is not a direct proportion between area of the physically-based model and data sets. Usually, smaller data sets are associated with a smaller size of the physically-based model. However, in some cases, large extent models (which is, larger than 10,000 km^2) are covered by a relatively small number of data (e.g., ref. [3]). There is not any recommendation in the reviewed papers about the density of samples which optimises the model performance. However, denser distributed training data allow achieving the best performance in temporal prediction scenarios. For example, the ANNs' ability to learn or generalise system behaviour is limited by the data with which it is trained. Machine learning models can fail to accurately predict GWL in areas where a scarce number of data for training is available, and results can be worse than those of numerical models.

4.6. Subset for Machine Learning Model Training, Validation and Testing

As explained in Section 2.2.1, the data available for modelling are subdivided into a training dataset (used during the learning phase of the machine learning model to produce a function representing the system behaviour) and into a testing dataset (used to evaluate the model's performance). Some authors subdivide data in three groups: training, testing, and validation; validation aims to check the model's prediction ability with a new input dataset.

There is not a specific rule for determining the optimum percent of data division for training, validation, and testing tasks. However, it can be noted that in all cases (except Sahoo et al., 2013 [102]) the dataset for training in the reviewed papers was always at least 60% (Figure 5), reaching 95%. In the majority of the papers (9 among 16), the percent of training dataset exceeded 80%. With regard to the testing dataset, authors use a percentage highly variable, between 4.5% and 40%. Only three of the reviewed papers used three subsets for training, for testing, and for validation, respectively. In these cases, the main subset was used for training (60%, 69%, and 52%), and the remaining data were equally distributed between testing and validation subsets or subdivided into 30% and 18% for testing and validation, respectively. In Banerjee et al. [96], the division into validation or testing sets was not mentioned, and the performance criteria were only mentioned for the training data, as already reported in ref. [19]. It can be concluded that a robust machine learning model should always be based on at least 60% of the training data, and 40% of the testing data.

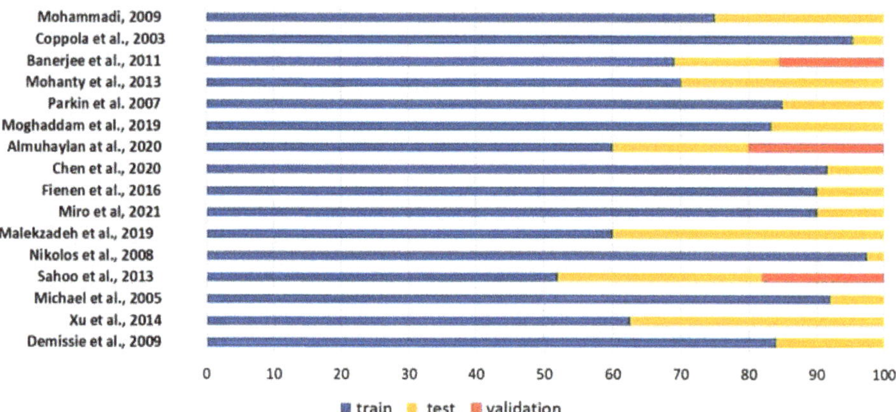

Figure 5. Percentage of the training and testing datasets used in machine learning modelling. Data from: Mohammadi, 2009 [59], Coppola et al., [3], Banerjee et al., 2011 [96], Mohanty et al., 2013 [97], Parkin et al., 2007 [98], Moghaddam et al., 2019 [76], Almuhaylan at al., 2020 [68], Chen et al., 2020 [63], Fienen et al., 2016 [95], Miro et al., 2021 [99], Malekzadeh et al., 2019 [100], Nikolos et al., 2008 [101], Sahoo et al., 2013 [102], Michael et al., 2005 [82], Xu et al., 2014 [79], Demissie et al., 2009 [85].

4.7. Used Software

Table 2 shows the number of times that each software was used to develop the ML models. It should be noted that in 11 cases, the software was not mentioned. Matlab is the most used software (nine times): when mentioned, the specific toolbox which adapts to the models' purpose is reported. In the hybrid approach of Michael et al. [82], the data-to-knowledge D2K software was used, a java-based data mining tool from the National Canter of Supercomputing Applications which allows for graphic data flows [116]. The results of this review indicated that Matlab can be easily used to implement the machine learning models; the variability and flexibility of its toolboxes clearly represent an advantage. However, the modellers can choose a range-free software with comparable skills.

Table 2. Software used for the machine learning models in the reviewed papers.

Machine Learning Model	Software	Commercial/Free	n of Times
ANN	Matlab R-neuralnet package LINGO not specified	c f c	3 1 1 7
RBF	Matlab	c	1
ANFIS	Matlab	c	1
TLRN	NeuroSolution	c	1
ELM, WA-ELM	Matlab, Matlab wavelet toolbox	c	1
BN	Hugin Lite 8.3; netica Software, CVNetica (for cv)	c c	1 1
IBW	Matlab Statistic Toolbox TM not specified	c	1 1
SVM	Matlab Statistic Toolbox TM not specified	c	2 2
DT	not specified		1
RF	R (randomForest package) not specified	f	1 1
GBRT	Phyton (scikit-learn library)	f	1

5. Specific Results

This section aims to furnish specific information to orient modellers choosing the appropriate machine learning approach based either on the properties of each of the examined model (e.g., the most used algorithms, model structure, tuning parameters: Section 5.1) or on advantages and disadvantages arising from the comparison between different machine learning techniques (Sections 5.2–5.4).

5.1. Properties of the Machine Learning Techniques Used in the Reviewed Papers

This section describes the results of the assessment of the machine learning techniques mostly used in the reviewed papers: ANNs, RBF, ELM, BN, SVM, DT.

Artificial Neural Networks

An assessment of the reviewed studies on ANNs revealed the following issues:

- Feed-forward multilayer perceptron with a backpropagation learning algorithm was the most used ANN technique in the reviewed papers.
- The training algorithms used in the reviewed papers were Levenberg Marquardt, Bayesian regularisation, scaled conjugate gradient, quick propagation algorithm, back-propagation algorithm, and resilient backpropagation. The most used were Levenberg Marquardt [60,117], which integrates the advantages of two training algorithms,

namely the steepest descent, and Gaussian–Newton methods, and searches for the global minima function to optimise the solution [68]; some authors point out that this is the less time-consuming algorithm.

- The transfer functions used for the hidden layer are: sigmoid, sine, hardlim, triangle basis, radial basis, hyperbolic tangent, linear, and logistic.
- The most common structure of ANN in the reviewed papers is a feed-forward ANN with a single hidden layer, with sigmoid transfer function in the hidden layer and linear transfer function in output layer. The best structure and number of hidden neurons are chosen by trial-and-error or cross-validation.
- The final structure of multilayer perceptron is usually chosen as the one resulting in minimum error and maximum efficiency during training.
- ANNs are capable of achieving substantially higher predictive accuracy at observation wells than the physically-based numerical model, with fewer inputs and lower developmental effort and cost. The choice of the appropriate training data size is a key issue; it should be evaluated considering many aspects, such as the required model accuracy, the number of connection weights, the complexity, and the level of noise in the system [3]. Moreover, it is important to find the optimal ANN topology ensuring satisfactory generalisation capability for any given problem. This is generally achieved by testing different topologies and transfer functions.

Radial Basis Function

Chen et al. [63] applied the Gaussian radial basis function to the neurons of the hidden layer and the linear transfer function in the output layer. RBF showed a better predictive performance and its computation cost in training and prediction stages were much less than those of numerical model in calibration and verification stages.

Extreme Learning Machine

In the work of Malekzadeh et al. [100], the number of hidden neurons for the ELM model was optimised by trial-and-error; results showed that model prediction was not significantly improved by increasing the number of hidden layer neurons. The sigmoid activation function provided higher simulation accuracy. The advantages of ELM with respect to other models are its modelling simplicity, easy coding, and quick computation for simulations in complex systems.

Bayesian Network

In the work of Moghaddam et al. [76], a BN structure was built, generating 108 possible states. The input parameters included rainfall, GWL in the previous month, average temperature, aquifer recharge, and discharge. The performance of BN models was evaluated by means of the R^2 and RMSE derived for all the observation points. In Fienen et al. [95], the BN was implemented with variables that were supposed to have the greatest influence on the source of water to wells: the distance to surface water, the surface water percent, the distance of 1st-order stream, and the percent of 1st-order stream. The continuous values of variables were discretised into bins; this permits performing predictions as discrete conditional probabilities without requiring a priori assumptions about distributions. Both the number of nodes and the number and ranges of bins were adjusted by 10-fold cross validation, and the set of parameters resulting in highest R^2 was selected as the optimal model.

Support Vector Machine

The most used kernel function with Support Vector Machine technique was the Gaussian Radial Basis Function, although several functions were tested (linear, radial bias, sigmoid). Cross-validation was the most used method for the optimisation of parameters (i.e., gamma value for the radial basis function and the regularisation coefficient), although Chen et al., 2020 [63] used Sequential minimal optimisation.

Decision Tree

Three types of decision trees were used in the reviewed papers: decision trees, random forest, and Gradient-Boosted Regression tree. In Demissie et al. [85], k-fold cross-validation was used to optimise the DT's pruning levels (used to reduce the complexity of the trees and reduce overfitting by removing redundant trees sections). Michael et al. [82] highlighted that decision trees have the ability to incorporate different data sources; thus, existing historical data can be combined with new surrogate or indicator data (such as rainfall) to detect whether the new data indicate potential problems that would warrant the collection of more traditional samples. To note, while DT provides the most accurate prediction improvement with updated data, IDW represents a good compromise between prediction accuracy and easy implementation. In the Random Forest model of Miro et al. [99], the parameters to optimise were the pruning levels, the learning rate and maximum tree depth, and the number of trees examined. Hyperparameters were adjusted generally by cross validation. Furthermore, the RF with the number of trees providing sufficient performance with a reasonable computational time was chosen as best model. The main advantage of using RF model is the reduced computational time with respect to numerical models, which allows incorporating it as a step of decision-making studies to speeds up the process. In the Gradient-Boosted Regression Trees [95] the parameters defining the individual trees included tree depth, shrinkage (a form of regularisation), learning rate, and maximum number of leaves on a tree. One advantage of GBRT is the possibility to use a variety of loss functions; Fienen et al. [95] used the HUBER loss function, an intermediate between squared difference and absolute difference. Hyper parameters were adjusted by cross validation with k = 10. The key tuning hyperparameters were the learning rate and maximum tree depth. The tradeoff curves of the best set of tuning parameters were explored for each technique; other metrics of skill/fit were calculated based on R^2 score.

5.2. Comparison between Machine Learning Techniques

The comparison between different machine learning techniques in the reviewed studies showed that:

- The performance of ANN with RBF as the activation function performed the best in simulating groundwater dynamics in arid basins, compared with ANN multilayer perceptron and SVM [63]. In detail, SVM performed the best in the training stage, while RBF in the verification stage; ANN's performance was lower than these two.
- Regarding ANFIS, no improvements are remarked with respect to ANNs, although greater performance with respect to the MODFLOW numerical model is documented [68].
- With respect to multilayer perceptron ANN, TLRNs can provide an appropriate tool for processing time-varying information. The main advantage is that TLRNs require a lower memory compared to multilayer perceptron, due to their lower network size. Furthermore, TLRNs have a low sensitivity to noise.
- Compared to simple ANN, ELM showed better performance, much less modelling time, less modelling error, and less weights norm [100].
- With respect to ANN, BN models provided easier implementation, higher prediction accuracy, and a greater ability to deal with missing or incomplete data [46]. It allows an uncertainty estimation more accurate than other machine learning models because the variables are modelled by means of probability distributions. When used as a metamodel, replacing a regional groundwater model to simulate the source of water-to-well [95], BN showed lower cross validation predictive skill compared with ANN and GBRT. However, the BN includes estimates of the uncertainty of predictions as part of the technique. GBRT required the least time with respect to BN and ANN. Thus, in this case, the choice between a statistical learning approach such as ANN or GBRT and the BN approach depends upon the preference of the modeller and the aims of the problem.
- When used to predict the annual change in GWL as effect of managed recharge, RF produced the most accurate average basin GWL representation respect to observations, compared with SVM and ANN [99].

5.3. Results of Testing Hybrid or Ensemble Models

The use of hybrid models and a combination of techniques for data pre-processing (described in Section 3.3) allowed a significant improvement in each modelling phase.

- ELM and WA-ELM were both used to simulate GWL in an arid basin [100]. However, the ELM model with the db2 mother wavelet for data pre-processing showed a better performance with a significant accuracy improvement compared with the physically-based models.
- The hybrid approach of Nikolos et al. [101] provides a fast way to integrate the physically-based models within an evolution-based optimisation procedure (DE algorithm) by replacing the calls of the PTC model with an ANN. The ANN provides a tool to perform an optimisation run with the DE algorithm with very short time, serving as a fast and accurate surrogate model.
- The hybrid modelling approach HANN [102] showed a high model structure strength since it integrated a robust data pre-processing and input variable selection techniques.
- Using machine learning models in hierarchical approach can significantly improve the results of physics-based models [82]; moreover, by that way, advantages and disadvantages of different machine learning models are identified and insights are provided into which data are most valuable to long-term monitoring objectives and which are not. In particular, Michael et al. [82] found that DT consistently provided the most accurate predictions of hydraulic head compared with IDW and ANN. However, when using all of the data across time, IDW showed substantial improvements. Given that IDW is simple to use and is widely accepted among practitioners, it could be considered as an optimum choice.
- The computational time of regional physically-based models can be substantially reduced by introducing an empirical (or statistical) representation of numerical models; this consists of machine learning models trained using numerical models inputs and outputs, which can be used to make predictions of variable of interest [95,99].

5.4. Results of Machine Learning Models Used to Reduce or Correct Errors in Physically-Based Models

This section summarises the main features of the machine learning models used for error correction and reduction (described in Section 3.4). IBW models were constructed to correct MODFLOW models by using the position of observation wells, calculated heads, evapotranspiration rates, and stress periods as inputs, and the residuals of MODFLOW model as outputs [79,82]. The parameters to optimise were the values of weighting function parameters and the number of neighbors n. Parameters of SVM models were already described in Section 5.1. When used to correct the error of physically-based models, both IBW and SVM have been shown to successfully reduce the magnitude and biasedness of the prediction error. Xu et al. [79] remarked that the popularity of SVM can be attributed to: (1) good generalisation performance; (2) always having a globally optimal solution (instead of local optima); (3) representation of the solution sparsely by a small subset of all training examples (Support Vectors) [87]. On the other hand, because IBW does not involve the training process and is less affected by the size of the training dataset, it is particularly recommended when the number of data is too small for other techniques to learn the spatial pattern of residuals [85]. In the case of spatial prediction, the simple IBW updates the future predictions better than DT and ANN and SVM; IBW models allow locally improving the results, and its degree of localisation and complexity can be adjusted flexibly. Thus, when groundwater model errors show local patterns, the application of IBW is advantageous. When considering both spatial and temporal prediction, IBW performed roughly as well as the more sophisticated SVM.

6. Discussion

Assessments of machine learning applications in GWL forecasting reveal that the performance of such methods is comparable to, or even more accurate than, that of numerical ones. Overall, the reviewed papers prove the capability of machine learning methods for capturing the nonlinear relation between groundwater and climate variables, especially where physically-based models would be difficult to implement. Machine learning models require a lesser number of input parameters and avoid the model building and parameter estimation stages typical of numerical models. Machine learning models can be a valid alternative for numerical models requiring long runtimes (i.e., complex regional models, models simulating many different processes, uncertainty analysis, sensitivity analysis), being capable of reducing computational times without sacrificing accuracy of detail in GWL forecasting. The very short time allows integrating machine learning models in decision-making processes when rapid and effective solutions for groundwater management need to be considered. Data-driven models are computationally efficient tools to correct head error prediction of numerical models; they work for error from multiple sources, and do not invoke assumptions on the error distribution [49]. Input data different from those used in the training stage can be included (e.g., pumping rate, boundary conditions, etc.); therefore, the data-driven models can be used to improve the prediction of physically-based models under scenarios that differ from the conditions used for calibration. Moreover, machine learning models can be applied successfully for modelling river–aquifer interactions.

Many studies exist concerning the use of machine learning models for groundwater simulation, developed on the basis of a limited number of observation points, without comparing results with numerical models. Conversely, the comparison of numerical and machine learning models is still a scarcely diffused task. In these comparative studies, each modeler uses the machine learning techniques for fixing a specific weakness of the numerical model, or to ameliorate poor fitting between simulated and observed values; in most cases, modellers explore different machine learning techniques to establish which one adapts better to its scopes. However, there are currently no well-defined procedures for the use of machine learning techniques to enhance results of numerical models, and this can limit the diffusion of the method. Another reason can be that the modeler should be familiar with both numerical and data driven models to correctly use both model types. Indeed, even if machine learning modelling does not consider the behaviour of the natural system, a certain degree of knowledge about the hydrological parameters and how they affect the results is required in order to avoid, for example, model overfitting (which means fitting the model to all the input parameters, preventing the generalisation ability of model, which is, in turn, given from the parameters effectively influencing groundwater level). In other words, the modeler should be able to manage both physically based data and statistical distributions of data, coupling different skills: those typical of hydrogeologists and those typical of statisticians/mathematicians. In many cases, a modeler (or a team of modellers) can meet both these requirements, but it is not so common. In addition, machine learning models are viewed with some skepticism by numerical modellers. Physically based represent the technique most widely diffused and used by local administrators for groundwater management. Usually, the results of a physically based model are improved by the integrating new observations (when available) or by tuning model parameters in order to modify the conceptual model. The machine learning approach, instead, aims at detecting the inherent mechanism, increasing prediction skills without deriving this from physical knowledge. This 'black box' nature, where no insight is gained into how the model generated the solution, is not widely accepted among numerical modellers and can prevent the use of machine learning models.

Regarding different machine learning methods to simulate the GWL when numerical models already exist, it can be said that from this review it is not possible to make a recommendation about one particular type of machine learning model for a specific problem. One advisable option could be testing different types of machine learning techniques in the different phases of the GWL modelling to detect the proper machine learning method in each stage and then couple them to achieve an optimum performance. However, hybrid modelling such as the combination of different techniques (e.g., data pre-processing such as time series decomposition or spatial clustering) and the hierarchical combination of machine learning models help to improve the accuracy of prediction. Moreover, some of the machine learning models appear to be suitable for updating numerical models previously calibrated, improving predictions as new data are collected (i.e., DT, [82]; IBW and SVM, [79]). Furthermore, when using machine learning models to correct the error of physically-based models, both IBW and SVM show better performance than DT and ANN. However, the simple IBW allows locally improving the results, and this suggests that it is suitable when errors show local patterns.

Some authors highlighted the main disadvantages of machine learning models with respect to numerical models:

- The numerical models are comparatively more reliable. While showing a lower prediction error than the physical models, machine learning models cannot return many of the outputs of a physical model, such as flux estimates or total water balance.
- Xu et al. [79] found that data-driven models are difficult to interpret physically. The updated head no longer conserved mass for the given model inputs, which can confound the physical interpretation of the results and prevent understanding errors in the conceptualisation of the groundwater system.
- Numerical models exhibit a higher generalisation ability than machine learning methods because they are based on the physics of the system [63]. Conversely, machine learning models are applicable to problems that require a high number of model runs without considering the physical system (e.g., optimisations, real-time models, sensitivity/uncertainty analysis).
- Usually, while the machine learning models may be more efficacious for predicting short-term GWL and reproducing highly localised flow impacts, numerical modelling is more appropriate for long-term projections, or in areas where field data are insufficient for the given problem. However, it should be remarked that Almuhaylan et al. [68] were able to use machine learning models to perform long-term prediction (up to 50 years), by training the ANN/ANFIS model for the prediction of changes in groundwater levels instead of the direct simulation of water levels.

Thus, each type of model (numerical or machine learning) is suitable for a specific type of problem. As suggested by many authors, numerical and machine learning models can be successfully used as complementary to each other as a powerful groundwater management tool:

- when few field data exist, the results of numerical models can be improved by training machine learning models, which allow to obtain accurate groundwater level forecasting at specific observation wells;
- machine learning models cannot substitute a numerical model as one single model, but can be used to simulate water table fluctuation at every individual observation well with reduced computational time;
- accurate results of machine learning models in specific test sites can be used to obtain the best GWL data required by the numerical model as input;
- the physical dynamics of the system must be sufficiently understood by the modeller in order to identify the important predictor input variables of machine learning models. Results of numerical models help to understand the physical system; this can help, in turn, choosing the input parameters for machine learning models. Coppola et al. [3] suggested using ANNs to perform a sensitivity analysis on the interrelationships between input and output variables;
- Numerical models can simulate different scenarios, allowing for detection areas requiring particular management strategies, thereby supporting the design of an effective monitoring network, which, in turn, may improve both machine learning predictive capability and performance.

Given the results of this review, one should evaluate the best machine learning technique based on:

- The aim of the work, for example: improvement of prediction at some well location, numerical model error correction, numerical model updating;

- the need to produce a probability distribution of the results and obtain uncertainty estimation within the model, (i.e., in areas with few data);
- the availability of data for training and testing (number and spatial-temporal distribution);
- the need to speed up decision making processes and reduce the computational time;
- the degree of expertise of the modeller, which should drive the searching for a good compromise between model complexity and prediction performance.

To note, this review only accounts for groundwater flow models; robust groundwater flow models are the basis for setting up groundwater solute transport models. The comparison between physically based and machine learning models focused on groundwater solute transport should be the subject of future research.

7. Conclusions

This study presents a review of 16 papers regarding the use of numerical models and machine learning techniques for the prediction of groundwater level, which were published in 10 international journals and 1 book from 2003 to 2020. Machine learning techniques are used to improve or speed the prediction process of physically-based models, which are developed with different codes and software, from regional to site scale, and with data collected over time windows spanning from one to hundreds of years. Machine learning methodologies, approximating the complex behavior and dynamics of physical systems, allow for the optimisation of predictions of a large number of scenarios within a short period of time, compared with the long computational time required for the corresponding simulation time using a numerical model. Machine learning models do not return many of the outputs of a physical model, such as flux estimates and residence time calculations, or total water balance. Thus, machine learning models cannot be used to substitute numerical models in large study areas, but are affordable tools to improve predictions at specific observation wells. Results of this review suggest that numerical and machine learning models can be successfully used as complementary to each other as a powerful groundwater management tool. The machine learning techniques can be used to improve the calibration of numerical models, whereas the results of numerical models allow understanding the physical system and selecting proper input variables for machine learning models. Among the machine learning techniques, the hybrid machine learning models show better results accuracy.

Funding: This research received no external funding.

Acknowledgments: Author is very grateful to the reviewers for their interesting comments, which allowed to improve the quality of the manuscript. A special thank also goes to Daniel Feinstein, for his precious suggestions and support.

Conflicts of Interest: The author declares no conflict of interest.

References

1. Daliakopoulos, I.N.; Tsanis, I.K. Comparison of an artificial neural network and a conceptual rainfall–runoff model in the simulation of ephemeral streamflow. *Hydrol. Sci. J.* **2016**, *61*, 2763–2774. [CrossRef]
2. Besaw, L.E.; Rizzo, D.M.; Bierman, P.R.; Hackett, W.R. Advances in ungauged streamflow prediction using artificial neural networks. *J. Hydrol.* **2010**, *386*, 27–37. [CrossRef]
3. Coppola, E., Jr.; Szidarovszky, F.; Poulton, M.; Charles, E. Artificial neural network approach for predicting transient water levels in a multilayered groundwater system under variable state, pumping, and climate conditions. *J. Hydrol. Eng.* **2003**, *8*, 348–360. [CrossRef]
4. Neuman, S.P.; Wierenga, P.J. *A Comprehensive Strategy of Hydrogeologic Modeling and Uncertainty Analysis for Nuclear Facilities and Sites (NUREG/CR-6805)*; Report prepared for US Nuclear Regulatory Commission: Washington, DC, USA, 2003; p. 309.
5. Cooley, R.L. A theory for modeling ground-water flow in heterogeneous media. In *US Geological Survey Professional Paper 1679*; U.S. Geological Survey: Reston, VA, USA, 2004; p. 220.
6. Doherty, J.; Christensen, S. Use of paired simple and complex models to reduce predictive bias and quantify uncertainty. *Water Resour. Res.* **2011**, *47*, 1–21. [CrossRef]
7. Refsgaard, J.C.; van der Sluijs, J.P.; Brown, J.; van der Keur, P. A framework for dealing with uncertainty due to model structure error. *Adv. Water Resour.* **2006**, *29*, 1586–1597. [CrossRef]

8. Hunt, R.J.; Welter, D.E. Taking account of "unknown unknowns". *GroundWater* **2010**, *48*, 477. [CrossRef]
9. Tiedeman, C.R.; Hill, M.C. Model calibration and issues related to validation, sensitivity analysis, post-audit, uncertainty evaluation and assessment of prediction data needs. In *Groundwater: Resource Evaluation, Augmentation, Contamination, Restoration, Modeling and Management*; Thangarajian, M., Ed.; Springer: New York, NY, USA, 2007; pp. 237–282.
10. Liu, Y.; Gupta, H.V. Uncertainty in hydrologic modeling: Toward an integrated data assimilation framework. *Water Resour. Res.* **2007**, *43*, 1–18. [CrossRef]
11. Vrugt, J.A.; Stauffer, P.H.; Wohling, T.; Robinson, B.A.; Vesselinov, V.V. Inverse modeling of subsurface flow and transport properties: A review with new developments. *Vadose Zone J.* **2008**, *7*, 843–864. [CrossRef]
12. Bierkens, M.F. Modeling water table fluctuations by means of a stochastic differential equation. *Water Resour. Res.* **1998**, *34*, 2485–2499. [CrossRef]
13. Bidwell, V.J. Realistic forecasting of groundwater level, based on the Eigenstructure of aquifer dynamics. *Math. Comput. Simul.* **2005**, *69*, 12–20. [CrossRef]
14. Maier, H.R.; Dandy, G.C. The use of artificial neural networks for the prediction of water quality parameters. *Water Resour. Res.* **1996**, *32*, 1013–1022. [CrossRef]
15. Maity, R.; Nagesh Kumar, D. Probabilistic prediction of hydroclimatic variables with nonparametric quantification of uncertainty. *J. Geophys. Res. Atmos.* **2008**, *113*, 1–12. [CrossRef]
16. Vellido, A.; Martín-Guerrero, J.D.; Lisboa, P.J. Making machine learning models interpretable. In Proceedings of the European Symposium on Artificial Neural Networks, Computational Intelligence and Machine Learning, Bruges, Belgium, 25–27 April 2012; pp. 163–172.
17. Abraham, A.; Pedregosa, F.; Eickenberg, M.; Gervais, P.; Mueller, A.; Kossaifi, J.; Gramfort, A.; Thirion, B.; Varoquaux, G. Machine learning for neuroimaging with scikit-learn. *Front. Neuroinform.* **2014**, *14*, 1–10.
18. Park, C.; Took, C.C.; Seong, J.K. Machine learning in biomedical engineering. *Biomed. Eng. Lett.* **2018**, *8*, 1–3. [CrossRef]
19. Reich, Y. Machine learning techniques for civil engineering problems. *Comput.-Aided Civ. Infrastruct. Eng.* **1997**, *12*, 295–310. [CrossRef]
20. Reich, Y.; Barai, S.V. Evaluating machine learning models for engineering problems. *Artif. Intell. Eng.* **1999**, *13*, 257–272. [CrossRef]
21. Vadyala, S.R.; Betgeri, S.N.; Matthews, J.C.; Matthews, E. A review of physics-based machine learning in civil engineering. *Results Eng.* **2021**, *13*, 100316. [CrossRef]
22. Zander, S.; Nguyen, T.; Armitage, G. Automated traffic classification and application identification using machine learning. In Proceedings of the IEEE Conference on Local Computer Networks 30th Anniversary, Sydney, Australia, 15–17 November 2005; (LCN'05) l. pp. 250–257.
23. Yu, H.; Wu, Z.; Wang, S.; Wang, Y.; Ma, X. Spatiotemporal recurrent convolutional networks for traffic prediction in transportation networks. *Sensors* **2017**, *17*, 1501. [CrossRef]
24. Nguyen, H.; Kieu, L.M.; Wen, T.; Cai, C. Deep learning methods in transportation domain: A review. *IET Intell. Transp. Syst.* **2018**, *12*, 998–1004. [CrossRef]
25. Tahmasebi, P.; Kamrava, S.; Bai, T.; Sahimi, M. Machine learning in geo-and environmental sciences: From small to large scale. *Adv. Water Resour.* **2020**, *142*, 103619. [CrossRef]
26. Sun, A.Y.; Scanlon, B.R. How can Big Data and machine learning benefit environment and water management: A survey of methods, applications, and future directions. *Environ. Res. Lett.* **2019**, *14*, 073001. [CrossRef]
27. Lary, D.J.; Alavi, A.H.; Gandomi, A.H.; Walker, A.L. Machine learning in geosciences and remote sensing. *Geosci. Front.* **2016**, *7*, 3–10. [CrossRef]
28. Mosavi, A.; Ozturk, P.; Chau, K.W. Flood prediction using machine learning models: Literature review. *Water* **2018**, *10*, 1536. [CrossRef]
29. Choubin, B.; Mosavi, A.; Alamdarloo, E.H.; Hosseini, F.S.; Shamshirband, S.; Dashtekian, K.; Ghamisi, P. Earth fissure hazard prediction using machine learning models. *Environ. Res.* **2019**, *179*, 108770. [CrossRef]
30. Elith, J.; Leathwick, J.R.; Hastie, T. A working guide to boosted regression trees. *J. Anim. Ecol.* **2008**, *77*, 802–813. [CrossRef]
31. Jeong, J.H.; Resop, J.P.; Mueller, N.D.; Fleisher, D.H.; Yun, K.; Butler, E.E.; Timlin, D.J.; Shim, K.-M.; Gerber, J.S.; Reddy, V.R.; et al. Random forests for global and regional crop yield predictions. *PLoS ONE* **2016**, *11*, e0156571. [CrossRef]
32. Lamorski, K.; Šimůnek, J.; Sławiński, C.; Lamorska, J. An estimation of the main wetting branch of the soil water retention curve based on its main drying branch using the machine learning method. *Water Resour. Res.* **2017**, *53*, 1539–1552. [CrossRef]
33. Povak, N.A.; Hessburg, P.F.; McDonnell, T.C.; Reynolds, K.M.; Sullivan, T.J.; Salter, R.B.; Cosby, B.J. Machine learning and linear regression models to predict catchment-level base cation weathering rates across the southern Appalachian Mountain region, USA. *Water Resour. Res.* **2014**, *50*, 2798–2814. [CrossRef]
34. Singh, N.; Chakrapani, G. ANN modelling of sediment concentration in the dynamic glacial environment of Gangotri in Himalaya. *Env. Monit Assess* **2015**, *187*, 1–14. [CrossRef]
35. Piotrowski, A.P.; Napiorkowski, J.J. A comparison of methods to avoid overfitting in neural networks training in the case of catchment runoff modelling. *J. Hydrol.* **2013**, *476*, 97–111. [CrossRef]
36. Kingston, G.B.; Maier, H.R.; Lambert, M.F. Calibration and validation of neural networks to ensure physically plausible hydrological modeling. *J. Hydrol.* **2005**, *314*, 158–176. [CrossRef]

37. Solomatine, D.P. Data-driven modeling and computational intelligence methods in hydrology. In *Encyclopedia of Hydrological Sciences*; Anderson, M., Ed.; Wiley: New York, NY, USA, 2005.
38. Deka, P.C. Support vector machine applications in the field of hydrology: A review. *Appl. Soft Comput.* **2014**, *19*, 372–386.
39. Lange, H.; Sippel, S. Machine learning applications in hydrology. In *Forest-Water Interactions*; Levia, D.F., Carlyle-Moses, D.E., Lida, S., Michalzik, B., Nanko, K., Tischer, A., Eds.; Ecological Studies; Springer Nature: Cham, Switzerland, 2020; Volume 240, pp. 233–257. [CrossRef]
40. Rasouli, K.; Hsieh, W.W.; Cannon, A.J. Daily streamflow forecasting by machine learning methods with weather and climate inputs. *J. Hydrol.* **2012**, *414*, 284–293. [CrossRef]
41. Xu, T.; Liang, F. Machine learning for hydrologic sciences: An introductory overview. *Wiley Interdiscip. Rev. Water* **2021**, *8*, e1533. [CrossRef]
42. Yaseen, Z.M.; Sulaiman, S.O.; Deo, R.C.; Chau, K.W. An enhanced extreme learning machine model for river flow forecasting: State-of-the-art, practical applications in water resource engineering area and future research direction. *J. Hydrol.* **2019**, *569*, 387–408. [CrossRef]
43. Wu, W.; Dandy, G.C.; Maier, H.R. Protocol for developing ANN models and its application to the assessment of the quality of the ANN model development process in drinking water quality modelling. *Environ. Model. Softw.* **2014**, *54*, 108–127. [CrossRef]
44. Fienen, M.N.; Nolan, B.T.; Kauffman, L.J.; Feinstein, D.T. Metamodeling for groundwater age forecasting in the Lake Michigan Basin. *Water Resour. Res.* **2018**, *54*, 4750–4766. [CrossRef]
45. Maier, H.R.; Jain, A.; Dandy, G.C.; Sudheer, K.P. Methods used for the development of neural networks for the prediction of water resource variables in river systems: Current status and future directions. *Environ. Model. Softw.* **2010**, *25*, 891–909. [CrossRef]
46. Solomatine, D.P. Applications of data-driven modelling and machine learning in control of water resources. In *Computational Intelligence in Control*; Mohammadian, M., Sarker, R.A., Yao, X., Eds.; Idea Group Publishing: Hershey, PA, USA, 2002; pp. 197–217.
47. Yan, J.; Jia, S.; Lv, A.; Zhu, W. Water resources assessment of China's transboundary river basins using a machine learning approach. *Water Resour. Res.* **2019**, *55*, 632–655. [CrossRef]
48. Nourani, V.; Baghanam, A.H.; Adamowski, J.; Kisi, O. Applications of hybrid wavelet–artificial intelligence models in hydrology: A review. *J. Hydrol.* **2014**, *514*, 358–377. [CrossRef]
49. Rajaee, T.; Ebrahimi, H.; Nourani, V. A review of the artificial intelligence methods in groundwater level modeling. *J. Hydrol.* **2019**, *572*, 336–351. [CrossRef]
50. McDonald, M.G.; Harbaugh, A.W. A modular three-dimensional finite-difference ground-water flow model. In *US Geological Survey Report 06-A1*; US Geological Survey: Reston, VA, USA, 1988; p. 586.
51. Harbaugh, A.W.; Banta, E.R.; Hill, M.C.; Mcdonald, M.G. MODFLOW-2000, the US geological survey modular ground-water model—User guide to modularization concepts and the ground-water flow process. In *US Geological Survey Open-File Report 00-92*; US Geological Survey: Reston, VA, USA, 2000; p. 121.
52. Winston, R.B. MODFLOW-related freeware and shareware resources on the internet. *Comput. Geosci.* **1999**, *25*, 377–382. [CrossRef]
53. Voss, C.I. A Finite-Element Simulation Model for Saturated–Unsaturated, Fluid-Density-dependent Ground-Water Flow with Energy Transport or Chemically Reactive Single-species. In *Water-Resources Investigations Report 84-4369*; US Geological Survey: Reston, VA, USA, 1984. [CrossRef]
54. Babu, D.K.; Pinder, G.F. A finite element–finite difference alternating direction algorithm for 3- dimensional groundwater transport. *Adv. Water Resour.* **1984**, *7*, 116–119. [CrossRef]
55. Bentley, L.R.; Kieper, G.M. Verification of the Princeton Transport Code (PTC). In *Engineering Hydrology, Proceedings of the Symposium Sponsored by the Hydraulics Division of the American Society of Civil Engineers, San Francisco, CA, USA, 25–30 July 1993*; American Society of Civil Engineers: New York, NY, USA, 1993; pp. 1037–1042.
56. Ewen, J.; Parkin, G.; O'Connell, P.E. SHETRAN: A coupled surface/subsurface modelling system for 3D water flow and sediment and solute transport in river basins. *ASCE J. Hydrol. Eng.* **2000**, *5*, 250–258. [CrossRef]
57. Hsu, K.L.; Gupta, H.V.; Sorooshian, S. Artificial neural network modeling of the rainfall-runoff process. *Water Resour. Res.* **1995**, *31*, 2517–2530. [CrossRef]
58. Schalkoff, R.J. *Artificial Neural Networks*; McGraw-Hill Higher Education: New York, NY, USA, 1997; p. 448.
59. Mohammadi, K. Groundwater Table Estimation Using MODFLOW and Artificial Neural Networks. In *Practical Hydroinformatics*; Abrahart, R.J., See, L.M., Solomatine, D.P., Eds.; Water Science and Technology Library: Springer: Berlin/Heidelberg, Germany, 2009; Volume 68. [CrossRef]
60. Samarasinghe, S. *Neural Networks for Applied Sciences and Engineering: From Fundamentals to Complex Pattern Recognition*; Auerbach Publications: New York, NY, USA, 2016; ISBN 0429115784.
61. Taormina, R.; Chau, K.-W.; Sethi, R. Artificial neural network simulation of hourly groundwater levels in a coastal aquifer system of the Venice lagoon. *Eng. Appl. Artifi. Intellig.* **2012**, *25*, 1670–1676. [CrossRef]
62. Wunsch, A.; Liesch, T.; Broda, S. Forecasting groundwater levels using nonlinear autoregressive networks with exogenous input (NARX). *J. Hydrol.* **2018**, *567*, 743–758. [CrossRef]
63. Chen, C.; He, W.; Zhou, H.; Xue, Y.; Zhu, M. A comparative study among machine learning and numerical models for simulating groundwater dynamics in the Heihe River Basin, northwestern China. *Sci. Rep.* **2020**, *10*, 1–13. [CrossRef]
64. Schwenker, F.; Kestler, H.A.; Palm, G. Three learning phases for radial-basis-function networks. *Neural Netw.* **2001**, *14*, 439–458. [CrossRef]

65. Buhmann, M.D. *Radial Basis Functions: Theory and Implementations*; Cambridge University Press: Cambridge, UK, 2003; p. 258.
66. Jang, J.S.R. ANFIS adaptive-network-based fuzzy inference systems. *IEEE Trans. Syst. Man. Cybern.* **1993**, *23*, 665–685. [CrossRef]
67. Kurtulus, B.; Razack, M. Modeling daily discharge responses of a large karstic aquifer using soft computing methods: Artificial neural network and neuro-fuzzy. *J. Hydrol.* **2010**, *381*, 101–111. [CrossRef]
68. Almuhaylan, M.R.; Ghumman, A.R.; Al-Salamah, I.S.; Ahmad, A.; Ghazaw, Y.M.; Haider, H.; Shafiquzzaman, M. Evaluating the Impacts of Pumping on Aquifer Depletion in Arid Regions Using MODFLOW, ANFIS and ANN. *Water* **2020**, *12*, 2297. [CrossRef]
69. Chen, S.H.; Lin, Y.H.; Chang, L.C.; Chang, F.J. The strategy of building a flood forecast model by neuro fuzzy network. *Hydr. Proc.* **2006**, *20*, 1525–1540. [CrossRef]
70. Haykin, S. *Communication Systems*, 2nd ed.; Wiley: New York, NY, USA, 1994; pp. 45–90.
71. Saharia, M.; Bhattacharjya, R.K. Geomorphology-based time-lagged recurrent neural networks for runoff forecasting. *KSCE J. Civ. Eng.* **2012**, *16*, 862–869. [CrossRef]
72. Sattari, M.; Taghi, K.Y.; Pal, M. Performance evaluation of artificial neural network approaches in forecasting reservoir inflow. *Appl. Math. Model.* **2012**, *36*, 2649–2657. [CrossRef]
73. Huang, G.B.; Chen, L.; Siew, C.K. Universal approximation using incremental constructive feedforward networks with random hidden nodes. *IEEE Trans. Neural Netw.* **2006**, *17*, 879–892. [CrossRef]
74. Huang, G.B.; Chen, L. Convex incremental extreme learning machine. *Neurocomputing* **2007**, *70*, 3056–3062. [CrossRef]
75. Huang, G.B.; Chen, L. Enhanced random search based incremental extreme learning machine. *Neurocomputing* **2008**, *71*, 3460–3468. [CrossRef]
76. Moghaddam, H.K.; Moghaddam, H.K.; Rahimzadeh Kivi, Z.; Bahreinimotlagh, M.; Javad Alizadeh, M. Developing comparative mathematic models, BN and ANN for forecasting of groundwater levels. *Groundw. Sustain. Dev.* **2019**, *9*, 100237. [CrossRef]
77. Cleary, J.G.; Trigg, L.E. K*: An instance-based learner using an entropic distance measure. In *Machine Learning, Proceedings of the Twelfth International Conference, San Francisco, CA, USA, 9–12 July 1995*; Morgan Kaufmann Publishers Inc.: San Francisco, CA, USA, 1995; pp. 108–114.
78. Smith, E.E.; Medin, D.L. *Categories and Concepts*; Harvard University Press: Cambridge, MA, USA, 1981; p. 203.
79. Xu, T.; Valocchi, A.J.; Choi, J.; Amir, E. Use of machine learning methods to reduce predictive error of groundwater models. *Groundwater* **2014**, *52*, 448–460. [CrossRef]
80. Aha, D.W. Feature Weighting for Lazy Learning algorithms. In *Feature Extraction, Construction and Selection: A Data Mining Perspective*; The American Statistical Association: Boston, MA, USA, 1998; Volume 1, p. 410.
81. Aha, D.W.; Kibler, D.; Albert, M.C. Instance-Based Learning Algorithms. *Mach. Learn.* **1991**, *6*, 37–66. [CrossRef]
82. Michael, W.J.; Minsker, B.S.; Tcheng, D.; Valocchi, A.J.; Quinn, J.J. Integrating data sources to improve hydraulic head predictions: A hierarchical machine learning approach. *Water Resour. Res.* **2005**, *41*, 1–14. [CrossRef]
83. Vapnik, V.N. *The Nature of Statistical Learning Theory*; Springer: New York, NY, USA, 1995; p. 314.
84. Gunn, S.R. Support vector machines for classification and regression. In *ISIS Technical Report*; University of Southampton: Southampton, UK, 1998; p. 66.
85. Demissie, Y.K.; Valocchi, A.J.; Minsker, B.S.; Bailey, B.A. Integrating a calibrated groundwater flow model with error-correcting data-driven models to improve predictions. *J. Hydrol.* **2009**, *364*, 257–271. [CrossRef]
86. Yoon, H.; Jun, S.-C.; Hyun, Y.; Bae, G.-O.; Lee, K.-K. A comparative study of artificial neural networks and support vector machines for predicting groundwater levels in a coastal aquifer. *J. Hydrol.* **2011**, *396*, 128–138. [CrossRef]
87. Cao, L.J.; Chua, K.S.; Chong, W.K.; Lee, H.P.; Gu, Q.M. A comparison of PCA, KPCA and ICA for dimensionality reduction in support vector machine. *Neurocomputing* **2003**, *55*, 321–336. [CrossRef]
88. Vapnik, V.N. *Statistical Learning Theory*; John Wiley & Sons: New York, NY, USA, 1998; p. 768.
89. Smola, A.J.; Sch¨olkopf, B. A tutorial on support vector regression. *Stat. Comput.* **2004**, *14*, 199–222. [CrossRef]
90. Quinlan, J.R. Induction of decision trees. *Mach. Learn.* **1986**, *1*, 81–106.
91. Breiman, L.; Friedman, J.H.; Olshen, R.A.; Stone, C.J. *Classification and Regression Trees*; Routhledge: New York, NY, USA, 1984; p. 368.
92. Anderton, S.P.; White, S.M.; Alvera, B. Evaluation of spatial variability of snow water equivalent in a high mountain catchment. *Hydrol. Processes* **2004**, *18*, 435–453. [CrossRef]
93. Breiman, L. Random forests. *Mach. Learn.* **2001**, *45*, 5–32. [CrossRef]
94. Aertsen, W.; Kint, V.; Van Orshoven, J.; Muys, B. Evaluation of Modelling Techniques for Forest Site Productivity Prediction in Contrasting Ecoregions Using Stochastic Multicriteria Acceptability Analysis (SMAA). *Environ. Model. Softw.* **2011**, *26*, 929–937. [CrossRef]
95. Fienen, M.N.; Nolan, B.T.; Feinstein, D.T. Evaluating the sources of water to wells: Three techniques for metamodeling of a groundwater flow model. *Environ. Model. Softw.* **2016**, *77*, 95–107. [CrossRef]
96. Banerjee, P.; Singh, V.S.; Chatttopadhyay, K.; Chandra, P.C.; Singh, B. Artificial neural network model as a potential alternative for groundwater salinity forecasting. *J. Hydrol.* **2011**, *398*, 212–220. [CrossRef]
97. Mohanty, S.; Jha, M.K.; Kumar, A.; Sudheer, K.P. Artificial neural network modeling for groundwater level forecasting in a river island of eastern India. *Water Resour. Manag.* **2010**, *24*, 1845–1865. [CrossRef]
98. Parkin, G.; Birkinshaw, S.J.; Younger, P.L.; Rao, Z.; Kirk, S. A numerical modelling and neural network approach to estimate the impact of groundwater abstractions on river flows. *J. Hydrol.* **2007**, *339*, 15–28. [CrossRef]

99. Aghlmand, R.; Abbasi, A. Application of MODFLOW with boundary conditions analyses based on limited available observations: A case study of Birjand plain in East Iran. *Water* **2019**, *11*, 1904. [CrossRef]
100. Feinstein, D.T.; Eaton, T.T.; Hart, D.J.; Krohelski, J.T.; Bradbury, K.R. Regional aquifer model for southeastern Wisconsin; Report 1: Data collection, conceptual model development, numerical model construction, and model calibration. In *US Geological Survey Techniques Report*; US Geological Survey: Reston, VA, USA, 2005.
101. Miro, M.E.; Groves, D.; Tincher, B.; Syme, J.; Tanverakul, S.; Catt, D. Adaptive water management in the face of uncertainty: Integrating machine learning, groundwater modeling and robust decision making. *Clim. Risk Manag.* **2021**, *34*, 100383. [CrossRef]
102. Malekzadeh, M.; Kardar, S.; Shabanlou, S. Simulation of groundwater level using MODFLOW, extreme learning machine and Wavelet-Extreme Learning Machine models. *Groundw. Sustain. Dev.* **2019**, *9*, 100279. [CrossRef]
103. Nikolos, I.K.; Stergiadi, M.; Papadopoulou, M.P.; Karatzas, G.P. Artificial neural networks as an alternative approach to groundwater numerical modelling and environmental design. *Hydrol. Processes Int. J.* **2008**, *22*, 3337–3348. [CrossRef]
104. Sahoo, S.; Jha, M.K. Groundwater-level prediction using multiple linear regression and artificial neural network techniques: A comparative assessment. *Hydrogeol. J.* **2013**, *21*, 1865–1887. [CrossRef]
105. Clark, B.R.; Hart, R.M.; Gurdak, J.J. Groundwater availability of the Mississippi Embayment. In *US Geological Survey Professional Paper 2011*; US Geological Survey: Reston, VA, USA, 1785; p. 62.
106. Luckey, R.R.; Becker, M.F. Hydrogeology, water use, and simulation of flow in the High Plains aquifer in northwestern Oklahoma, southeastern Colorado, southwestern Kansas, northeastern New Mexico, and northwestern Texas. In *US Geological Survey Water Resources Investment Report 99-4104*; US Geological Survey: Reston, VA, USA, 1999; p. 73.
107. Quinn, J.J.; Negri, M.C.; Hinchman, R.R.; Moos, L.P.; Wozniak, J.B.; Gatliff, E.G. Predicting the effect of deep-rooted hybrid poplars on the groundwater flow system at a large-scale phytoremediation site. *Int. J. Phytoremediation* **2001**, *3*, 41–60. [CrossRef]
108. Lefebvre, C.; Principe, J. *NeuroSolutions User's Guide*; Neurodimension Inc.: Gainesville, FL, USA, 1998; p. 786.
109. Uusitalo, L. Advantages and challenges of Bayesian networks in environmental modelling. *Ecol. Model.* **2007**, *203*, 312–318. [CrossRef]
110. Liu, Z.; Malone, B.; Yuan, C. Empirical evaluation of scoring functions for Bayesian network model selection. *BMC Bioinform.* **2012**, *13*, 1–16. [CrossRef] [PubMed]
111. Misiti, M.; Misiti, Y.; Oppenheim, G.; Poggi, J.M. *Wavelet Toolbox for Use with Matlab*; The Mathworks, Inc.: Natick, MA, USA, 1996; p. 1030.
112. Nikolos, I.K. Inverse design of aerodynamic shapes using differential evolution coupled with artificial neural network. In Proceedings of the ERCOFTAC Conference in Design Optimization: Methods and Applications, Athens, Greece, 31 March–2 April 2004.
113. Feinstein, D.T.; Hunt, R.; Reeves, H. Regional groundwater-flow model of the Lake Michigan Basin in support of Great Lakes Basin water availability and use studies. In *Scientific Investigations Report 2010–5109*; United States Geological Survey: Reston, VA, USA, 2010.
114. Fienen, M.N.; Nolan, B.T.; Feinstein, D.T.; Starn, J.J. *Metamodels to Bridge the Gap between Modeling and Decision Support*; United States Geological Survey: Reston, VA, USA, 2015; p. 860.
115. Republican River Compact Administration (RRCA). Appendix A: Groundwater Model for 1918–2000 (June 30, 2003). Available online: https://www.republicanrivercompact.org/v12p/html/ch01.html (accessed on 4 April 2022).
116. Welge, M.; Auvil, L.; Shirk, A.; Bushell, C.; Bajcsy, P.; Cai, D.; Redman, T.; Clutter, D.; Aydt, R.; Tcheng, D. *Data to Knowledge, Technical Report*; Automated Learning Group, National Center for Supercomputing Applications: Urbana, IL, USA, 2003.
117. Djurovic, N.; Domazet, M.; Stricevic, R.; Pocuca, V.; Spalevic, V.; Pivic, R.; Gregoric, E.; Domazet, U. Comparison of Groundwater Level Models Based on Artificial Neural Networks and ANFIS. *Sci. World J.* **2015**, *13*, 1–13. [CrossRef] [PubMed]

MDPI
St. Alban-Anlage 66
4052 Basel
Switzerland
Tel. +41 61 683 77 34
Fax +41 61 302 89 18
www.mdpi.com

Water Editorial Office
E-mail: water@mdpi.com
www.mdpi.com/journal/water

www.ingramcontent.com/pod-product-compliance
Lightning Source LLC
LaVergne TN
LVHW070129170726
843515LV00007B/1773

* 9 7 8 3 0 3 6 5 7 1 4 2 3 *